Fundamental Biomedical Technologies

Series Editor

Mauro Ferrari
Department of Biomedical Engineering
The University of Texas
Houston, TX, USA

More information about this series at http://www.springer.com/series/7045

Dan Ohtan Wang • Daniel Packwood
Editors

Cell-Inspired Materials and Engineering

Springer

Editors
Dan Ohtan Wang
Kyoto University iCeMS
Sakyo-ku, Kyoto, Japan

Daniel Packwood
Kyoto University iCeMS
Sakyo-ku, Kyoto, Japan

ISSN 1559-7083 ISSN 2626-8655 (electronic)
Fundamental Biomedical Technologies
ISBN 978-3-030-55923-6 ISBN 978-3-030-55924-3 (eBook)
https://doi.org/10.1007/978-3-030-55924-3

This Springer imprint is published by the registered company Springer Nature Switzerland AG
The registered company address is: Gewerbestrasse 11, 6330 Cham, Switzerland

Preface

It is a curious fact that, despite starting in quite different places, different scientific fields often converge as they develop over time. Prominent examples include the convergence of mathematics and physics some 300 years ago and of physics and chemistry some 100 years ago. As researchers from these areas can attest, such "interdisciplinary convergence" can profoundly enrich the fields involved and can also create new avenues for fundamental research and applications.

While perhaps not widely recognized, materials science and cell biology are also heading towards a point of convergence. Over the last decade, materials scientists, particularly those with a chemistry bent, have been gradually shifting attention away from their traditional domain of atomic scales. Inspired by advances in coordination and supramolecular chemistry, considerable efforts have been made to use molecules to build materials with structural features (such as pores and voids) on nanometer and larger scales. Similarly, demands for regenerative medicine and novel disease treatments have inspired great efforts by cell biologists to control cellular processes by manipulation of biological molecules and signaling pathways. These efforts naturally shift the attention of cell biologists from their traditional domain of micrometer scales towards smaller scales. These opposing trends, namely "from smaller to larger" in materials science versus "from larger to smaller" in cell biology, suggest that the two fields are converging towards a point of contact where expertise can be exchanged.

iCeMS (Institute for Integrated Cell-Material Sciences) at Kyoto University is an institute which is uniquely located at this point of contact between materials science and cell biology. Our institute is home to a rich mixture of materials chemistry and cell biology research groups, each of which aspires to be world leading in their own field while actively developing an awareness of other fields. By cultivating scientists with a broad vision and strong interdisciplinary communication skills, and supplying them with high-end facilities such as advanced microscopy and single-molecule imaging, we aim to accelerate the convergence of materials science and cell biology and ultimately achieve the following three missions: to develop materials to understand cell functions, to produce materials to control cell processes, and to create functional cell-inspired materials.

This book, entitled *Cell-Inspired Materials and Engineering*, highlights some of the breakthroughs that have emerged from iCeMS over the last few years. Each of the eleven chapters is authored by past and present iCeMS scientists and their collaborators. The chapters are divided into five parts. The first two parts deal with fundamental cell biology and materials chemistry, whereas the later parts showcase examples of "interdisciplinary convergence" in the form of materials and methods for cell control. While the diversity of topics presented may appear large at first, the attentive reader will spot common themes between the chapters and gain a sense of how the two fields are converging. They will also learn about the potential new research directions and applications that this convergence promises to bring.

All authors are sincerely thanked for their contributions and extraordinary patience with publication. All staff at Springer who coordinated with publication, particularly Merry Stuber, Sanjana Sundaram, and Vishnu Prakash, are sincerely thanked for their effort, advice, and patience. Bruna Corradetti is sincerely thanked for her constructive comments and advice as well.

Kyoto, Japan
Kyoto, Japan
17 June 2020

Dan Ohtan Wang
Daniel Packwood

Contents

Part I
Fundamental Stem Cell Science for Organ Engineering

Generation of Hepatocytes from Human ES/iPS Cells for Regenerative Medicine

Tomoki Yamashita, Kazuo Takayama, and Hiroyuki Mizuguchi

1 Introduction

The liver plays important roles in a range of functions, including glycogen storage, bile synthesis, and drug metabolism. Liver disease is thus a serious threat to human health. In recent years, liver disease, including hereditary hepatic diseases, fulminant hepatitis, hepatocirrhosis, and liver cancer, has been one of the main causes of global deaths. Although liver transplantation is the most effective treatment for these liver diseases, many patients are denied this life-saving approach due to a shortage of donors [1]. In the USA, there are more than 20,000 patients awaiting liver transplantation. Over a 1-year period, fewer than 1/3 of these patients will receive liver transplantation [2]. To overcome this donor shortage, various alternative regenerative medicine technologies have been developed, such as hepatocyte transplantation, liver tissue engineering, and bio-artificial liver devices [3]. However, all of these methods require the preparation of large amounts of hepatocytes. In recent years, the wide use of human embryonic stem (ES) cells [4] and human

T. Yamashita · K. Takayama (✉)
Laboratory of Biochemistry and Molecular Biology, Graduate School of Pharmaceutical Sciences, Osaka University, Osaka, Japan
e-mail: takayama@phs.osaka-u.ac.jp

H. Mizuguchi
Laboratory of Biochemistry and Molecular Biology, Graduate School of Pharmaceutical Sciences, Osaka University, Osaka, Japan

Laboratory of Hepatocyte Regulation, National Institute of Biomedical Innovation, Health and Nutrition, Osaka, Japan

Global Center for Medical Engineering and Informatics, Osaka University, Osaka, Japan

Integrated Frontier Research for Medical Science Division, Institute for Open and Transdisciplinary Research Initiatives (OTRI), Osaka University, Osaka, Japan

D. O. Wang, D. Packwood (eds.), *Cell-Inspired Materials and Engineering*, Fundamental Biomedical Technologies, https://doi.org/10.1007/978-3-030-55924-3_1

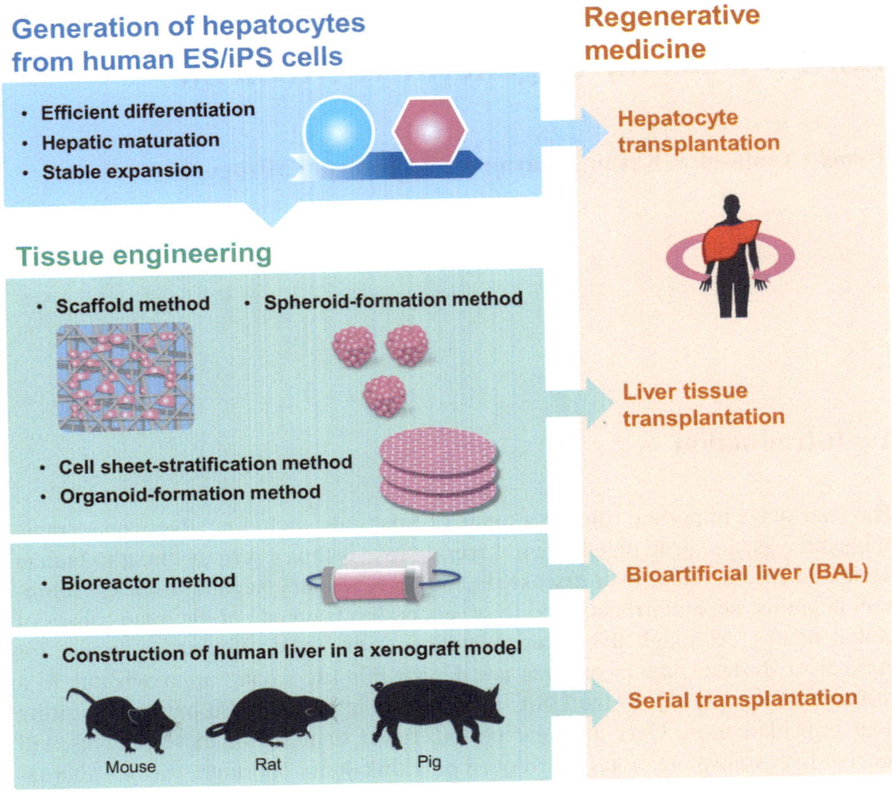

Fig. 1 Schematic overview of the contents described in this review

induced pluripotent stem (iPS) cells [5] as hepatocyte sources has been anticipated. These cells are expected to be ideal cell sources of hepatocytes because of their infinite self-renewal potential and pluripotency. In this review, we will provide an overview of the current approaches to human ES/iPS cell-based regenerative medicine for the treatment of liver diseases (Fig. 1).

2 Cell Sources of Hepatocytes

It is difficult to obtain an immunologically compatible liver for transplantation within the necessary timeframe. The difficulty of the timing is compounded by the fact that the transplantable liver can only be preserved for a few hours [6]. To resolve these problems, it is expected that human hepatocytes will be increasingly used for regenerative medicine. However, because human hepatocytes do not readily proliferate in vitro [7, 8], new approaches are needed to supply the abundant numbers of human hepatocytes required for regenerative medicine. In one such attempt, Shan et al. identified a small molecule, FPH1, that can facilitate in vitro proliferation of human hepatocytes [9]. They found that FPH1 treatment enabled a tenfold increase in the proliferation of human hepatocytes, while the increase was limited to around twofold under the standard hepatocyte culture condition. Recently, Katsuda et al. also found that a set of three small molecules, Y-27632, A-83-01, and CHIR99021, could convert mouse hepatocytes into expandable liver progenitor cells (LPCs) that, in turn, could self-replicate and differentiate into mature hepatocyte-like cells (HLCs) [10]. If this technology could be applied to human hepatocytes as well as mouse hepatocytes, these three molecule-treated human hepatocytes would be a usable cell source for regenerative medicine. At the present time, however, there is no technology that enables infinite proliferation of human hepatocytes. Therefore, in order to supply the abundance of human hepatocytes needed for regenerative medicine, it is expected that certain types of stem cells that have higher proliferation potential than human hepatocytes, such as human hepatic stem cells and human ES/iPS cells, will be utilized as alternative cell sources.

Human hepatic stem cells (human HSCs) have been obtained from human fetal liver at around 20 weeks of gestational age by sorting epithelial cell adhesion molecule-positive (EpCAM+) cells [11]. In other studies, human HSCs exhibited around 1400-fold expansion in Kubota's medium, a serum-free medium [12, 13]. Moreover, human HSCs can be differentiated into mature HLCs by embedding them in a mixture of 40% hyaluronan and 60% type II collagen and supplementing with glucagon, epidermal growth factor (EGF), and hepatocyte growth factor (HGF) [13]. Nonetheless, there are ethical limitations on the use of human fetal liver.

Human ES/iPS cells can proliferate infinitely and differentiate into most of the cell types in the human body, including hepatocytes. In addition, human iPS cells can avoid immune rejection when they are established directly from the patient. Further, HLCs derived from genetic error-corrected human iPS cells via genome-editing technology might contribute to the treatment of hereditary hepatic diseases [14–16]. In consideration of all these findings, human ES/iPS cells are expected to be an ideal hepatocyte source for regenerative medicine.

3 Human ES/iPS Cell-Derived Hepatocyte-Like Cells

3.1 Differentiation and Expansion Technology

Human ES/iPS cells can differentiate into HLCs through definitive endoderm cells and hepatoblast-like cells [17] (Fig. 2). Most of the hepatocyte differentiation protocols are divided in three steps: a definitive endoderm differentiation step, a hepatoblast differentiation step, and a hepatocyte differentiation step. Each of these differentiation steps requires certain growth factors and cytokines known to be important for liver development.

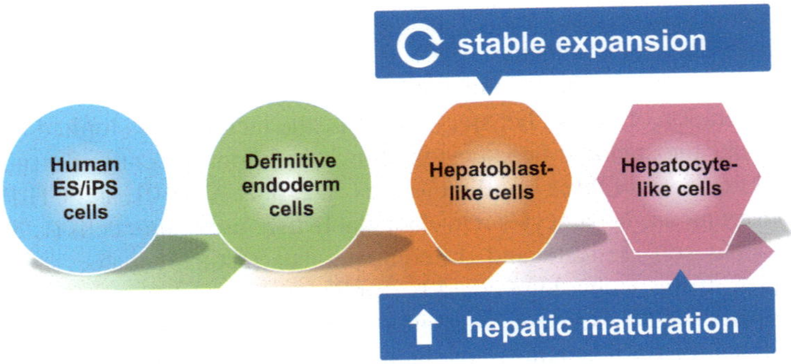

Fig. 2 Hepatocyte differentiation from human ES/iPS cells

For the definitive endoderm differentiation step, activin A is used in most of the available protocols. D'Amour et al. demonstrated that activin A could efficiently promote the differentiation of definitive endoderm from human ES cells in the presence of a low concentration of fetal bovine serum (FBS) [18]. Their immunostaining analysis showed that the percentage of SRY-related HMG-box 17 (SOX17, a definitive endoderm marker)-positive cells was higher than 80% at 5 days of differentiation in the presence of 100 ng/ml activin A and 0.5% FBS. Consistently, fluorescence-activated cell sorting (FACS) analysis showed that the percentage of C-X-C chemokine receptor type 4 (CXCR4, definitive endoderm marker)-positive cells was approximately 90%. It is also known that combination treatment with activin A and Wnt3a can promote more efficient definitive endoderm differentiation than treatment with activin A alone [19, 20].

For the hepatoblast differentiation step, fibroblast growth factor (FGF) and bone morphogenetic protein (BMP), which play important roles in liver specification, are often used. Cai et al. have demonstrated that combination treatment with FGF4 and BMP2 could efficiently promote hepatoblast differentiation from human ES cells [21]. Their immunostaining analysis showed that the percentage of alpha-fetoprotein (AFP, a hepatoblast marker)-positive cells in the human ES cell-derived hepatoblast-like cells was approximately 70%. In addition, dimethyl sulfoxide (DMSO), which exhibits histone deacetylase inhibitor activity, is sometimes used in the hepatoblast differentiation step [22, 23].

For the hepatocyte differentiation step, HGF and oncostatin M (OsM), which are known to play important roles in liver development [24, 25], are widely used. Hay et al. demonstrated that HGF and OsM treatment could efficiently promote hepatocyte differentiation from human ES cells [23]. Their western blotting analysis showed that the expression level of mature hepatocyte markers, such as albumin (ALB) and alpha-1-antitrypsin (AAT), was increased by the hepatocyte differentiation process. The immunostaining analysis showed that most of the human ES cell-derived HLCs ware positive for CK18, which is also a hepatocyte marker. In addition, the periodic-acid Schiff (PAS) staining analysis showed that the human ES cell-derived HLCs have a glycogen storage function, which is one of the functions of mature hepatocytes. Although it is possible to generate functional HLCs from human ES/iPS cells, the hepatocyte differentiation process takes a long time (generally more than 20 days) [26–28].

Hepatocyte differentiation from human ES/iPS cells is time-consuming. It is thus expected that the establishment of an expandable source of hepatoblast-like cells, which are the progenitor of HLCs, would be effective to shorten the hepatocyte differentiation process. To obtain an expandable source of hepatoblast-like cells, hepatoblast marker-positive cells were sorted in various studies. Zhao et al. reported that human ES cell-derived N-cadherin-positive hepatoblast-like cells were proliferative and could be maintained for more than 100 days on STO feeder cells [29]. Yanagida et al. reported that both CD13 and CD133-positive hepatoblast-like cells generated from human iPS cells were proliferative and could be maintained for more than 1 month on MEF feeder cells [30]. Kido et al. have also shown that human iPS cell-derived carboxypeptidase M-positive hepatoblast-like cells were proliferative and could be passaged more than five times on MEF feeder cells [31]. We also found that human ES/iPS cell-derived hepatoblast-like cells were proliferative and could be maintained for more than 3 months on recombinant human laminin-111 [32]. From these reports, hepatoblast expansion technology could make it easier than ever before to obtain an abundant supply of hepatocytes for regenerative medicine.

3.2 Maturation Technology

Human ES/iPS cell-derived HLCs (human ES/iPS-HLCs) show various hepatic functions [21, 23, 28, 33]. However, the level of these functions in human ES/iPS-HLCs is still lower than that of human hepatocytes. Also, human ES/iPS-HLCs resemble fetal hepatocytes rather than adult hepatocytes [34, 35]. To enhance the therapeutic effect of human ES/iPS-HLCs, the maturation technology of human ES/iPS-HLCs is considered to be important.

Several studies have reported that the small molecule-compound treatment could promote the maturation of human ES/iPS-HLCs [9, 36]. Ogawa et al. showed that bromo-cAMP-treatment promoted the maturation of human ES/iPS-HLCs [36]. Shan et al. demonstrated that the small molecule compounds FPH1 and FH1 respectively increased and decreased the percentage of CYP3A4- and AFP-positive cells [9]. This result suggests that FPH1 and FH1 can promote the maturation of human ES/iPS-HLCs. In addition, Siller et al. demonstrated that hepatocyte differentiation can be performed by using small molecule compounds alone, without the need for growth factors [37]. Therefore, small molecule compound-based differentiation methods are a cost-effective way to promote the maturation of human ES/iPS-HLCs.

The coculture method is also an effective strategy to generate mature human ES/iPS-HLCs, because human hepatocytes in vivo coexist with many types of cells, such as hepatic stellate cells, Kupffer cells, liver sinusoidal endothelial cells (LSECs), and biliary epithelial cells. Shiraki et al. have reported that the hepatocyte differentiation from human ES cells could be promoted by coculturing with M15 cells, which is a mouse mesoderm-derived cell line [38]. Similarly, Berger et al. reported that the hepatocyte differentiation from human iPS cells could be promoted by coculturing with Swiss 3T3-J2 cells, which are mouse embryonic fibroblasts [39]. We also showed that the maturation of human ES/iPS-HLCs could be promoted by the coculture with Swiss 3T3 cells [40]. Although these methods might be effective, use of xenogeneic cells would be problematic in a clinical setting. Recently, Koui et al. demonstrated that the hepatocyte differentiation from human iPS cells could be promoted by coculturing with hepatic stellate cells and LSECs, which were both generated from human iPS cells [41]. Taken together, these results show that hepatocyte differentiation could be promoted by coculturing with various types of cells, such as fibroblasts, hepatic stellate cells, and LSECs.

Gene transfer technology is also a potent strategy to generate mature human ES/iPS-HLCs [42–48]. Because hepatic transcription factors are known to play important roles in liver development, we have attempted to use a fiber-modified adenovirus (Ad) vector to overexpress hepatic transcription factors during hepatocyte differentiation [49, 50]. In other studies, we demonstrated that the definitive endoderm, hepatoblast, and hepatocyte differentiation steps were promoted by overexpression of SOX17 [43], hematopoietically expressed homeobox (HEX) [44], and hepatocyte nuclear factor (HNF) 4α [45], respectively. Moreover, we determined that the combination of two transcription factors, FOXA2 and HNF1α, could more strongly promote the hepatocyte differentiation than the previously used protocol [46]. Indeed, the gene expression levels of drug metabolic enzymes in FOXA2- and HNF1α-overexpressed human ES/iPS-HLCs were close to those in human hepatocytes. Similarly, Sasaki et al. showed that HNF6 overexpression enhanced CYP3A4 expression levels in human iPS-HLCs [48]. From these reports, the overexpression of hepatic transcription factors is an effective method to generate mature human ES/iPS-HLCs.

3.3 Transplantation Technology

Types of Liver-Failure Model Mouse

To conduct the basic research of human ES/iPS-HLC transplantation, several small animal models that reproduce the pathology of human liver failure are widely used. There are mainly four types of the liver-failure model mice. First one is the partial hepatectomy model mouse. Second one is the chemically induced model mice, such as carbon tetrachloride (CCl_4)-, D-galactosamine-, and Fas ligand agonist-treated mice. Third one is the mice model with radiation-induced liver injury. Fourth one is the genetically manipulated mice, such as fumarylacetoacetate hydrolase deficient (Fah−/−) mouse [51], transgenic mouse with liver-specific expression of urokinase-type plasminogen activator (uPA) [52, 53], and transgenic mouse with liver-specific expression of herpes simplex virus type 1 thymidine kinase (HSVtk) [54]. In many reports, to avoid the immune rejection of transplanted cells, these liver-failure model mice are combined with immunodeficient mice.

Engraftment Efficiency

In this decade, a number of different groups have performed human ES/iPS-HLC transplantation into liver-failure model mice. To the best of our knowledge, the 2007 report of Cai et al. was the first to examine whether human ES-HLCs could be engrafted into mouse liver [21]. The immunostaining analysis in that report revealed the presence of human nuclei- and human AAT-positive cells in the liver sections of a transplanted mouse. However, the level of engraftment efficiency was low (not quantified). In 2008, Agarwal et al. transplanted human ES cell-derived definitive endoderm cells, which are progenitor cells of hepatoblast-like cells, into CCl_4-treated immunodeficient mice [55]. The immunostaining analysis showed that the percentage of human AAT-, CD26-, or mitochondria-positive cells was lower than 1%. In 2009, Haridass et al. compared the engraftment efficiencies among human hepatocytes, human fetal LPCs, and human ES cell-derived hepatic precursor cells (human ES-HPCs) by transplanting these cells into Alb-uPA+/− immunodeficient mice [56]. The immunostaining analysis showed that the percentages of human ALB-positive cells were around 10%, 3.9%, and 0% in mice transplanted with human hepatocytes, human fetal LPCs, and human ES-HPCs, respectively. Thus the results suggested that immature human ES-HLCs were not readily engrafted into the recipient liver.

To enhance the engraftment efficiency, the maturation of human ES/iPS-HLCs has been considered to be important. Basma et al. enriched mature human ES-HLCs by sorting asialoglycoprotein receptor 1 (ASGR1)-positive cells, then transplanted the enriched mature human ES-HLCs into Alb-uPA SCID mice [57]. In the liver sections of the transplanted mice, small colonies of human CK18-expressing cells were observed (not quantified). In 2010, Si-Tayeb et al. succeeded in generating functional hepatocytes from human ES/iPS cells with high efficiency, and transplanting the human ES/iPS-HLCs into mice [58]. The immunostaining analysis showed that clusters of human ALB-positive cells were detected in the liver sections of transplanted mice. Touboul et al. have also succeeded in the efficient generation of human ES-HLCs under a chemically defined condition, and transplanted the human ES-HLCs into uPA immunodeficient mice [59]. They observed both small and large clusters of human AAT- or human ALB-positive cells in the liver sections of the transplanted mice. Collectively, these results indicate that the nearly homogeneous and functional human ES/iPS-HLCs can be engrafted into recipient liver to a certain extent.

3.4 Therapeutic Effects

It is important to evaluate the therapeutic effects of human ES/iPS-HLC transplantation as well as the engraftment efficiency. Chen et al. transplanted human iPS-HLCs into a mouse model of acute liver-failure induced by CCl_4 [60]. The survival rate was significantly improved by human iPS-HLC transplantation (from 0% to 71%). Consistently, several liver disorder markers in mouse serum, such as glutamyl oxaloacetic aminotransferase (GOT), glutamyl pyruvic aminotransferase (GPT), lactate dehydrogenase (LDH), and total bilirubin were decreased by human iPS-HLC transplantation. Woo et al. used laser microdissection to separate mature human ES-HLCs, which can abundantly uptake indocyanine green (ICG). The separated mature human ES-HLCs were then transplanted into CCl_4-treated nude mice [61]. The number of bromodeoxyuridine (BrdU)-positive mouse hepatocytes was increased by the human ES-HLC transplantation, suggesting that human ES-HLCs have the potential to promote the proliferation of host hepatocytes. In addition, platelet endothelial cell adhesion molecule (PECAM)-positive cells in the mouse liver sections were increased by the human ES-HLC transplantation, suggesting that the vascularization was promoted in the host liver. In addition, the fibrin-positive area in the mouse liver section was decreased by the transplantation of human ES-HLCs, suggesting that the human ES-HLCs would have a curative effect on liver fibrosis. Finally, the mouse serum alanine aminotransferase (ALT) level was decreased by the human ES-HLC transplantation. These results suggested that human ES/iPS-HLC transplantation would have therapeutic effects against both acute and chronic liver injury.

To further enhance the engraftment efficiency or the therapeutic effect, gene expression manipulation methods have been applied to hepatocyte transplantation. We examined whether the overexpression of FNK (a hyperactive mutant of an anti-apoptotic gene, Bcl-xL) in human iPS-HLCs could improve the efficiency of engraftment into the uPA/SCID mouse liver [62]. Our immunostaining analysis showed that the percentage of human AAT-positive cells in the liver section was upregulated by FNK transduction (from 3% to 19%), indicating that FNK transduction into human iPS-HLCs did indeed enhance the engraftment efficiency. Similarly, Möbus et al. examined whether inhibition of miR-199a-5p, which is one of the important microRNAs for hepatocyte differentiation, in human ES-HLCs could enhance the engraftment efficiency and the therapeutic effect in Fah−/− immunodeficient mice [63]. The immunostaining analysis showed that human ALB-positive cell-clusters in the mouse liver section were enlarged by the miR-199a-5p inhibition, indicating that the engraftment efficiency was enhanced by miR-199a-5p inhibition. Additionally, the mouse serum ALT and aspartate aminotransferase (AST) levels were both decreased by the miR-199a-5p inhibition, indicating that miR-199a-5p inhibition could also enhance the therapeutic effects of human ES-HLCs. The reports of human ES/iPS-HLC transplantation are summarized in Table 1. Based on their findings, gene expression manipulation might be a powerful tool for hepatocyte transplantation.

Table 1 Summary of the reports on human ES/iPS-HLC transplantation into mice

Year	Reference			Mouse	Cell type	Cell number (cells)	Transplantation method (cell suspension)	Observation (after transplantation)	Engraftment efficiency	Theraeutic effect
2007	Hepatology	Cai et al.	[21]	SCID, CCl4, 10 weeks old	Human ES-HLCs	1×10^6	Intrasplenic injection	8 weeks	Low efficienccy (not quantified)	Not evaluated
2008	Stem cells	Agarwal et al.	[55]	NOD-SCID, retrorsine, CCl4, 10 weeks old	Human ES-DE cells	1×10^6	Injected into the portal vein	4 weeks	<1%	Not evaluated
2009	Am J Pathol	Haridass et al.	[56]	Alb-uPA+/− Rag2−/−γc−/−	PHHs	1×10^6	Intrasplenic injection	3 months	10%	Not evaluated
					Human fetal LPCs				3.9%	
					Human ES-HPCs				No engraftment	
2009	Gastroenterology	Basma et al.	[57]	Alb-uPA SCID, 3 weeks old	ASGR1-sorted human ES-HLCs	1×10^6	Intrasplenic injection	At death	Small clusters of engrafted cells (not quantified)	Not evaluated
2010	Hepatology	Si-Tayeb et al.	[58]	CD1 neonate	Human ES/iPS-HLCs	3×10^5	Injected transcutaneously into the right lobe of the liver	1 week	Claster formation (not quantified)	Not evaluated

(continued)

Table 1 (continued)

Year	Reference			Mouse	Cell type	Cell number (cells)	Transplantation method (cell suspension)	Observation (after transplantation)	Engraftment efficiency	Theraeutic effect
2010	Hepatology	Touboul et al.	[59]	uPAxRag2−/− γc−/−, 5 days old	Human ES-HLCs	5×10^5	Injected into the liver	8 weeks	Small and large clusters (not quantified)	Not evaluated
2012	Hepatology	Chen et al.	[60]	NOD-SCID, CCl4, 5–8 weeks old	Human iPS-HLCs	4×10^7 (cells/kg body weight)	Intrasplenic injection	28 days	Could detect human cells (not quantified)	5 of 7 mice rescued from acute liver failure, decrease serum glutamyl oxaloacetic aminotransferase, decrease serum glutamyl pyruvic aminotransferase, decrease lactate dehydrogenase, decrease total bilirubinin
2012	Gastroenterology	Woo et al.	[61]	BALB/c nude, CCl4, 5–6 weeks old	ICG-sorted human ES-HLCs	2×10^6	Intrasplenic injection	3 days	20.2–23%	Promote endogenous host hepatocytes proliferation, promote endogenous host vascularization, promote fibrin clearance in injured nectrotic region, decrease serum ALT
								35 days	10.2%	Not evaluated

2015	Cell transplant	Nagamoto et al.	[62]	uPA+/+/SCID, 2–4 weeks old	Human iPS-HLCs	1×10^6	Intrasplenic injection	4 weeks	19%	Not evaluated
2015	J Hepatol	Möbus et al.	[63]	Fah−/−/Rag2−/−/Il2rg−/−, 1/3 partial hepatectomy 3-weeks after transplantation	miR-199a-5p inhibited human ES-HLCs	5×10^5	Intrasplenic injection	3 weeks (without hepatectomy)	Small clusters (not quantified)	Not evaluated
								7 weeks	Lerge clusters (not quantified)	Decrease serum ALT and AST level

4 Liver Tissue Engineering Using Human ES/iPS Cells

In several studies related to human ES/iPS-HLC transplantation, low engraftment efficiency and uncontrollability of the engraftment site have been reported [57, 61, 64]. Therefore, the human iPS-HLC transplantation method must be improved before its clinical use. Tissue engineering is a potent technology in cell transplantation therapy because of its potential for delivering a large number of cells to the desired organ without loss of transplanted cells. Therefore, many groups have applied the tissue engineering technology to generate liver tissue from human ES/iPS cells.

4.1 Scaffold Method

Scaffolds are porous materials for cell culture, which have been engineered to mimic the in vivo microenvironment (Fig. 3). Scaffolds have uniformly sized pore structures that can regulate cellular attachment, proliferation, differentiation, and cell–cell interaction. Several groups have applied scaffolds to promote the hepatocyte differentiation from human ES/iPS cells.

Baharvand et al. performed hepatocyte differentiation from human ES cells in 24-well tissue culture dishes containing a 3D scaffold, Vitrogen (type I collagen) (2 mm height) [65]. Under the 3D culture condition using Vitrogen, ALB and G6P were detected earlier and higher levels of urea and AFP were produced than under the 2D culture condition. Ramasamy et al. also performed hepatocyte differentiation from human ES cells in 24-well tissue culture dishes containing a 3D scaffold, AlgiMatrix (alginate from brown seaweed) (3–4 mm height) [66]. Under the 3D culture condition using AlgiMatrix, the CYP3A4 activity and urea production levels

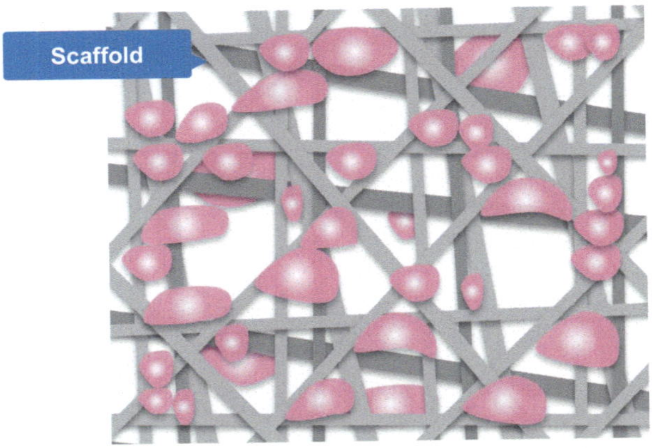

Fig. 3 Graphical image of a cell culture using the scaffold method

were higher than those under the 2D culture condition. Recently, Gieseck et al. performed hepatocyte differentiation from human iPS cells in a 6-well plate containing a 3D type I collagen scaffold concentrated by a RAFT system [67]. This system enables the concentration of cell-containing collagen hydrogel by removing interstitial fluid. In the 3D culture condition created by the RAFT system, CD26- and ABCC2-positive canalicular structures were observed, while they were not observed under the 2D culture condition. Additionally, the CYP3A4 activity under the 3D culture condition created by the RAFT system was detected for a longer period of time (over 75 days) than that under the 2D culture (45 days). From these results, the scaffolds might be useful materials for efficient hepatocyte differentiation from human ES/iPS cells.

In recent years, liver decellularization technology has been expected to provide a more suitable scaffold for hepatocytes. Liver decellularization is accomplished mainly by detergent-based perfusion [68–71]. Yagi et al. performed the decellularization of porcine livers by perfusing sodium dodecyl sulfate (SDS) and Triton X-100 for 72–96 h and 30 min, respectively [72]. After the perfusion procedure, the histological analysis showed that no DAPI-positive cells were present in the porcine liver scaffolds, indicating that the porcine livers were successfully decellularized. To perform recellularization, isolated porcine hepatocytes were infused into the liver scaffolds via the portal vein. The hematoxylin and eosin (H&E) staining demonstrated that the engraftment efficiency of attached cells was 74 ± 13%. In addition, Mirmalek-Sani et al. have transplanted pig or rat liver scaffolds into rats [73]. The numbers of white blood cells, lymphocytes, and monocytes were not significantly different between the transplanted rats and sham-operated rats, suggesting that the immunogenicity of the liver scaffolds is low. Therefore, an immune response might not be seen in either allogeneic (rat-into-rat) or xenogeneic (pig-into-rat) transplants. In the near future, it is expected that a human liver-like structure consisting of animal liver scaffolds and human ES/iPS-HLCs will be transplantable into patients without immune rejection.

4.2 Spheroid-Formation Method

Spheroids are three-dimensial cell agregates (Fig. 4). They can faithfully reproduce in vivo cell conditions rather than 2D culture conditions in terms of cellular contact. Several groups have applied the spheroid-formation method to promote hepatocyte differentiation from human ES/iPS cells.

Subramanian et al. generated hepatocyte spheroids from human ES cells by using ultralow-attachment round-bottomed 96-well plates [74]. After the plates were centrifuged for 4 min at 400 × *g*, the seeded cells settled to the bottom of the well and formed spheroids over the subsequent 2–3 days. By adapting the spheroid culture method, the percentage of ASGR1-positive cells was increased as compared with that by a 2D culture method (from 16% to 60%). Similarly, Sengupta et al. generated hepatocyte spheroids from human ES cells by using ultralow-attachment

4

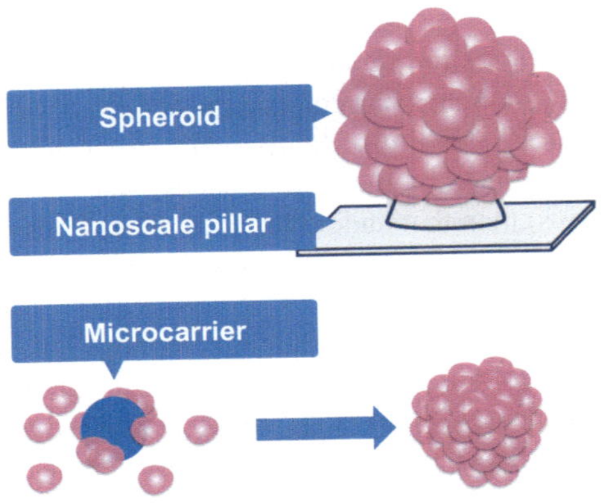

Fig. 4 Graphical image of spheroid-formation using a nanopillar plate and microcarrier

12-well cell culture plates [75]. The plates were set in a nutator (rocking mixer, at 24 rpm) to allow the added cells to form spheroids in suspension. The CYP1A2 and CYP3A4 induction potencies in hepatocyte spheroids generated from human ES cells were higher than those in HLCs cultured under a 2D condition. Although these spheroid-formation methods promoted the maturation of human ES/iPS-HLCs, necrosis might have occurred in the central part of the large spheroids. An earlier study suggested that such necrosis, if present, would have been caused mainly by the hypoxic conditions in the central part of the large spheroid (>100 μm in diameter) [76]. Therefore, to regulate the diameter of the spheroids, we generated hepatocyte spheroids from human ES/iPS cells using a nanopillar plate [77]. The Nanopillar Plate has an array of μm-scale holes at the bottom of each well, with nanoscale pillars set in the middle of each. The seeded cells aggregate on the nanoscale pillars, which promotes their formation into uniform spheroids within the holes. The size of the hepatocyte spheroids created by the Nanopillar Plate was considered to be optimal (approximately 100 μm in diameter). Importantly, more than 90% of the hepatocytes in the spheroids were alive. Further, by using the nanopillar plate, the ALB and urea secretion levels were increased as compared with those under the 2D culture condition. These results suggest that functional hepatocytes could be generated from human ES/iPS cells by using the spheroid-formation method.

In the spheroid-formation method described above, the scale of the cell culture is limited because of the need for multiwell plates. To resolve this problem, Park et al. generated hepatocyte spheroids from human ES cells in a 100 ml spinner flask by using Cytodex microcarriers (alginate beads) [78]. The culture medium was stirred at 20–25 rpm on a magnetic stir plate. The CYP3A4 activity level in hepatocyte spheroids was higher than that in HLCs generated under the 2D culture condition, suggesting that the larger-scale spheroid-formation method can also promote the maturation of human ES/iPS-HLCs. Taken together, these results suggest that the spheroid-formation method might be useful to provide abundant hepatocytes for regenerative medicine.

4.3 Cell Sheet-Stratification Method

Cell sheets can be obtained by using culture dishes made from a thermoresponsive polymer, poly(N-isopropylacrylamide) (PIPAAm) [79]. Because the recovery of the cell sheet can be accomplished only by changing the temperature, the cell sheets are recovered without losing their extracellular matrix (ECM). The stratified cell sheets have sufficient physical strength, and thus they can be easily handled in applications such as cell sheet transplantation (Fig. 5). Several groups have reported the construction of various types of cell sheets by using this thermoresponsive polymer [80–82]. Ohashi et al. succeeded in constructing human hepatocyte sheets by using PIPAAm, and demonstrated that efficient and controlled engraftment could be accomplished by the sheet transplantation [83]. We also performed the construction of human iPS-HLC sheets by using PIPAAm and transplantation into liver-failure model mice [84]. The immunostaining analysis and semiquantitative PCR analysis showed that the engraftment site in the mice with transplanted human iPS-HLC sheets was closely controlled. Moreover, as compared with the single-cell human iPS-HLC transplantation, the human iPS-HLC sheet transplantation successfully improved the survival rate of acute liver injury model mice (from 33.3% to 63.2%). From these results, it is considered that the cell sheet-stratification method would be effective for use in the treatment of liver injury.

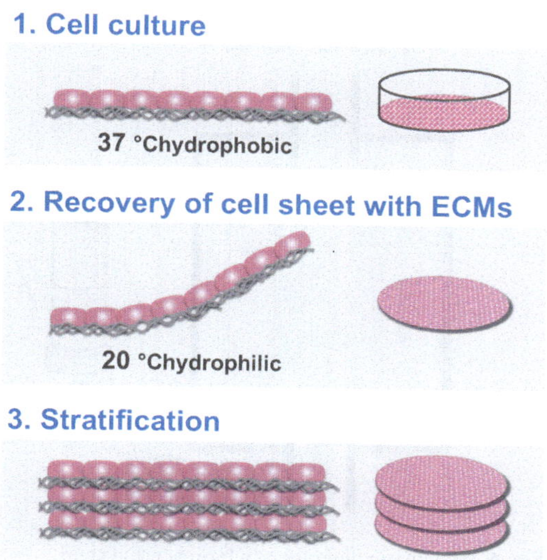

Fig. 5 Schematic diagram of the cell sheet-stratification protocol

4.4 Bioreactor Method

A bioreactor is a device that is continuously perfused with culture medium to supply a biologically active environment for cells (Fig. 6). The continuous supply of fresh medium improves the nutritional status of cells and thereby enhances their maturation. Miki et al. performed hepatocyte differentiation from human ES cells under a 3D culture condition in a hollow fiber-based bioreactor [85]. They found that the ALB and urea secretion levels were increased as compared with those under a 2D culture condition, suggesting that the bioreactor method would be effective for generating mature human ES/iPS-HLCs. In addition, bioreactors containing functional mature human ES/iPS-HLCs are expected to be utilized as extracorporeal bioartificial livers, supporting the hepatic functions of liver-failure patients ex vivo without the risk of immune rejection [86].

4.5 Organoid-Formation Method

Organoids are an organotypic 3D structure containing multiple cell types. Because organoids contain stem cells, they can spontaneously form the organ structure. Huch et al. successfully generated a liver organoid from Lgr5-positive mouse liver stem cells [87]. They also succeeded in the generation of a liver organoid from EpCAM-positive human ductal cells [88]. Takebe et al. showed that liver organoids (which they call liver buds) could be generated from a mixture of human iPS cell-derived hepatic endodermal cells, HUVECs, and human MSCs [89–91]. As compared with a sham operation, their liver-bud transplantation method improved the survival rate in a mouse model of drug-induced lethal liver failure (from 23% to

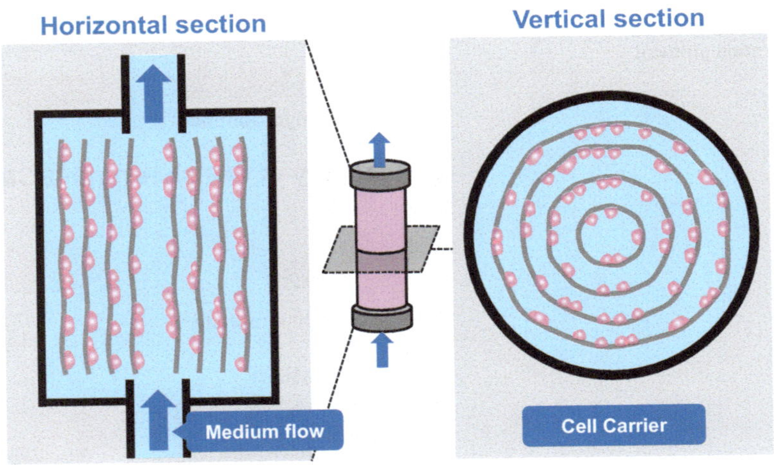

Fig. 6 Schematic diagram of a bioreactor system

94%) [89]. These results suggest that the organoid-formation method can provide expandable functional liver tissues for regenerative medicine.

4.6 Construction of Human Liver in a Xenograft Model

The construction of a human liver in an animal model is now eagerly anticipated. If a functional human liver could be created in an animal body, it could be used for serial transplantation into patients. Such a human liver might be also used as an external bio-artificial liver device to support the hepatic functions of a patient ex vivo. To develop the technology for human liver construction in an animal model, several groups have studied the generation of chimeric mice with humanized livers. There are well-known two types of chimeric mice.

The first type of chimeric mice employs *Fah−/−Rag2−/−Il2rg−/−* mice (FRG mice). FRG mice suffer from severe liver damage due to the accumulation of fumarylacetoacetic acid, which is metabolized by Fah. This damage can be blocked by administration of NTBC or by a low tyrosine diet. Transplanted human hepatocytes can be engrafted into FRG mouse livers. Azuma et al. succeeded in the generation of chimeric mice with humanized livers by transplanting human hepatocytes into the livers of FRG mice [92]. The transplanted human hepatocytes could repopulate up to 90% of the FRG mouse livers. The chimeric FRG mice with humanized livers are now commercially available from Yecuris Corporation. It is also known that chimeric FRG mice with humanized livers can be generated by transplantation of human ES/iPS cells [93].

The second type is based on uPA/SCID mice. uPA mice suffer from severe liver damage due to the expression of a urokinase-type plasminogen activator gene under the control of the *ALB* promoter. Tateno et al. have succeeded in the generation of chimeric mice with humanized livers by the transplantation of human hepatocytes into uPA/SCID mouse livers [94]. The transplanted human hepatocytes could repopulate up to 96% of the uPA/SCID mice livers. The chimeric uPA/SCID mice with humanized livers are commercially available from the PhoenixBio Company. In addition, just as for the model using FRG mice, chimeric uPA/SCID mice with humanized livers can be generated by the transplantation of human ES/iPS cells [57, 62].

Recently, Yamaguchi et al. have shown that a mouse pancreas could be created in a rat body by using blastocyst complementation technology [95]. In the same study, they demonstrated that the blood glucose levels of diabetes model mice could be normalized and maintained by the serial transplantation of mouse islets isolated from chimeric rats. In an earlier work, they found that the blastocyst complementation technology could be applied in pigs as well as mice and rats [96]. Finally, Wu et al. reported the chimeric contribution and differentiation of human iPS cells in pig embryos [97]. The Fah−/− pig was recently established [98, 99]. Taken together, these investigations suggest that a chimeric Fah−/− pig with a humanized liver might be created in the near future.

5 Conclusion

To realize the safe and therapeutically effective transplantation of human ES/iPS-HLCs, further hepatic maturation and safety assessments of human ES/iPS-HLCs are needed. To advance the hepatic maturation, it will be important to attempt different combinations of the technologies and methods described in this review. For example, the microfabricated coculture is accomplished by combining coculture technology with the scaffold method [39, 100, 101]. In addition, to ensure the safety of human ES/iPS-HLC transplantation, the risk of teratoma formation and the genetic instability of human ES/iPS-HLCs should be eliminated. If we can meet these remaining challenges, human ES/iPS-HLC transplantation will be rescuing patients with severe liver damage in the near future.

Acknowledgments We thank Dr. Eiri Ono (K-CONNEX, Kyoto University) for drawing the figures for this review. *Competing financial interests*: The authors declare no competing financial interests.

References

1. Ioannou GN, Perkins JD, Carithers RL (2008) Liver transplantation for hepatocellular carcinoma: impact of the MELD allocation system and predictors of survival. Gastroenterology 134:1342–1351. https://doi.org/10.1053/j.gastro.2008.02.013
2. Strom SC, Bruzzone P, Cai H et al (2006) Hepatocyte transplantation: clinical experience and potential for future use. Cell Transplant 15:105–110. https://doi.org/10.3727/000000006783982395
3. Allen J, Hassanein T, Bhatia SN (2001) Advances in bioartificial liver devices. Hepatology 34:447–455. https://doi.org/10.1053/jhep.2001.26753
4. Thomson JA, Itskovitz-Eldor J, Shapiro SS et al (1998) Embryonic stem cell lines derived from human blastocysts. Science (80) 282:1145–1147. https://doi.org/10.1126/science.282.5391.1145
5. Takahashi K, Tanabe K, Ohnuki M et al (2007) Induction of pluripotent stem cells from adult human fibroblasts by defined factors. Cell 131:861–872. https://doi.org/10.1016/j.cell.2007.11.019
6. Totsuka E, Fung JJ, Lee M-C et al (2002) Influence of cold ischemia time and graft transport distance on postoperative outcome in human liver transplantation. Surg Today 32:792–799. https://doi.org/10.1007/s005950200152
7. Kidambi S, Yarmush RS, Novik E et al (2009) Oxygen-mediated enhancement of primary hepatocyte metabolism, functional polarization, gene expression, and drug clearance. Proc Natl Acad Sci U S A 106:15714–15719. https://doi.org/10.1073/pnas.0906820106
8. Nahmias Y, Berthiaume F, Yarmush ML (2007) Integration of technologies for hepatic tissue engineering. In: Lee K, Kaplan D (eds) Tissue engineering. II Basics of tissue engineering. Springer, Berlin, pp 309–329
9. Shan J, Schwartz RE, Ross NT et al (2013) Identification of small molecules for human hepatocyte expansion and iPS differentiation. Nat Chem Biol 9:514–520. https://doi.org/10.1038/nchembio.1270

10. Katsuda T, Kawamata M, Hagiwara K et al (2017) Conversion of terminally committed hepatocytes to culturable bipotent progenitor cells with regenerative capacity. Cell Stem Cell 20:41–55. https://doi.org/10.1016/j.stem.2016.10.007
11. Schmelzer E, Zhang L, Bruce A et al (2007) Human hepatic stem cells from fetal and postnatal donors. J Exp Med 204:1973–1987. https://doi.org/10.1084/jem.20061603
12. Kubota H, Reid LM (2000) Clonogenic hepatoblasts, common precursors for hepatocytic and biliary lineages, are lacking classical major histocompatibility complex class I antigen. Proc Natl Acad Sci U S A 97:12132–12137. https://doi.org/10.1073/pnas.97.22.12132
13. Wang Y, Yao H, Cui C et al (2010) Paracrine signals from mesenchymal cell populations govern expansion and differentiation of human hepatic stem cells to adult liver fates. Hepatology 52:1443–1454. https://doi.org/10.1002/hep.23829
14. Takayama K, Igai K, Hagihara Y et al (2017) Highly efficient biallelic genome editing of human ES/iPS cells using a CRISPR/Cas9 or TALEN system. Nucleic Acids Res 45:5198–5207. https://doi.org/10.1093/nar/gkx130
15. Ding Q, Regan SN, Xia Y et al (2013) Enhanced efficiency of human pluripotent stem cell genome editing through replacing TALENs with CRISPRs. Cell Stem Cell 12:393–394. https://doi.org/10.1016/J.STEM.2013.03.006
16. Yu C, Liu Y, Ma T et al (2015) Small molecules enhance CRISPR genome editing in pluripotent stem cells. Cell Stem Cell 16:142–147. https://doi.org/10.1016/j.stem.2015.01.003
17. Takayama K, Mizuguchi H (2017) Generation of human pluripotent stem cell-derived hepatocyte-like cells for drug toxicity screening. Drug Metab Pharmacokinet 32:12–20. https://doi.org/10.1016/J.DMPK.2016.10.408
18. D'Amour KA, Agulnick AD, Eliazer S et al (2005) Efficient differentiation of human embryonic stem cells to definitive endoderm. Nat Biotechnol 23:1534–1541. https://doi.org/10.1038/nbt1163
19. Hay DC, Fletcher J, Payne C et al (2008) Highly efficient differentiation of hESCs to functional hepatic endoderm requires ActivinA and Wnt3a signaling. Proc Natl Acad Sci U S A 105:12301–12306. https://doi.org/10.1073/pnas.0806522105
20. Toivonen S, Lundin K, Balboa D et al (2013) Activin a and Wnt-dependent specification of human definitive endoderm cells. Exp Cell Res 319:2535–2544. https://doi.org/10.1016/j.yexcr.2013.07.007
21. Cai J, Zhao Y, Liu Y et al (2007) Directed differentiation of human embryonic stem cells into functional hepatic cells. Hepatology 45:1229–1239. https://doi.org/10.1002/hep.21582
22. Kajiwara M, Aoi T, Okita K et al (2012) Donor-dependent variations in hepatic differentiation from human-induced pluripotent stem cells. Proc Natl Acad Sci U S A 109:12538–12543. https://doi.org/10.1073/pnas.1209979109
23. Hay DC, Zhao D, Fletcher J et al (2008) Efficient differentiation of hepatocytes from human embryonic stem cells exhibiting markers recapitulating liver development in vivo. Stem Cells 26:894–902. https://doi.org/10.1634/stemcells.2007-0718
24. Kamiya A, Kinoshita T, Ito Y et al (1999) Fetal liver development requires a paracrine action of oncostatin M through the gp130 signal transducer. EMBO J 18:2127–2136. https://doi.org/10.1093/emboj/18.8.2127
25. Suzuki A, Iwama A, Miyashita H et al (2003) Role for growth factors and extracellular matrix in controlling differentiation of prospectively isolated hepatic stem cells. Development 130:2513–2524. https://doi.org/10.1242/dev.00459
26. Mallanna SK, Duncan SA (2013) Differentiation of hepatocytes from pluripotent stem cells. Curr Protoc Stem Cell Biol 1:1–15. https://doi.org/10.1002/9780470151808.sc01g04s26
27. Hannan NRF, Segeritz C-P, Touboul T, Vallier L (2013) Production of hepatocyte-like cells from human pluripotent stem cells. Nat Protoc 8:430–437. https://doi.org/10.1038/nprot.2012.153
28. Takayama K, Morisaki Y, Kuno S et al (2014) Prediction of interindividual differences in hepatic functions and drug sensitivity by using human iPS-derived hepatocytes. Proc Natl Acad Sci U S A 111:16772–16777. https://doi.org/10.1073/pnas.1413481111

29. Zhao D, Chen S, Cai J et al (2009) Derivation and characterization of hepatic progenitor cells from human embryonic stem cells. PLoS One 4:e6468. https://doi.org/10.1371/journal.pone.0006468
30. Yanagida A, Ito K, Chikada H et al (2013) An in vitro expansion system for generation of human iPS cell-derived hepatic progenitor-like cells exhibiting a Bipotent differentiation potential. PLoS One 8:e67541. https://doi.org/10.1371/journal.pone.0067541
31. Kido T, Koui Y, Suzuki K et al (2015) CPM is a useful cell surface marker to isolate expandable bi-potential liver progenitor cells derived from human iPS cells. Stem Cell Rep 5:508–515. https://doi.org/10.1016/j.stemcr.2015.08.008
32. Takayama K, Nagamoto Y, Mimura N et al (2013) Long-term self-renewal of human ES/iPS-derived hepatoblast-like cells on human laminin 111-coated dishes. Stem Cell Rep 1:322–335. https://doi.org/10.1016/j.stemcr.2013.08.006
33. Sakurai F, Kunito T, Takayama K et al (2017) Hepatitis C virus-induced innate immune responses in human iPS cell-derived hepatocyte-like cells. Virus Res 242:7–15. https://doi.org/10.1016/J.VIRUSRES.2017.09.004
34. Yu Y, Liu H, Ikeda Y et al (2012) Hepatocyte-like cells differentiated from human induced pluripotent stem cells: relevance to cellular therapies. Stem Cell Res 9:196–207. https://doi.org/10.1016/j.scr.2012.06.004
35. Baxter M, Withey S, Harrison S et al (2015) Phenotypic and functional analyses show stem cell-derived hepatocyte-like cells better mimic fetal rather than adult hepatocytes. J Hepatol 62:581–589. https://doi.org/10.1016/j.jhep.2014.10.016
36. Ogawa S, Surapisitchat J, Virtanen C et al (2013) Three-dimensional culture and cAMP signaling promote the maturation of human pluripotent stem cell-derived hepatocytes. Development 140:3285–3296. https://doi.org/10.1242/dev.090266
37. Siller R, Greenhough S, Naumovska E, Sullivan GJ (2015) Small-molecule-driven hepatocyte differentiation of human pluripotent stem cells. Stem Cell Rep 4:939–952. https://doi.org/10.1016/j.stemcr.2015.04.001
38. Shiraki N, Umeda K, Sakashita N et al (2008) Differentiation of mouse and human embryonic stem cells into hepatic lineages. Genes Cells 13:731–746. https://doi.org/10.1111/j.1365-2443.2008.01201.x
39. Berger DR, Ware BR, Davidson MD et al (2015) Enhancing the functional maturity of induced pluripotent stem cell-derived human hepatocytes by controlled presentation of cell-cell interactions in vitro. Hepatology 61:1370–1381. https://doi.org/10.1002/hep.27621
40. Nagamoto Y, Tashiro K, Takayama K et al (2012) The promotion of hepatic maturation of human pluripotent stem cells in 3D co-culture using type I collagen and Swiss 3T3 cell sheets. Biomaterials 33:4526–4534. https://doi.org/10.1016/j.biomaterials.2012.03.011
41. Koui Y, Kido T, Ito T et al (2017) An in vitro human liver model by iPSC-derived parenchymal and non-parenchymal cells. Stem Cell Rep 9:490–498. https://doi.org/10.1016/j.stemcr.2017.06.010
42. Mitani S, Takayama K, Nagamoto Y et al (2017) Human ESC/iPSC-derived hepatocyte-like cells achieve zone-specific hepatic properties by modulation of WNT signaling. Mol Ther 25:1420–1433. https://doi.org/10.1016/J.YMTHE.2017.04.006
43. Takayama K, Inamura M, Kawabata K et al (2011) Efficient and directive generation of two distinct endoderm lineages from human ESCs and iPSCs by differentiation stage-specific SOX17 transduction. PLoS One 6:e21780. https://doi.org/10.1371/journal.pone.0021780
44. Inamura M, Kawabata K, Takayama K et al (2011) Efficient generation of hepatoblasts from human ES cells and iPS cells by transient overexpression of Homeobox gene HEX. Mol Ther 19:400–407. https://doi.org/10.1038/mt.2010.241
45. Takayama K, Inamura M, Kawabata K et al (2012) Efficient generation of functional hepatocytes from human embryonic stem cells and induced pluripotent stem cells by HNF4α transduction. Mol Ther 20:127–137. https://doi.org/10.1038/mt.2011.234

46. Takayama K, Inamura M, Kawabata K et al (2012) Generation of metabolically functioning hepatocytes from human pluripotent stem cells by FOXA2 and HNF1α transduction. J Hepatol 57:628–636. https://doi.org/10.1016/j.jhep.2012.04.038
47. Nakamori D, Takayama K, Nagamoto Y et al (2016) Hepatic maturation of human iPS cell-derived hepatocyte-like cells by ATF5, c/EBPα, and PROX1 transduction. Biochem Biophys Res Commun 469:424–429. https://doi.org/10.1016/j.bbrc.2015.12.007
48. Sasaki T, Takahashi S, Numata Y et al (2013) Hepatocyte nuclear factor 6 activates the transcription of CYP3A4 in hepatocyte-like cells differentiated from human induced pluripotent stem cells. Drug Metab Pharmacokinet 28:250–259. https://doi.org/10.2133/dmpk.DMPK-12-RG-132
49. Koizumi N, Mizuguchi H, Utoguchi N et al (2003) Generation of fiber-modified adenovirus vectors containing heterologous peptides in both the HI loop and C terminus of the fiber knob. J Gene Med 5:267–276. https://doi.org/10.1002/jgm.348
50. Tashiro K, Kawabata K, Inamura M et al (2010) Adenovirus vector-mediated efficient transduction into human embryonic and induced pluripotent stem cells. Cell Reprogram 12:501–507. https://doi.org/10.1089/cell.2010.0023
51. Grompe M, Al-Dhalimy M, Finegold M et al (1993) Loss of fumarylacetoacetate hydrolase is responsible for the neonatal hepatic dysfunction phenotype of lethal albino mice. Genes Dev 7:2298–2307
52. Heckel JL, Sandgren EP, Degen JL et al (1990) Neonatal bleeding in transgenic mice expressing urokinase-type plasminogen activator. Cell 62:447–456. https://doi.org/10.1016/0092-8674(90)90010-C
53. Sandgren EP, Palmiter RD, Heckel JL et al (1991) Complete hepatic regeneration after somatic deletion of an albumin-plasminogen activator transgene. Cell 66:245–256. https://doi.org/10.1016/0092-8674(91)90615-6
54. Hasegawa M, Kawai K, Mitsui T et al (2011) The reconstituted "humanized liver" in TK-NOG mice is mature and functional. Biochem Biophys Res Commun 405:405. https://doi.org/10.1016/j.bbrc.2011.01.042
55. Agarwal S, Holton KL, Lanza R (2008) Efficient differentiation of functional hepatocytes from human embryonic stem cells. Stem Cells 26:1117–1127. https://doi.org/10.1634/stemcells.2007-1102
56. Haridass D, Yuan Q, Becker PD et al (2009) Repopulation efficiencies of adult hepatocytes, fetal liver progenitor cells, and embryonic stem cell-derived hepatic cells in albumin-promoter-enhancer urokinase-type plasminogen activator mice. Am J Pathol 175:1483–1492. https://doi.org/10.2353/ajpath.2009.090117
57. Basma H, Soto-Gutiérrez A, Yannam GR et al (2009) Differentiation and transplantation of human embryonic stem cell-derived hepatocytes. Gastroenterology 136:990–999.e4. https://doi.org/10.1053/j.gastro.2008.10.047
58. Si-Tayeb K, Noto FK, Nagaoka M et al (2010) Highly efficient generation of human hepatocyte-like cells from induced pluripotent stem cells. Hepatology 51:297–305. https://doi.org/10.1002/hep.23354
59. Touboul T, Hannan NRF, Corbineau S et al (2010) Generation of functional hepatocytes from human embryonic stem cells under chemically defined conditions that recapitulate liver development. Hepatology 51:1754–1765. https://doi.org/10.1002/hep.23506
60. Chen Y-F, Tseng C-Y, Wang H-W et al (2012) Rapid generation of mature hepatocyte-like cells from human induced pluripotent stem cells by an efficient three-step protocol. Hepatology 55:1193–1203. https://doi.org/10.1002/hep.24790
61. Woo D, Kim S, Lim H et al (2012) Direct and indirect contribution of human embryonic stem cell-derived hepatocyte-like cells to liver repair in mice. Gastroenterology 142:602–611. https://doi.org/10.1053/j.gastro.2011.11.030
62. Nagamoto Y, Takayama K, Tashiro K et al (2015) Efficient engraftment of human induced pluripotent stem cell-derived hepatocyte-like cells in uPA/SCID mice by overexpression of

FNK, a Bcl-xL mutant gene. Cell Transplant 24:1127–1138. https://doi.org/10.3727/096368914X681702
63. Möbus S, Yang D, Yuan Q et al (2015) MicroRNA-199a-5p inhibition enhances the liver repopulation ability of human embryonic stem cell-derived hepatic cells. J Hepatol 62:101–110. https://doi.org/10.1016/j.jhep.2014.08.016
64. Liu H, Kim Y, Sharkis S et al (2011) In vivo liver regeneration potential of human induced pluripotent stem cells from diverse origins. Sci Transl Med 3:82ra39. https://doi.org/10.1126/scitranslmed.3002376
65. Baharvand H, Hashemi SM, Kazemi Ashtiani S, Farrokhi A (2006) Differentiation of human embryonic stem cells into hepatocytes in 2D and 3D culture systems in vitro. Int J Dev Biol 50:645–652. https://doi.org/10.1387/ijdb.052072hb
66. Ramasamy TS, Yu JSL, Selden C et al (2012) Application of three-dimensional culture conditions to human embryonic stem cell-derived definitive endoderm cells enhances hepatocyte differentiation and functionality. Tissue Eng Part A 19:360–367. https://doi.org/10.1089/ten.tea.2012.0190
67. Gieseck RL III, Hannan NRF, Bort R et al (2014) Maturation of induced pluripotent stem cell derived hepatocytes by 3D-culture. PLoS One 9:e86372. https://doi.org/10.1371/journal.pone.0086372
68. Bao J, Shi Y, Sun H et al (2011) Construction of a portal implantable functional tissue-engineered liver using perfusion-decellularized matrix and hepatocytes in rats. Cell Transplant 20:753–766. https://doi.org/10.3727/096368910X536572
69. Baptista PM, Siddiqui MM, Lozier G et al (2011) The use of whole organ decellularization for the generation of a vascularized liver organoid. Hepatology 53:604–617. https://doi.org/10.1002/hep.24067
70. Soto-Gutierrez A, Zhang L, Medberry C et al (2011) A whole-organ regenerative medicine approach for liver replacement. Tissue Eng Part C Methods 17:677–686. https://doi.org/10.1089/ten.tec.2010.0698
71. Uygun BE, Soto-Gutierrez A, Yagi H et al (2010) Organ reengineering through development of a transplantable recellularized liver graft using decellularized liver matrix. Nat Med 16:814–820. https://doi.org/10.1038/nm.2170
72. Yagi H, Fukumitsu K, Fukuda K et al (2013) Human-scale whole-organ bioengineering for liver transplantation: a regenerative medicine approach. Cell Transplant 22:231–242. https://doi.org/10.3727/096368912X654939
73. Mirmalek-Sani S-H, Sullivan DC, Zimmerman C et al (2013) Immunogenicity of decellularized porcine liver for bioengineered hepatic tissue. Am J Pathol 183:558–565. https://doi.org/10.1016/j.ajpath.2013.05.002
74. Subramanian K, Owens DJ, Raju R et al (2014) Spheroid culture for enhanced differentiation of human embryonic stem cells to hepatocyte-like cells. Stem Cells Dev 23:124–131. https://doi.org/10.1089/scd.2013.0097
75. Sengupta S, Johnson BP, Swanson SA et al (2014) Aggregate culture of human embryonic stem cell-derived hepatocytes in suspension are an improved in vitro model for drug metabolism and toxicity testing. Toxicol Sci 140:236–245. https://doi.org/10.1093/toxsci/kfu069
76. Glicklis R, Merchuk JC, Cohen S (2004) Modeling mass transfer in hepatocyte spheroids via cell viability, spheroid size, and hepatocellular functions. Biotechnol Bioeng 86:672–680. https://doi.org/10.1002/bit.20086
77. Takayama K, Kawabata K, Nagamoto Y et al (2013) 3D spheroid culture of hESC/hiPSC-derived hepatocyte-like cells for drug toxicity testing. Biomaterials 34:1781–1789. https://doi.org/10.1016/j.biomaterials.2012.11.029
78. Park Y, Chen Y, Ordovas L, Verfaillie CM (2014) Hepatic differentiation of human embryonic stem cells on microcarriers. J Biotechnol 174:39–48. https://doi.org/10.1016/j.jbiotec.2014.01.025

79. Yamato M, Utsumi M, Kushida A et al (2001) Thermo-responsive culture dishes allow the intact harvest of multilayered keratinocyte sheets without dispase by reducing temperature. Tissue Eng 7:473–480. https://doi.org/10.1089/10763270152436517
80. Shimizu T, Yamato M, Isoi Y et al (2002) Fabrication of pulsatile cardiac tissue grafts using a novel 3-dimensional cell sheet manipulation technique and temperature-responsive cell culture surfaces. Circ Res 90:e40. https://doi.org/10.1161/hh0302.105722
81. Nishida K, Yamato M, Hayashida Y et al (2004) Functional bioengineered corneal epithelial sheet grafts from corneal stem cells expanded ex vivo on a temperature-responsive cell culture surface. Transplantation 77:379–385. https://doi.org/10.1097/01.TP.0000110320.45678.30
82. Shiroyanagi Y, Yamato M, Yamazaki Y et al (2004) Urothelium regeneration using viable cultured urothelial cell sheets grafted on demucosalized gastric flaps. BJU Int 93:1069–1075. https://doi.org/10.1111/j.1464-410X.2004.04783.x
83. Ohashi K, Yokoyama T, Yamato M et al (2007) Engineering functional two- and three-dimensional liver systems in vivo using hepatic tissue sheets. Nat Med 13:880–885. https://doi.org/10.1038/nm1576
84. Nagamoto Y, Takayama K, Ohashi K et al (2016) Transplantation of a human iPSC-derived hepatocyte sheet increases survival in mice with acute liver failure. J Hepatol 64:1068–1075. https://doi.org/10.1016/j.jhep.2016.01.004
85. Miki T, Ring A, Gerlach J (2011) Hepatic differentiation of human embryonic stem cells is promoted by three-dimensional dynamic perfusion culture conditions. Tissue Eng Part C Methods 17:557–568. https://doi.org/10.1089/ten.tec.2010.0437
86. Sakiyama R, Blau BJ, Miki T (2017) Clinical translation of bioartificial liver support systems with human pluripotent stem cell-derived hepatic cells. World J Gastroenterol 23:1974–1979. https://doi.org/10.3748/wjg.v23.i11.1974
87. Huch M, Dorrell C, Boj SF et al (2013) In vitro expansion of single Lgr5+ liver stem cells induced by Wnt-driven regeneration. Nature 494:247–250. https://doi.org/10.1038/nature11826
88. Huch M, Gehart H, van Boxtel R et al (2015) Long-term culture of genome-stable bipotent stem cells from adult human liver. Cell 160:299–312. https://doi.org/10.1016/j.cell.2014.11.050
89. Takebe T, Sekine K, Enomura M et al (2013) Vascularized and functional human liver from an iPSC-derived organ bud transplant. Nature 499:481–484. https://doi.org/10.1038/nature12271
90. Takebe T, Zhang R-R, Koike H et al (2014) Generation of a vascularized and functional human liver from an iPSC-derived organ bud transplant. Nat Protoc 9:396–409. https://doi.org/10.1038/nprot.2014.020
91. Camp JG, Sekine K, Gerber T et al (2017) Multilineage communication regulates human liver bud development from pluripotency. Nature 546:533–538. https://doi.org/10.1038/nature22796
92. Azuma H, Paulk N, Ranade A et al (2007) Robust expansion of human hepatocytes in fah−/−/Rag2−/−/Il2rg−/− mice. Nat Biotechnol 25:903–910. https://doi.org/10.1038/nbt1326
93. Carpentier A, Tesfaye A, Chu V et al (2014) Engrafted human stem cell–derived hepatocytes establish an infectious HCV murine model. J Clin Invest 124:4953–4964. https://doi.org/10.1172/JCI75456
94. Tateno C, Yoshizane Y, Saito N et al (2004) Near completely humanized liver in mice shows human-type metabolic responses to drugs. Am J Pathol 165:901–912. https://doi.org/10.1016/S0002-9440(10)63352-4
95. Yamaguchi T, Sato H, Kato-Itoh M et al (2017) Interspecies organogenesis generates autologous functional islets. Nature 542:191–196. https://doi.org/10.1038/nature21070
96. Matsunari H, Nagashima H, Watanabe M et al (2013) Blastocyst complementation generates exogenic pancreas in vivo in apancreatic cloned pigs. Proc Natl Acad Sci U S A 110:4557–4562. https://doi.org/10.1073/pnas.1222902110

97. Wu J, Platero-Luengo A, Sakurai M et al (2017) Interspecies chimerism with mammalian pluripotent stem cells. Cell 168:473–486.e15. https://doi.org/10.1016/j.cell.2016.12.036
98. Hickey RD, Lillegard JB, Fisher JE et al (2011) Efficient production of fah-null heterozygote pigs by chimeric adeno-associated virus-mediated gene knockout and somatic cell nuclear transfer. Hepatology 54:1351–1359. https://doi.org/10.1002/hep.24490
99. Hickey RD, Mao SA, Glorioso J et al (2014) Fumarylacetoacetate hydrolase deficient pigs are a novel large animal model of metabolic liver disease. Stem Cell Res 13:144–153. https://doi.org/10.1016/j.scr.2014.05.003
100. Ma X, Qu X, Zhu W et al (2016) Deterministically patterned biomimetic human iPSC-derived hepatic model via rapid 3D bioprinting. Proc Natl Acad Sci U S A 113:2206–2211. https://doi.org/10.1073/pnas.1524510113
101. Du C, Narayanan K, Leong MF, Wan ACA (2014) Induced pluripotent stem cell-derived hepatocytes and endothelial cells in multi-component hydrogel fibers for liver tissue engineering. Biomaterials 35:6006–6014. https://doi.org/10.1016/j.biomaterials.2014.04.011

Part II
Fundamental Chemistry for Cell-Inspired Materials and Imaging

Construction of Multistep Catalytic Systems in Protein Assemblies

Hiroyasu Tabe and Takafumi Ueno

1 Introduction

In nature, bionanoreactors are used for multistep reactions in catabolism, anabolism, and energy transfer. The bacterial microcompartments (BMCs) found in diverse microorganisms represent one example of well-known natural bionanoreactors [1–13]. The carboxysomes of cyanobacteria are the best studied example of BMCs [9–13]. Carboxysomes are the core component of the carbon-concentrating mechanism used for cellular uptake of inorganic carbon and the accumulation of hydrogen carbonate (HCO_3^-). Structural studies of carboxysomes by electron microscopy and single-crystal X-ray structure analyses have indicated that they consist of a scaffold composed of a few thousand protein subunits and multiple catalytic components such as ribulose-1,5-bisphosphate carboxylase-oxygenase (RuBisCO), carbonic anhydrase and several coenzymes. In this context, the main challenge in the development of artificial bionanoreactors is controllable immobilization of multiple catalytic components within a single scaffold.

Immobilization of catalytic components in proteins has been the focus of much attention in preparation of artificial enzymes. Since Whitesides and coworkers immobilized Rh^I-biotin complex in avidin for development of enantioselective hydrogenation reactions, various methods have been developed for immobilization of catalytic components within proteins for construction of artificial enzymes [14–22]. Recently, artificial enzymes have been used in a solution containing other catalysts to develop cascade reactions [23–29]. For example, Ward and coworkers used

H. Tabe
Research Center for Artificial Photosynthesis (ReCAP), Osaka City University, Osaka, Japan

T. Ueno (✉)
School of Life Science and Technology, Tokyo Institute of Technology, Yokohama, Japan
e-mail: tueno@bio.titech.ac.jp

D. O. Wang, D. Packwood (eds.), *Cell-Inspired Materials and Engineering*, Fundamental Biomedical Technologies, https://doi.org/10.1007/978-3-030-55924-3_2

Ir^{III}-modified streptavidin, monoamine oxidase, and catalase in a cascade reaction for amine deracemization coupled to formate oxidation and water reduction [28]. Pordea and coworkers also reported a cascade reaction for enantioselective reduction of ketones using NADPH as an electron donor coupled to $NADH^+$ reduction by formate ion using a mixture of unmodified and Rh^{III}-modified alcohol dehydrogenase. In this system, the same protein scaffold is used for the different reactions [29]. However, immobilization of multiple catalytic components within a single protein remains problematic due to synthetic challenges, size limits of nanospaces and mutual inactivation via unintended interactions between the catalytic components.

Protein assemblies are promising for use as scaffolds for multistep reactions such as the reactions catalyzed within natural BMCs, because multiple catalytic components can be immobilized in large interior nanospaces among periodically aligned protein subunits. Recently, the techniques used in preparation of artificial enzymes have been applied to protein assemblies such as protein cages and porous protein crystals with interior nanospaces [30–32]. Several protein cages such as ferritin (Fr) and porous protein crystals such as hen egg white lysozyme (HEWL) crystals can be routinely and abundantly obtained and functionalized due to their high stability (Fig. 1) [33–37]. Moreover, theoretical methods have predicted that the physical properties of interior nanospaces provide unique chemical environments for selective catalysis. In this chapter, we review the techniques developed for modification of protein assemblies by incorporation of single or multiple catalytic chemical components with a focus on recent progress in development of multistep catalysis within protein assemblies.

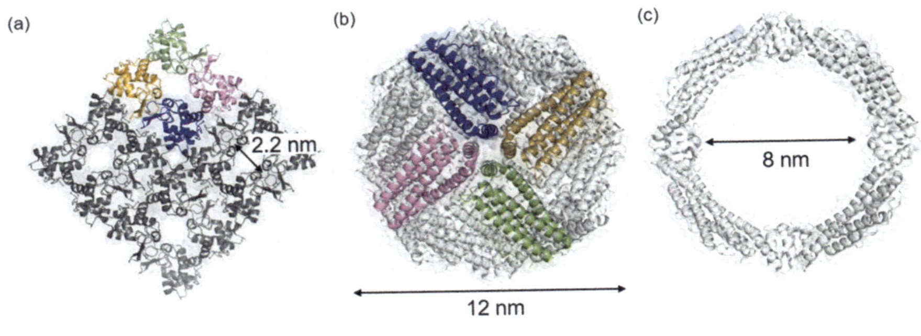

Fig. 1 The overall structure of (**a**) a porous hen egg white lysozyme (HEWL) crystal in the tetragonal form (PDB ID: 193L). (**b**) The whole structure and (**c**) the interior view of a cage of apoferritin from horse liver (apo-rHLFr) (PDB ID: 1DAT)

2 Immobilization of Exogenous Molecules and Nanoparticles Within Protein Assemblies

Modification techniques established for protein monomers are applicable to protein assemblies. Most of the modification techniques include immobilization of exogeneous molecules by forming linkages to side chains of amino acid residues exposed on the interior nanospaces of protein assemblies. The side chains used for molecular immobilization should be carefully chosen, because the exogenous molecules tend to disrupt inter-subunit interactions of the protein assemblies.

Covalent modification of amino acid residues on the interior nanospaces of porous protein crystals has been demonstrated using myoglobin (Mb) crystals, which provide a pore with a diameter of 40 Å [38]. Analysis of the structure of the Mb crystal led to replacement of several amino acids exposed on the interior nanospaces with cysteine. The cysteine residues were then modified to produce several maleimide derivatives via a cysteine-maleimide coupling reaction. The modified myoglobin was successfully crystallized retaining the original *P*6 lattice, resulting in 100% modification yield. Another technique for covalently modifying porous protein crystals with precrystallized proteins modified by maleimide derivatives, was first established for protein crystals known as polyhedra, which are naturally obtained in insect cells infected by cypovirus [39]. Cysteine residues were introduced on the outer surface of polyhedra by site-directed mutagenesis. The polyhedra mutant was immersed in a solution containing propargylmaleimide. The distribution and density of the alkyne moiety per surface area can be tuned by selecting amino acids for replacement with cysteine. The obtained alkyne sites were used for the display of sugar molecules via Cu^{I}-catalyzed azide–alkyne cycloaddition.

Metal ions and complexes are immobilized in protein assemblies via direct coordination of the side chains of amino acids. The imidazole group of histidine residues, the thiolate group of cysteine residues, and the carboxylate group of aspartate and glutamate often act as the metal-binding sites. A typical technique for immobilization of metal complexes is replacement of the native heme cofactor of hemoproteins [40–45]. Because hemoproteins have a cavity to accommodate the presence of heme, removal of heme to provide an apo-hemoprotein provides an empty cavity suitable for immobilization of various macrocyclic complexes. Another example is seen in immobilization of metal ions and complexes in interior nanospaces of crystalline or caged protein assemblies [46–57]. Interestingly, immobilization of metal ions is initiated by the cooperative dynamics of side chains and water molecules participating in hydrogen bond networks around the metal ions, as observed in the successive accumulation of Rh^{III}- and Au^{III}-ion in a porous hen egg white lysozyme (HEWL) crystal and an apoferritin cage, respectively [50, 51]. Coimmobilization of multiple metal ions has been reported. For example, Co^{II} ions and Pt^{II} ions were coimmobilized in a single nanospace in HEWL crystals because each metal ion is coordinated by different amino residues [57].

The immobilized metal ions and complexes have been converted to inorganic nanomaterials such as metal or metal oxide nanoparticles [57–71]. Immobilization of nanoparticles in a single protein is usually difficult due to the absence of nanospaces. However, the nanoparticles can be immobilized in protein assemblies where larger interior nanospaces are available. The nanoparticles precisely synthesized in protein assemblies exhibit various functional properties such as magnetism, electrical conductivity, and photoluminescence. These properties are relatively rare in protein assemblies synthesized via conventional methods.

3 Electron Transfer Reactions in Porous Protein Crystals

A well-known class of reactions promoted by multiple functional molecules is a photoinduced electron-transfer involving a photosensitizer, an electron mediator, and an electron acceptor. Protein engineering has been employed to provide artificial electron transfer systems in efforts elucidate the general mechanism of photoinduced electron-transfer and to develop an artificial photosynthesis system. For example, a photoinduced electron-transfer system with long-lived charge separation was constructed in the interior nanospace of porous myoglobin (Mb) crystals [72]. In this system, zinc porphyrin (ZnP) acts as a photosensitizer, $[Ru_3(\mu_3\text{-}O)(\mu\text{-}CH_3COO)_6(H_2O)_3]$ (Ru_3O) acts as an electron acceptor and methyl viologen (MV) acts as an electron mediator (Fig. 2a). ZnP-substituted myoglobin (ZnMb) was obtained by the cofactor replacement method. ZnMb was successfully crystallized and confirmed to retain the original Mb *P*6 lattice with a pore diameter of 40 Å. Then, the ZnMb crystals were soaked in a buffer solution containing Ru_3O^+. Single-crystal X-ray structure analysis indicated that the ZnMb monomer has two Ru_3O^+-binding sites at Nεs of His12 and His81 located in crevices formed by inter-

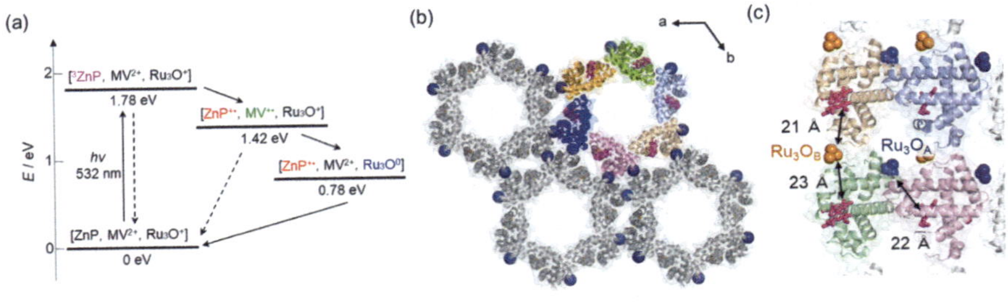

Fig. 2 (**a**) Reaction and energy diagram for a photoinduced electron-transfer system using zinc porphyrin (ZnP) as a photosensitizer, $[Ru_3(\mu_3\text{-}O)(\mu\text{-}CH_3COO)_6(H_2O)_3]$ (Ru_3O) as an electron acceptor, and methyl viologen (MV) as an electron mediator. (**b**) The overall structure of a ZnMb crystal containing Ru_3O^+ (PDB ID: 3OHQ). The position of Ru_3O bound to His12 (Ru_3O_A) is indicated by a dark blue sphere and the position of His81 (Ru_3O_B) is indicated by an orange sphere. (**c**) Magnifications of distances between Ru_3O moieties and ZnP in the crystal packing

molecular associations within the ZnMb crystal (Ru_3O_A and Ru_3O_B, respectively, Fig. 2b). An electron-transfer reaction was examined by photoirradiation of the ZnMb crystal ($\lambda = 532$ nm) in a buffer solution containing MV^{2+}. The ZnMb crystal has a charge-separated state with a half-life up to 2800 times longer than that of the model system in an organic solution. Furthermore, electron transfer was not observed in a buffer solution containing uncrystallized ZnMb, Ru_3O^+ and MV^{2+}. This is because of (1) the site-selective immobilization of ZnP and Ru_3O^+ with a distance of ca. 2 nm, (2) the efficient electron transfer by the fast diffusion of MV and (3) the low-level reorganization energy of Ru_3O^+ in the ZnMb crystals (Fig. 2c).

4 Catalytic Reactions in Porous Protein Crystals

4.1 *Selective Catalytic Reactions in Porous Protein Crystals*

Crystals of hen egg white lysozyme (HEWL), which can be routinely and abundantly crystallized to obtain yields of several hundred milligrams, are suitable for use as templates for construction of heterogeneous catalysts. Lu and coworkers reported heterogeneous catalysis of porous HEWL crystals containing gold nanoparticles for reduction of *p*-nitrophenol in the presence of sodium borohydride ($NaBH_4$) and as a sacrificial reductant [73]. The gold nanoparticles were synthesized via disproportionation of gold(I) ions immobilized in the HEWL crystals and characterized by transition electron microscopy. Enhanced catalytic activity was observed with precise tuning of the size of the gold nanoparticles. Silver or palladium/gold alloy nanoparticles synthesized in HEWL crystals have also been found to act as efficient reduction catalysts with catalytic reactivity comparable to that of gold nanoparticles [74–78].

The first example of selective heterogeneous catalysis using HEWL crystals is enantioselective hydrogenation of acetophenone derivatives by an immobilized organoruthenium(II) complex [79]. HEWL crystals in the tetragonal form (T-HEWL) and in the orthorhombic form (O-HEWL) were stabilized by a cross-linking treatment using glutaraldehyde (CL-T-HEWL and CL-O-HEWL, respectively). The crystals were immersed in a buffer solution containing $[Ru^{II}(benzene)Cl_2]_2$ to prepare **Ru^{II}(benzene)•CL-T-HEWL** and **Ru^{II}(benzene)•CL-O-HEWL**. The crystal structures of **Ru^{II}(benzene)•CL-HEWLs** were determined by single-crystal X-ray structure analysis (Fig. 3). Mononucleated $Ru^{II}(benzene)Cl_2$ is bound to Nε of His15 with an Ru–Nε bond length of 2.5 Å and 2.2 Å for **Ru^{II}(benzene)•CL-T-HEWL** and **Ru^{II}(benzene)•CL-O-HEWL**, respectively. Interestingly, the Ru^{II}(benzene) Cl_2–histidine moiety is stabilized by participating in hydrogen-bonding networks with surrounding amino acid residues in the solvent channels. Heterogeneous catalysis of the HEWL crystals was investigated in a transfer hydrogenation of acetophenone derivatives in the presence of sodium formate as a hydride donor. The catalytic conversion and enantioexcess (*ee*) values in the reactions using **Ru^{II}(benzene)•CL-**

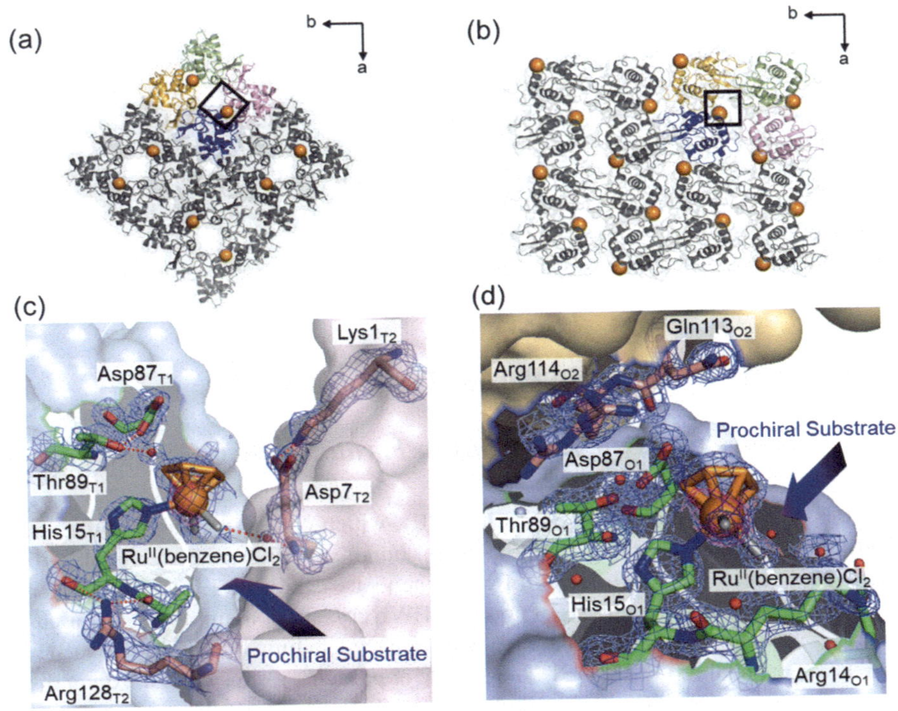

Fig. 3 (**a**, **b**) The overall structure of (**a**) Ru^{II}(benzene)Cl_2•CL-T-HEWL and (**b**) Ru^{II}(benzene) Cl_2•CL-O-HEWL (PDB ID: 3W6A and 4J7V, respectively). The positions of the Ru atoms are indicated by orange spheres. (**c**, **d**) Magnifications of Ru binding sites indicated by the squares in (**a**) and (**b**), respectively. Anomalous difference Fourier maps at 3.0 σ indicate the positions of individual Ru atoms (magenta). Selected 2Fo–Fc electron-density maps at 1.0 σ are indicated by the dark blue mesh

HEWLs as heterogeneous catalysts were generally found to be higher than the *ee* values obtained when the Ru^{II}(benzene)Cl_2-HEWL complex is in solution because the hydrogen bonding networks stabilizing the Ru^{II}(benzene)Cl_2–histidine moiety are absent in solution. Surprisingly, **Ru^{II}(benzene)•CL-T-HEWL** and **Ru^{II}(benzene)•CL-O-HEWL** exhibit the opposite enantioselectivity for several substrates. The different structures of hydrogen-bonding networks around the Ru^{II} complex affect the active site–substrate interactions, leading to altered enantioselectivity based on the crystal forms.

Confined nanospaces of porous protein crystals are also suitable for other selective reactions. For example, heterogeneous catalysis for selective oxidation of olefins has been reported using a cross-linked periplasmic nickel-binding protein crystal immobilizing the derivatives of an Fe-EDDA complex (EDDA = *N*, *N′*-

ethylenediaminediacetate) [80]. The complexes are immobilized in the nickel-binding protein crystal by hydrogen bonding networks between an arginine residue and π-stacking interactions between tryptophan and tyrosine residues [81]. The engineered crystals catalyze the selective oxidation of a series of styrene derivatives to aldehydes or ketones without forming epoxides, diols, or carboxylic acids. The crystals are stable heterogeneous catalysts exhibiting a turnover number of at least 30,000 without any loss of activity. The engineered cross-linked nickel-binding protein crystals can also catalyze chemo- and regioselective hydroxychlorination of a series of styrene derivatives.

4.2 Photocatalytic Systems Constructed in Porous Protein Crystals

Coimmobilization of multiple molecules by linkage to different functional groups in solvent channels of CL-HEWLs is essential for construction of a heterogeneous catalytic system promoted by several catalytic components. A heterogeneous catalyst for photocatalytic hydrogen (H_2) evolution was constructed using CL-HEWLs, in which rose bengal functions as an organic photosensitizer and platinum nanoparticles (PtNPs) act as a hydrogen evolution catalyst (Fig. 4a) [82]. CL-T-HEWL was immersed in a buffer solution containing rose bengal and hexachloroplatinic(IV) acid (H_2PtCl_6), which is a precursor of PtNPs. Single-crystal X-ray crystallography suggests that the amine and imidazole groups derived from Lys1 and His15, respectively, coordinate to the chloroplatinates (Fig. 4b–e). Although unambiguous positions of rose bengal molecules were not identified accurately, they are expected to be immobilized at positively charged areas with guanidyl groups derived from Arg5, Arg125 and Arg21 (Fig. 4c). The distance between the binding sites for rose bengal molecules and chloroplatinates are close enough to allow electron transfer (ca. 2 nm). Photocatalytic H_2 evolution was performed by visible-light irradiation of a buffer solution containing CL-T-HEWL immobilizing rose bengal, H_2PtCl_6 and β-nicotinamide adenine dinucleotide (NADH) as a sacrificial reductant. Prior to H_2 evolution, PtNPs were deposited by reduction of chloroplatinates in the vicinity of rose bengal, resulting in the enhancement of H_2 evolution due to the efficient electron transfer. A homogeneous catalytic system using a buffer solution containing PtNPs, rose bengal and NADH without CL-T-HEWL resulted in lower H_2 yield than the same combination with CL-T-HEWL. Moreover, the homogeneous mixture of PtNPs, rose bengal, NADH, and CL-HEWL in a buffer solution also resulted in a lower H_2 yield. These results suggest that the immobilization of each component in CL-T-HEWL enhances the efficiency of electron transfer and the stability of the catalytic system.

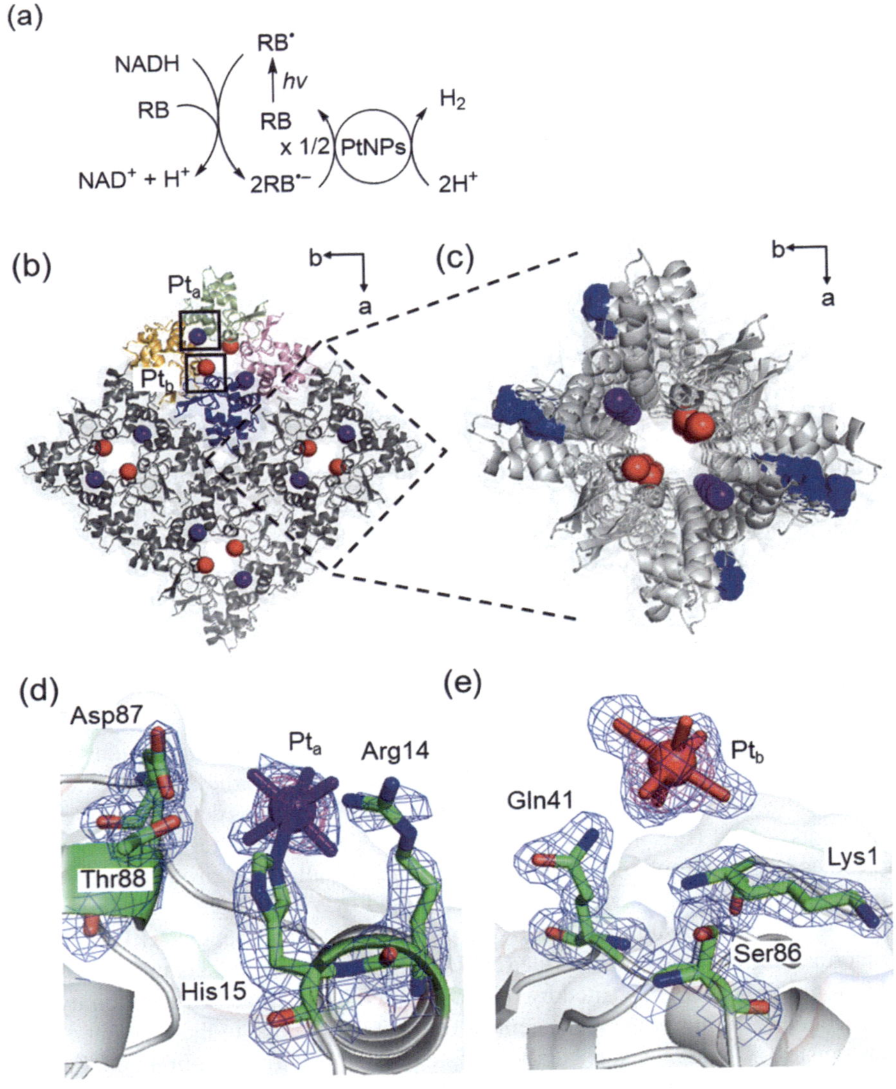

Fig. 4 (**a**) Overall photocatalytic cycle of hydrogen (H_2) evolution. *RB* rose bengal, *PtNPs* platinum nanoparticles, *NADH* β-nicotinamide adenine dinucleotide. (**b**) The overall structure of CL-T-HEWL immobilizing RB and chloroplatinate(IV) ions (PDB ID: 5YKY). The positions of the Pt atoms bound to His15 (Pt_a) and Lys1 (Pt_b) are indicated by purple and magenta spheres, respectively. (**c**) Magnifications of a pore indicated by a dashed square in (**b**). Blue residues indicate positively charged areas on the surfaces of the solvent channels. (**d, e**) Magnifications of two Pt binding sites are indicated by the black squares in (**b**). Anomalous difference Fourier maps at 3.0 σ indicate the positions of individual Ru atoms (magenta). Selected 2Fo–Fc lectron-density maps at 1.0 σ are indicated by the dark blue mesh

5 Tandem Catalytic Reactions in Protein Cages

Protein cages possess nanospaces where various exogenous molecules can be immobilized [51–56]. Recently, protein cages such as heat-shock protein, tobacco mosaic virus, and ferritin have been used as templates for development of integrated catalytic systems, because the inner surfaces of their cage-like structures can be conveniently modified [83–91]. These nanospaces are attractive for immobilizing multiple catalytic molecules including metal complexes, metal nanoparticles and enzymes for tandem reactions. The first example of tandem catalysis in protein cages was reported by Douglas and coworkers, who genetically encapsulated two different enzymes in a single nanospace of a protein cage known as the virus-like particle [88]. Virus-like particles derived from the capsid of bacteriophage P22 are uniform cages, each composed of 420 monomers of coat proteins. The assembly has an interior nanospace with a diameter of 50 nm filled by 100–300 copies of scaffolding proteins. Coexpression of the scaffold proteins and foreign protein enables encapsulation of the foreign proteins in the interior nanospaces of virus-like particles. A tandem reaction catalyzed by galactokinase (GALK), glucokinase (GLUK) and β-glucosidase (CelB) enzymes encapsulated in virus-like particles has been reported. The three enzymes are linked together with flexible peptide linkers, which allow quantitative encapsulation of the enzymes with proper folding structures. This system converts lactose to galactose-1-phosphate and glucose-6-phosphate via the formation of galactose and glucose, respectively. Other examples of coencapsulation in a protein cage have also been reported by using cowpea chlorotic mottle virus (CCMV) and the bacteriophage MS2 capsid protein [89].

Heterogeneous catalysis of the immobilized enzymes has been reported by preparing a hierarchical assembly based on bacteriophage P22 [90, 91]. The assembly of bacteriophage P22 with two- or three-dimensional structure is driven by electrostatic interactions in the presence of the polyamidoamine dendrimer. Thus, fabrication of a coassembly composed of P22s containing ketoisovalerate decarboxylase (**P22-KivD**) or alcohol dehydrogenase A (**P22-AdhA**) is a promising method for preparing a heterogeneous catalyst for a tandem reaction. The structure of the P22 assembly (**P22-KivD/P22-AdhA**) was observed by transition electron microscopy and small-angle X-ray scattering (SAXS) measurements. The assembly catalyzes a tandem reaction in which α-ketoisovalerate is first converted to isobutyraldehyde by KivD and then to isobutanol by AdhA in the presence of thiamine pyrophosphate and NADH as coenzymes. The overall yield of the product in the reaction system containing the assembly of **P22-KivD/P22-AdhA** was found to be higher than the yield produced by a system containing a physical mixture of **P22-KivD** and **P22-AdhA**. Interestingly, the assembly could be easily recovered and recycled with gentle centrifugation and used as the catalyst for repetitive runs.

Encapsulation of multiple metal-complex catalysts has also been reported. Coencapsulation of pentamethylcyclopentadienyliridium(III) ion ($Ir^{III}Cp^*$) and allylpalladium(II) ion (Pd^{II}(allyl)) was demonstrated using the recombinant L-chain apoferritin from horse liver (apo-rHLFr) as a scaffold [92]. First, $Ir^{III}Cp^*$ was

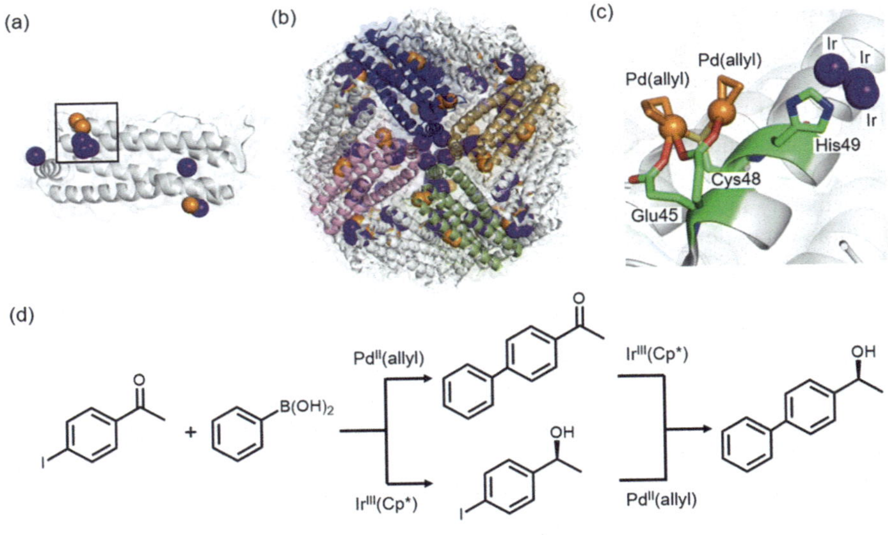

Fig. 5 (**a**) A subunit and (**b**) the overall structure of Ir^{III}Cp*/Pd^{II}(allyl)·apo-rHLFr (PDB ID: 5HQO). The positions of the Pd atoms are indicated by orange spheres. The positions of the Ir atoms are represented by the purple spheres. (**c**) Magnifications of metal-binding sites indicated by the black square in (**a**). (**d**) Scheme of a tandem reaction of 4-iodoacetophenone and phenylboronic acid

encapsulated by the reactions of pentamethylcyclopentadienyliridium(III) dichloride dimer ([Ir^{III}Cp*Cl$_2$]$_2$) and apo-rHLFr (**Ir^{III}Cp*·apo-rHLFr**). Then, allylpalladium(II) chloride dimer ([Pd^{II}(allyl)Cl]$_2$) was reacted with **Ir^{III}Cp*·apo-rHLFr** to obtain **Ir^{III}Cp*/Pd^{II}(allyl)·apo-rHLFr**. Single-crystal X-ray structure analyses suggested that Ir^{III}Cp* and Pd^{II}(allyl) can be simultaneously immobilized at the different positions within a single apo-rHLFr (Fig. 5a, b). The distance between Pd^{II}(allyl) bound to Nε of His49 and Ir^{III}Cp* bound to Sγ of Cys48 and Oδ of Glu45 in **Ir^{III}Cp*/Pd^{II}(allyl)·apo-rHLFr** is enough close (ca. 8 Å) to promote an efficient tandem reaction (Fig. 5c). Surprisingly, a reversed procedure, in which the Pd^{II}(allyl) encapsulation in apo-rHLFr (**Pd^{II}(allyl)·apo-rHLFr**) followed by the Ir^{III}Cp* encapsulation in apo-rHLFr, was found to be unsuccessful because of the occupation of binding sites for Ir^{III}Cp* by Pd^{II}(allyl). This suggests that a particular reaction sequence is important for co-encapsulation of multiple metal complexes in apo-rHLFr. A tandem reaction was performed by adding 4-iodoacetophenone and phenylboronic acid to a buffer solution containing **Ir^{III}Cp*/Pd^{II}(allyl)·apo-rHLFr** (Fig. 5d). The product of the tandem reaction, 1-(4-phenyl)phenylethanol, was obtained with moderate enantioselectivity. On the other hand, 4-phenylacetophenone was obtained in the reaction system using [Ir^{III}(Cp*)Cl$_2$]$_2$ and [Pd^{II}(allyl)Cl]$_2$ as the catalysts, suggesting that encapsulation of both metal complexes within a single ferritin molecule is essential for the tandem reaction. The reaction using a mixture of **Ir^{III}Cp*·apo-rHLFr** and **Pd^{II}(allyl)·apo-rHLFr** as the catalysts resulted in the formation of 1-(4-phenyl)phenylethanol with comparable yield and selectivity as **Ir^{III}Cp*/Pd^{II}(allyl)·apo-rHLFr**. However, the composition of intermediates was found to be slightly different. This result indicates that the reaction mechanism varies among the systems using **Ir^{III}Cp*/Pd^{II}(allyl)·apo-rHLFr** and a mixture of **Ir^{III}Cp*·apo-rHLFr** and **Pd^{II}(allyl)·apo-rHLFr** as the catalysts.

6 Conclusions

We have described catalytic systems constructed in interior nanospaces of protein assemblies. Immobilization of multiple molecules or nanoparticles in the interior nanospaces enables complex reactions such as electron transfer reactions, photocatalytic reactions and tandem reactions. These reactions are promoted by precise coimmobilization of the catalytic components with specifically arranged molecular interactions in the internal nanospaces of protein assemblies. More sophisticated catalytic systems can be provided by employing protein assemblies and artificial bacterial microcompartments as scaffolds for multistep reactions.

References

1. Kerfeld CA, Aussignargues C, Zarzycki J, Cai F, Sutter M (2018) Bacterial microcompartments. Nat Rev Microbiol 16:277–290. https://doi.org/10.1038/nrmicro.2018.10
2. Bernhardsgrütter I, Vögeli B, Wagner T, Peter DM, Cortina NS, Kahnt J, Bange G, Engilberge S, Girard E, Riobé F, Maury O, Shima S, Zarzycki J, Erb TJ (2018) The multicatalytic compartment of propionyl-CoA synthase sequesters a toxic metabolite. Nat Chem Biol 14:1127–1132. https://doi.org/10.1038/s41589-018-0153-x
3. Sutter M, Greber B, Aussignargues C, Kerfeld CA (2017) Assembly principles and structure of a 6.5-MDa bacterial microcompartment shell. Science 356:1293–1297. https://doi.org/10.1126/science.aan3289
4. Pieters BJGE, van Eldijk MB, Nolte RJM, Mecinović J (2016) Natural supramolecular protein assemblies. Chem Soc Rev 45:24–39. https://doi.org/10.1039/c5cs00157a
5. Tanaka S, Sawaya MR, Yeates TF (2010) Structure and mechanisms of a protein-based organelle in *Escherichia coli*. Science 327:81–84. https://doi.org/10.1126/science.1179513
6. Savage DF, Afonso B, Chen AH, Silver PA (2010) Spatially ordered dynamics of the bacterial carbon fixation machinery. Science 327:1258–1261. https://doi.org/10.1126/science.1186090
7. Crowley CS, Sawaya MR, Bobik TA, Yeates TF (2008) Structure of the PduU shell protein from the Pdu microcompartment of salmonella. Structure 16:1324–1332. https://doi.org/10.1016/j.str.2008.05.013
8. Kerfeld CA, Sawaya MR, Tanaka S, Nguyen CV, Phillips M, Beeby M, Yeates TF (2005) Protein structures forming the shell of primitive bacterial organelles. Science 309:936–938. https://doi.org/10.1126/science.1113397
9. Cameron JC, Wilson SC, Bernstein SL, Kerfeld CA (2013) Biogenesis of a bacterial organelle: the carboxysome assembly pathway. Cell 155:1131–1140. https://doi.org/10.1016/j.cell.2013.10.044

10. Tanaka S, Sawaya MR, Phillips M, Yeates TF (2009) Insights from multiple structures of the shell proteins from the beta-carboxysome. Protein Sci 18:108–120. https://doi.org/10.1002/pro.14
11. Yeates TF, Kerfeld CA, Heinhorst S, Cannon GC, Shively JM (2008) Protein-based organelles in bacteria: carboxysomes and related microcompartments. Nat Rev Microbiol 6:681–691. https://doi.org/10.1038/nrmicro1913
12. Tanaka S, Kerfeld CA, Sawaya MR, Cai F, Heinhorst S, Cannon GC, Yeates TF (2008) Atomic-level models of the bacterial carboxysome shell. Science 319:1083–1086. https://doi.org/10.1126/science.1151458
13. Yeates TF, Tsai Y, Tanaka S, Sawaya MR, Kerfeld CA (2007) Self-assembly in the carboxysome: a viral capsid-like protein shell in bacterial cells. Biochem Soc Trans 35:508–511. https://doi.org/10.1042/bst0350508
14. Watanabe Y, Aiba Y, Ariyasu S, Abe S (2020) Molecular design and regulation of metalloenzyme activities through two novel approaches: ferritin and P450s. Bull Chem Soc Jpn 93:379–392. https://doi.org/10.1246/bcsj.20190305
15. Reetz MT (2019) Directed evolution of artificial metalloenzymes: a universal means to tune the selectivity of transition metal catalysts? Acc Chem Res 52:336–344. https://doi.org/10.1021/acs.accounts.8b00582
16. Yu Y, Hu C, Xia L, Wang JY (2018) Artificial metalloenzyme design with unnatural amino acids and non-native cofactors. ACS Catal 8:1851–1863. https://doi.org/10.1021/acscatal.7b03754
17. Matsuo T, Miyake T, Hirota S (2019) Recent developments on creation of artificial metalloenzymes. Tetrahedron Lett 60:151226. https://doi.org/10.1016/j.tetlet.2019.151226
18. Markel U, Sauer DF, Schiffels J, Okuda J, Schwaneberg U (2019) Towards the evolution of artificial metalloenzymes-a protein engineer's perspective. Angew Chem Int Ed 58:4454–4464. https://doi.org/10.1002/anie.201811042
19. Roelfes G (2019) LmrR: a privileged scaffold for artificial metalloenzymes. Acc Chem Res 52:545–556. https://doi.org/10.1021/acs.accounts.9b00004
20. Oohora K, Onoda A, Hayashi T (2019) Hemoproteins reconstituted with artificial metal complexes as biohybrid catalysts. Acc Chem Res 52:945–954. https://doi.org/10.1021/acs.accounts.8b00676
21. Mirts EN, Bhagi-Damodaran A, Lu Y (2019) Understanding and modulating metalloenzymes with unnatural amino acids, non-native metal ions, and non-native metallocofactors. Acc Chem Res 52:935–944. https://doi.org/10.1021/acs.accounts.9b00011
22. Schwizer F, Okamoto Y, Heinisch T, Gu YF, Pellizzoni MM, Lebrun V, Reuter R, Köhler V, Lewis JC, Ward TR (2018) Artificial metalloenzymes: reaction scope and optimization strategies. Chem Rev 118:142–231. https://doi.org/10.1021/acs.chemrev.7b00014
23. Vázquez-González M, Wang C, Willner I (2020) Biocatalytic cascades operating on macromolecular scaffolds and in confined environments. Nat Catal 3:256–273. https://doi.org/10.1038/s41929-020-0433-1
24. Zhang GQ, Quin MB, Schmidt-Dannert C (2018) Self-assembling protein scaffold system for easy in vitro coimmobilization of biocatalytic cascade enzymes. ACS Catal 8:5611–5620. https://doi.org/10.1021/acscatal.8b00986
25. Schmidt S, Castiglione K, Kourist R (2018) Overcoming the incompatibility challenge in chemoenzymatic and multi-catalytic cascade reactions. Chem Eur J 24:1755–1768. https://doi.org/10.1002/chem.201703353
26. Rudroff F, Mihovilovic MD, Groger H, Snajdrova R, Iding H, Bornscheuer UT (2018) Opportunities and challenges for combining chemo- and biocatalysis. Nat Catal 1:12–22. https://doi.org/10.1038/s41929-017-0010-4
27. Denard CA, Huang H, Bartlett MJ, Lu L, Tan YC, Zhao HM, Hartwig JF (2014) Cooperative tandem catalysis by an organometallic complex and a metalloenzyme. Angew Chem Int Ed 53:465–469. https://doi.org/10.1002/anie.201305778

28. Köhler V, Wilson YM, Durrenberger M, Ghislieri D, Churakova E, Quinto T, Knörr L, Häussinger D, Hollmann F, Turner NJ, Ward TR (2013) Synthetic cascades are enabled by combining biocatalysts with artificial metalloenzymes. Nat Chem 5:93–99. https://doi.org/10.1038/Nchem.1498
29. Morra S, Pordea A (2018) Biocatalyst-artificial metalloenzyme cascade based on alcohol dehydrogenase. Chem Sci 9:9. https://doi.org/10.1039/c8sc02371a
30. Maity B, Hishikawa Y, Lu DN, Ueno T (2019) Recent progresses in the accumulation of metal ions into the apo-ferritin cage: experimental and theoretical perspectives. Polyhedron 172:104–111. https://doi.org/10.1016/j.poly.2019.03.048
31. Abe S, Maity B, Ueno T (2018) Functionalization of protein crystals with metal ions, complexes and nanoparticles. Curr Opin Chem Biol 43:68–76. https://doi.org/10.1016/j.cbpa.2017.11.015
32. Abe S, Maity B, Ueno T (2016) Design of a confined environment using protein cages and crystals for the development of biohybrid materials. Chem Commun 52:6496–6512. https://doi.org/10.1039/c6cc01355d
33. Cristie-Dayid AS, Chen J, Nowak DB, Bondy AL, Sun K, Park S, Holl MMB, Su M, Marsh ENG (2019) Coiled-coil-mediated assembly of an icosahedral protein cage with extremely high thermal and chemical stability. J Am Chem Soc 141:9207–9216. https://doi.org/10.1021/jacs.8b13604
34. Aumiller WM, Uchida M, Douglas T (2018) Protein cage assembly across multiple length scales. Chem Soc Rev 47:3433–3469. https://doi.org/10.1039/c7cs00818j
35. Sasaki E, Bohringer D, van de Waterbeemd M, Leibundgut M, Zschoche R, Heck AJR, Ban N, Hilvert D (2017) Structure and assembly of scalable porous protein cages. Nat Commun 8:671. https://doi.org/10.1038/ncomms14663
36. Liljestrom V, Ora A, Hassinen J, Rekola HT, Nonappa HM, Hynninen V, Joensuu JJ, Ras RHA, Torma P, Ikkala O, Kostiainen MA (2017) Cooperative colloidal self-assembly of metal-protein superlattice wires. Nat Commun 8:671. https://doi.org/10.1038/s41467-017-00697-z
37. Luo Q, Hou CX, Bai YS, Wang RB, Liu JQ (2016) Protein assembly: versatile approaches to construct highly ordered nanostructures. Chem Rev 116:13571–13632. https://doi.org/10.1021/acs.chemrev.6b00228
38. Koshiyama T, Kawaba N, Hikage T, Shirai M, Miura Y, Huang CY, Tanaka K, Watanabe Y, Ueno T (2010) Modification of porous protein crystals in development of biohybrid materials. Bioconjug Chem 21:264–269. https://doi.org/10.1021/Bc9003052
39. Abe S, Tokura Y, Pal R, Komura N, Imamura A, Matsumoto K, Ijiri H, Sanghamitra NJM, Tabe H, Ando H, Kiso M, Mori H, Kitagawa S, Ueno T (2015) Surface functionalization of protein crystals with carbohydrate using site-selective bioconjugation. Chem Lett 44:29–31. https://doi.org/10.1246/cl.140865
40. Nastri F, Chino M, Maglio O, Bhagi-Damodaran A, Lu Y, Lombardi A (2016) Design and engineering of artificial oxygen-activating metalloenzymes. Chem Soc Rev 45:5020–5054. https://doi.org/10.1039/c5cs00923e
41. Hematian S, Garcia-Bosch I, Karlin KD (2015) Synthetic heme/copper assemblies: toward an understanding of cytochrome c oxidase interactions with dioxygen and nitrogen oxides. Acc Chem Res 48:2462–2474. https://doi.org/10.1021/acs.accounts.5b00265
42. Hayashi T, Sano Y, Onoda A (2015) Generation of new artificial metalloproteins by cofactor modification of native hemoproteins. Isr J Chem 55:76–84. https://doi.org/10.1002/ijch.201400123
43. Lin YW, Sawyer EB, Wang JY (2013) Rational heme protein design: all roads lead to Rome. Chem Asian J 8:2534–2544. https://doi.org/10.1002/asia.201300291
44. Watanabe Y, Nakajima H, Ueno T (2007) Reactivities of oxo and peroxo intermediates studied by hemoprotein mutants. Acc Chem Res 40:554–562. https://doi.org/10.1021/ar600046a
45. Ozaki S, Roach MP, Matsui T, Watanabe Y (2001) Investigations of the roles of the distal heme environment and the proximal heme iron ligand in peroxide activation by heme enzymes

via molecular engineering of myoglobin. Acc Chem Res 34:818–825. https://doi.org/10.1021/ar9502590
46. Abe S, Atsumi K, Yamashita K, Hirata K, Mori H, Ueno T (2018) Structure of in cell protein crystals containing organometallic complexes. Phys Chem Chem Phys 20:2986–2989. https://doi.org/10.1039/c7cp06651a
47. Tabe H, Shimoi T, Boudes M, Abe S, Coulibaly F, Kitagawa S, Mori H, Ueno T (2016) Photoactivatable CO release from engineered protein crystals to modulate NF-kappa B activation. Chem Commun 52:4545–4548. https://doi.org/10.1039/c5cc10440h
48. Tabe H, Shimoi T, Fujita K, Abe S, Ijiri H, Tsujimoto M, Kuchimaru T, Kizaka-Kondo S, Mori H, Kitagawa S, Ueno T (2015) Design of a CO-releasing extracellular scaffold using in vivo protein crystals. Chem Lett 44:342–344. https://doi.org/10.1246/cl.141035
49. Tabe H, Fujita K, Abe S, Tsujimoto M, Kuchimaru T, Kizaka-Kondoh S, Takano M, Kitagawa S, Ueno T (2015) Preparation of a cross-linked porous protein crystal containing Ru carbonyl complexes as a CO-releasing extracellular scaffold. Inorg Chem 54:215–220. https://doi.org/10.1021/ic502159x
50. Ueno T, Abe S, Koshiyama T, Ohki T, Hikage T, Watanabe Y (2010) Elucidation of metal-ion accumulation induced by hydrogen bonds on protein surfaces by using porous lysozyme crystals containing Rh^{III} ions as the model surfaces. Chem Eur J 16:2730–2740. https://doi.org/10.1002/chem.200903269
51. Ueno T, Abe M, Hirata K, Abe S, Suzuki M, Shimizu N, Yamamoto M, Takata M, Watanabe Y (2009) Process of accumulation of metal ions on the interior surface of apo-ferritin: crystal structures of a series of apo-ferritins containing variable quantities of Pd(II) ions. J Am Chem Soc 131:5094–5100. https://doi.org/10.1021/ja806688s
52. Fujita K, Tanaka Y, Sho T, Ozeki S, Abe S, Hikage T, Kuchimaru T, Kizaka-Kondoh S, Ueno T (2014) Intracellular CO release from composite of ferritin and ruthenium carbonyl complexes. J Am Chem Soc 136:16902–16908. https://doi.org/10.1021/ja508938f
53. Fujita K, Tanaka Y, Abe S, Ueno T (2016) A photoactive carbon-monoxide-releasing protein cage for dose-regulated delivery in living cells. Angew Chem Inter Ed 55:1056–1060. https://doi.org/10.1002/anie.201506738
54. Wang Z, Takezawa Y, Aoyagi H, Abe S, Hikage T, Watanabe Y, Kitagawa S, Ueno T (2011) Definite coordination arrangement of organometallic palladium complexes accumulated on the designed interior surface of apo-ferritin. Chem Commun 47:170–172. https://doi.org/10.1039/c0cc02221g
55. Takezawa Y, Bockmann P, Sugi N, Wang ZY, Abe S, Murakami T, Hikage T, Erker G, Watanabe Y, Kitagawa S, Ueno T (2011) Incorporation of organometallic Ru complexes into apo-ferritin cage. Dalton Trans 40:2190–2195. https://doi.org/10.1039/c0dt00955e
56. Abe S, Hikage T, Watanabe Y, Kitagawa S, Ueno T (2010) Mechanism of accumulation and incorporation of organometallic Pd complexes into the protein nanocage of apo-ferritin. Inorg Chem 49:6967–6973. https://doi.org/10.1021/ic1003758
57. Abe S, Tsujimoto M, Yoneda K, Ohba M, Hikage T, Takano M, Kitagawa S, Ueno T (2012) Porous protein crystals as reaction vessels for controlling magnetic properties of nanoparticles. Small 8:1314–1319. https://doi.org/10.1002/smll.201101866
58. Künzle M, Lach M, Beck T (2018) Crystalline protein scaffolds as a defined environment for the synthesis of bioinorganic materials. Dalton Trans 47:10382–10387. https://doi.org/10.1039/c8dt01192c
59. Guli MN, Li M, Hu Z, Yao JX, Mann S (2015) Fabrication of photoluminescent hybrid protein crystals of lysozyme and bromophenol blue. J Mater Sci 50:7026–7030. https://doi.org/10.1007/s10853-015-9255-y
60. Muskens OL, England MW, Danos L, Li M, Mann S (2013) Plasmonic response of Ag- and Au-infiltrated cross-linked lysozyme crystals. Adv Funct Mater 23:281–290. https://doi.org/10.1002/adfm.201201718

61. Guli M, Lambert EM, Li M, Mann S (2010) Template-directed synthesis of nanoplasmonic arrays by intracrystalline metalization of cross-linked lysozyme crystals. Angew Chem Int Ed 49:520–523. https://doi.org/10.1002/anie.200905070
62. Suzuki M, Abe M, Ueno T, Abe S, Goto T, Toda Y, Akita T, Yamadae Y, Watanabe Y (2009) Preparation and catalytic reaction of Au/Pd bimetallic nanoparticles in apo-ferritin. Chem Commun 2009:4871–4873. https://doi.org/10.1039/b908742g
63. Kang S, Jolley CC, Liepold LO, Young M, Douglas T (2009) From metal binding to nanoparticle formation: monitoring biomimetic iron oxide synthesis within protein cages using mass spectrometry. Angew Chem Inter Ed 48:4772–4776. https://doi.org/10.1002/anie.200900437
64. Klem MT, Mosolf J, Young M, Douglas T (2008) Photochemical mineralization of europium, titanium, and iron oxyhydroxide nanoparticles in the ferritin protein cage. Inorg Chem 47:2237–2239. https://doi.org/10.1021/ic701740q
65. Uchida M, Klem MT, Allen M, Suci P, Flenniken M, Gillitzer E, Varpness Z, Liepold LO, Young M, Douglas T (2007) Biological containers: protein cages as multifunctional nanoplatforms. Adv Mater 19:1025–1042. https://doi.org/10.1002/adma.200601168
66. Klem MT, Willits D, Solis DJ, Belcher AM, Young M, Douglas T (2005) Bio-inspired synthesis of protein-encapsulated CoPt nanoparticles. Adv Funct Mater 15:1489–1494. https://doi.org/10.1002/adfm.200400453
67. Ueno T, Suzuki M, Goto T, Matsumoto T, Nagayama K, Watanabe Y (2004) Size-selective olefin hydrogenation by a Pd nanocluster provided in an apo-ferritin cage. Angew Chem Int Ed 43:2527–2530. https://doi.org/10.1002/anie.200353436
68. Okuda M, Iwahori K, Yamashita I, Yoshimura H (2003) Fabrication of nickel and chromium nanoparticles using the protein cage of apoferritin. Biotechnol Bioeng 84:187–194. https://doi.org/10.1002/bit.10748
69. Allen M, Willits D, Young M, Douglas T (2003) Constrained synthesis of cobalt oxide nanomaterials in the 12-subunit protein cage from Listeria innocua. Inorg Chem 42:6300–6305. https://doi.org/10.1021/ic0343657
70. Allen M, Willits D, Mosolf J, Young M, Douglas T (2002) Protein cage constrained synthesis of ferrimagnetic iron oxide nanoparticles. Adv Mater 14:1562–1565. https://doi.org/10.1002/1521-4095(20021104)14:21<1562::Aid-Adma1562>3.0.Co;2-D
71. Maity B, Abe S, Ueno T (2017) Observation of gold sub-nanocluster nucleation within a crystalline protein cage. Nat Commun 8:14820. https://doi.org/10.1038/ncomms14820
72. Koshiyama T, Shirai M, Hikage T, Tabe H, Tanaka K, Kitagawa S, Ueno T (2011) Post-crystal engineering of zinc-substituted myoglobin to construct a long-lived photoinduced charge-separation system. Angew Chem Inter Ed 50:4849–4852. https://doi.org/10.1002/anie.201008004
73. Wei H, Lu Y (2012) Catalysis of gold nanoparticles within lysozyme single crystals. Chem Asian J 7:680–683. https://doi.org/10.1002/asia.201100942
74. Engstrom K, Johnston EV, Verho O, Gustafson KPJ, Shakeri M, Tai CW, Backvall JE (2013) Co-immobilization of an enzyme and a metal into the compartments of mesoporous silica for cooperative tandem catalysis: an artificial metalloenzyme. Angew Chem Int Ed 52:14006–14010. https://doi.org/10.1002/anie.201306487
75. Liang M, Wang LB, Liu X, Qi W, Su RX, Huang RL, Yu YJ, He ZM (2013) Cross-linked lysozyme crystal templated synthesis of Au nanoparticles as high-performance recyclable catalysts. Nanotechnology 24:245601. https://doi.org/10.1088/0957-4484/24/24/245601
76. Liang M, Wang LB, Su RX, Qi W, Wang MF, Yu YJ, He ZM (2013) Synthesis of silver nanoparticles within cross-linked lysozyme crystals as recyclable catalysts for 4-nitrophenol reduction. Cat Sci Technol 3:1910–1914. https://doi.org/10.1039/c3cy00157a
77. Liang M, Su RX, Qi W, Yu YJ, Wang LB (2014) Synthesis of well-dispersed Ag nanoparticles on eggshell membrane for catalytic reduction of 4-nitrophenol. J Mater Sci 49:1639–1647. https://doi.org/10.1007/s10853-013-7847-y

78. Liu MY, Wang LB, Huang RL, Yu YJ, Su RX, Qi W, He ZM (2016) Superior catalytic performance of gold nanoparticles within small cross-linked lysozyme crystals. Langmuir 32:10895–10904. https://doi.org/10.1021/acs.langmuir.6b02544
79. Tabe H, Abe S, Hikage T, Kitagawa S, Ueno T (2014) Porous protein crystals as catalytic vessels for organometallic complexes. Chem Asian J 9:1373–1378. https://doi.org/10.1002/asia.201301347
80. Lopez S, Rondot L, Lepretre C, Marchi-Delapierre C, Menage S, Cavazza C (2017) Cross-linked artificial enzyme crystals as heterogeneous catalysts for oxidation reactions. J Am Chem Soc 139:17994–18002. https://doi.org/10.1021/jacs.7b09343
81. Cavazza C, Bochot C, Rousselot-Pailley P, Carpentier P, Cherrier MV, Martin L, Marchi-Delapierre C, Fontecilla-Camps JC, Menage S (2010) Crystallographic snapshots of the reaction of aromatic C-H with O_2 catalysed by a protein-bound iron complex. Nat Chem 2:1069–1076. https://doi.org/10.1038/Nchem.841
82. Tabe H, Takahashi H, Shimoi T, Abe S, Ueno T, Yamada Y (2018) Photocatalytic hydrogen evolution systems constructed in cross-linked porous protein crystals. Appl Catal B 237:1124–1129. https://doi.org/10.1016/j.apcatb.2018.01.046
83. Chakraborti S, Lin TY, Glatt S, Heddle JG (2020) Enzyme encapsulation by protein cages. RSC Adv 10:13293–13301. https://doi.org/10.1039/c9ra10983h
84. Uchida M, McCoy K, Fukuto M, Yang L, Yoshimura H, Miettinen HM, LaFrance B, Patterson DP, Schwarz B, Karty JA, Prevelige PE, Lee B, Douglas T (2018) Modular self-assembly of protein cage lattices for multistep catalysis. ACS Nano 12:942–953. https://doi.org/10.1021/acsnano.7b06049
85. McCoy K, Uchida M, Lee B, Douglas T (2018) Templated assembly of a functional ordered protein macromolecular framework from P22 virus-like particles. ACS Nano 12:3541–3550. https://doi.org/10.1021/acsnano.8b00528
86. McCoy K, Selivanovitch E, Luque D, Lee B, Edwards E, Castón JR, Douglas T (2018) Cargo retention inside P22 virus-like particles. Biomacromolecules 19:3738–3746. https://doi.org/10.1021/acs.biomac.8b00867
87. Brasch M, Putri RM, de Ruiter MV, Luque D, Koay MST, Castón JR, Cornelissen JJLM (2017) Assembling enzymatic cascade pathways inside virus-based nanocages using dual-tasking nucleic acid tags. J Am Chem Soc 139:1512–1519. https://doi.org/10.1021/jacs.6b10948
88. Giessen TW, Silver PA (2016) A catalytic nanoreactor based on in vivo encapsulation of multiple enzymes in an engineered protein nanocompartment. Chembiochem 17:1931–1935. https://doi.org/10.1002/cbic.201600431
89. Frey R, Mantri S, Rocca M, Hilvert D (2016) Bottom-up construction of a primordial carboxysome mimic. J Am Chem Soc 138:10072–10075. https://doi.org/10.1021/jacs.6b04744
90. Azuma Y, Zschoche R, Tinzl M, Hilvert D (2016) Quantitative packaging of active enzymes into a protein cage. Angew Chem Int Ed 55:1531–1534. https://doi.org/10.1002/anie.201508414
91. Patterson DP, Schwarz B, Waters RS, Gedeon T, Douglas T (2014) Encapsulation of an enzyme cascade within the bacteriophage P22 virus-like particle. ACS Chem Biol 9:359–365. https://doi.org/10.1021/cb4006529
92. Maity B, Fukumori K, Abe S, Ueno T (2016) Immobilization of two organometallic complexes into a single cage to construct protein-based microcompartments. Chem Commun 52:5463–5466. https://doi.org/10.1039/c6cc00679e

Machine Learning and Monte Carlo Methods for Surface-Assisted Molecular Self-Assembly

Daniel Packwood

1 Introduction

Anybody who has ever been trapped in an elevator with a materials scientist or cell biologist would certainly have heard about molecular self-assembly. Molecular self-assembly, which refers to the spontaneous formation of functional structures from molecule precursors, is a major pathway by which materials scientists create nanomaterials and cells create organelles. Owning to its broad importance, molecular self-assembly has been the subject of numerous investigations and an enormous literature on the subject exists (for a seminal review, see Ref. [1]). However, in order to create new nanomaterials or artificial biomaterials via molecular self-assembly, our ability to predict the outcome of the self-assembly process must be dramatically improved [2].

Predictions of this kind are a major focus for the field of computational physics. While the calculation of physical properties for a given self-assembled structure is possible in principle, the prediction of self-assembled structure itself remains an enormous challenge. Broadly speaking, the standard methodologies of computational physics, such as molecular dynamics simulation and first-principles calculations, are plainly inadequate for making such predictions. The huge numbers of atoms involved and the enormous, microsecond-exceeding time-scales involved in the self-assembly process, are a major computational burden for these methods. Notwithstanding some significant developments (mostly via a clever use of classical Monte Carlo simulation [3–5]), theoretical breakthroughs are necessary before computational predictions for molecular self-assembly processes become routine.

D. Packwood (✉)
Kyoto University iCeMS, Kyoto, Japan
e-mail: dpackwood@icems.kyoto-u.ac.jp

D. O. Wang, D. Packwood (eds.), *Cell-Inspired Materials and Engineering*, Fundamental Biomedical Technologies, https://doi.org/10.1007/978-3-030-55924-3_3

In this chapter, we review our own theoretical work on the molecular self-assembly process, and highlight the potential of applied mathematics and information science in this area. In particular, this chapter discusses (1) the use of machine learning methods to develop simplified, yet accurate, models for molecular self-assembly, and (2) the use of nonclassical Markov chain Monte Carlo sampling methods to quickly predict the outcome of the molecular self-assembly process from these models. In addition, we also discuss (3) the use of machine learning methods to accelerate first-principles predictions for the molecular self-assembly process. Our exposition is nontechnical, and all technical details can be found in the relevant references. We focus on the specific case of organic molecule self-assembly on crystalline metal surfaces, which are both sufficiently simple for theoretical study and highly relevant to materials science. However, it should be noted that the methods presented here could, in principle, be applied to molecular self-assembly under a range of conditions.

2 Self-Assembly of Organic Precursors on Metal Surfaces

Before discussing any mathematical techniques, we first provide an overview of organic molecule self-assembly on metal surfaces and explain why the typical methods of computational physics are poorly suited for predicting the outcome of this process.

Small aromatic organic molecules interact strongly with metal surfaces and often undergo self-assembly on the surface without desorption back into the gas phase. Such systems are ideal for the study of molecular self-assembly for several reasons. Firstly, these experiments can be performed on highly crystalline metal surfaces under ultrahigh vacuum conditions, which means that molecular self-assembly occurs under well-defined conditions with minimal interference from unknown influences. In contrast, in experiments involving solvents or cells, the presence of solvent molecules or other solutes significantly complicates the self-assembly process. Secondly, the structure formed by the molecular self-assembly process can be examined in detail using a scanning tunneling microscope (STM), which is a standard (if expensive) apparatus in surface science. Finally, self-assembly of organic precursors on metal surfaces often results in interesting low-dimensional structures with novel electrical and optical properties. Molecular self-assembly of organic precursors on metal surfaces is therefore an topic of study [6], and a quick glance at any recent issue of *ACS Nano* will show just how actively this field is pursued [7].

Figure 1 shows a representative example of organic molecule self-assembly on a metal surface, as reported by Han and coworkers [8, 9]. In this experiment, Br_2BA (=10,10′-dibromo-9,9′-bianthracene, Fig. 1a) precursor molecules were deposited onto a copper surface (Cu(111)). STM imaging unveiled the presence of Br_2BA chains on the copper surface (Fig. 1b), resulting from Br_2BA self-assembly. Analysis of these images showed that the Br_2BA molecules in these chains align so that that stabilizing π-stacking interactions between adjacent molecules in the chain can

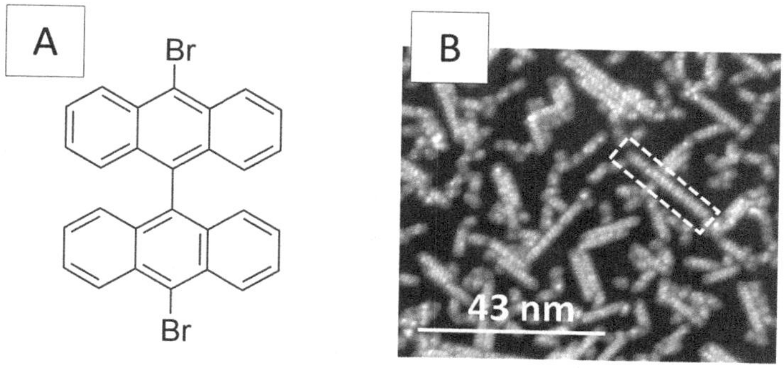

Fig. 1 (**a**) 10,10′-dibromo-9,9′-bianthracene (=Br_2BA) precursor molecule. (**b**) Scanning tunneling microscopy (STM) image of a copper surface (Cu(111)) following deposition of Br_2BA molecules. A single chain-shaped structure, formed by Br_2BA self-assembly, is highlighted by the dotted white box. STM image obtained by Patrick Han and colleagues (Tohoku University). STM conditions: Sample bias voltage = 1.1 V, tunneling current = 10 pA, STM imaging temperature = 5.6 K, annealing temperature = 400 °C

occur (Fig. 1b, insert). Most importantly, Han et al. showed that upon heating, these chains would undergo a chemical reaction to form graphene nanoribbons, a highly coveted nanomaterial which possesses a variety of remarkable electronic, optical, and mechanical properties (see [10, 11] for reviews on graphene nanoribbons). Interesting, in their STM imaging, Han et al. also observed the presence of "bent" graphene nanoribbons, and traced their appearance to the formation of "bent" chains following Br_2BA self-assembly [9]. These results demonstrate the use of organic molecule self-assembly in nanomaterials synthesis, and reinforces the importance of achieve controlled molecular self-assembly in order to control nanomaterial shape.

Of all the standard methods from computational physics, *density functional theory* (DFT) is arguably the method of choice for systems involving metal surfaces [12, 13]. DFT is a method for computing the ground-state properties of a system from the basic equations of quantum mechanics. Starting with the coordinates of the metal atoms and organic precursor molecules, DFT can estimate the energy of the system to reasonable accuracy within a reasonable amount of time. By arranging the precursor molecules in various ways on the metal surface and computing the energy of the system for each case with DFT, we can, in principle, identify the most energetically stable arrangement of precursor molecules on the surface. However, in practice this is an ineffective strategy for predicting the result of self-assembly processes for several reasons. Firstly, the number of atoms involved in each calculation can be very large, and exponentially increases as the number of molecules on the surface increases. Even though DFT is relatively fast compared to other types of quantum mechanical calculations, calculation times exceeding 24 h are common for systems of this type, even when modern computer clusters are used (problem A). Secondly, the number of possible ways to arrange the precursor molecules on the surface is typically very large and increases exponentially as the number of precursor molecules is increased. Given that a day of computation could be required to compute the energy of a single arrangement of molecules, years of computation would be needed to identify the most energetically stable arrangement of molecules (problem B). Finally, even if such computational constraints were not an issue, the most energetically stable arrangement of precursor molecules (as predicted by DFT) will not necessarily correspond to the outcome of the molecular self-assembly process. In general, the outcome of the self-assembly process is determined by the laws of thermodynamics, and may not necessarily correspond to the energy-minimizing configuration of molecules (problem C). Unfortunately, it is difficult to account for thermodynamics in DFT at present.

In response to the above drawbacks, one might be tempted use classical simulations based upon Newtonian mechanics in place of DFT for simulating precursor self-assembly. While some promising progress along these lines has been made [14], in general such simulations are inappropriate for simulating precursor self-assembly on metal surfaces, because the interaction between organic precursor molecules and the surface typically involves extensive charge transfer. The effects of this charge transfer on molecular conformation cannot be adequately simulated by classical methods at present. In summary, it is generally not feasible to study organic molecule self-assembly via any of the standard methods of computational physics at present.

3 Kernelized Machine Learning for Model Construction

In order to predict the outcome of organic precursor self-assembly on a metal surface, we must develop new methods which can properly account for the quantum mechanical nature of the system and are efficient enough to be simulated within a reasonable time frame. Recall from the previous section the three major difficulties in studying molecular self-assembly via computational methods.

(A) The calculation of the energy of any given arrangement of the precursor molecules, with proper account of the surface-molecule interaction, can be in the order of days.
(B) The number of ways of arranging the precursor molecules on the surface is extremely large. It is not possible to compute the energy for every possibility within a reasonable time frame.
(C) Even if the lowest energy way of arranging the precursor molecules on the surface is identified, it will not necessarily correspond to the true outcome of the molecular self-assembly process, because the latter is determined by the principles of thermodynamics.

In this section, we discuss how *machine learning* can be used to overcome problem (A). Machine learning is a branch of information science which is concerned with classification and prediction problems. This field is closely related to the field of statistics. However, whereas statistics is mainly concerned with understanding data sets, machine learning is usually more concerned with the accuracy and computational cost of making predictions from data sets. With the easy availability of large data sets and cheap computational resources, machine learning is rapidly becoming a defining part of contemporary science.

In Ref. [15], we report on a strategy for solving problem (A) via machine learning. Briefly, this strategy proceeds as follows. In Step I, we create a database containing a sample of precursor–precursor interactions and their corresponding interaction energies. These interaction energies are computed accurately using DFT methods. In Step II, we use this database of create an efficient scheme which predicts the energy of an arrangement of precursor molecules on the surface, without the need to perform any additional DFT calculations.

3.1 Generation of Database

Because the quality of machine learning predictions is dependent on the sample data, great care must be made when generating the database in Step I. In the present case, involving Br_2BA molecules adsorbed to Cu(111), each precursor–precursor interaction in our sample must follow the known physics of the system. To this end, we ensure that each precursor–precursor interaction in the sample obeys the following assumption. This assumption is justified by the fact that aromatic molecules such as Br_2BA undergo very strong van der Waals interactions with metal surfaces.

(Assumption) The conformation and orientation of each precursor molecule is determined entirely by the interaction between the precursor molecule and the surface, and is not affected by the intermolecular interaction.

By making this assumption, we can keep each molecule in the same conformation during the entire run of the simulation. The specific conformation to use can be found by performing standard DFT calculations for a single Br_2BA molecule adsorbed to the surface (Fig. 2a). These same DFT calculations also show that the Br_2BA molecule has a strong energetic preference to align with one of the three atomic planes of the Cu(111) surface (Fig. 2b), and to be positioned such that its center-of-mass lies directly above one of several points (adsorption sites) in the unit cell of the surface (Fig. 2c). Using the conformation for the Br_2BA molecule shown in Fig. 2a, we can generate a single precursor–precursor interaction by randomly choosing a pair of orientations from Fig. 2b and a pair of adsorption sites from Fig. 2c. By repeating this procedure many times, we can generate a database of precursor–precursor interactions. For each interaction in the sample, the interaction energy is then calculated via DFT.

In this work, we ignored the Cu atoms when calculating the interaction energies for the database. While this dramatically reduces the time required to generate the database, it means that the effect of the surface electrons on the precursor–precursor interactions is not accounted for. While additional DFT calculations suggest that this effect is not dramatic in the present system [15], this is not expected to hold for other types of molecules [20, 21]. We return to this point in Sect. 5. With the Cu

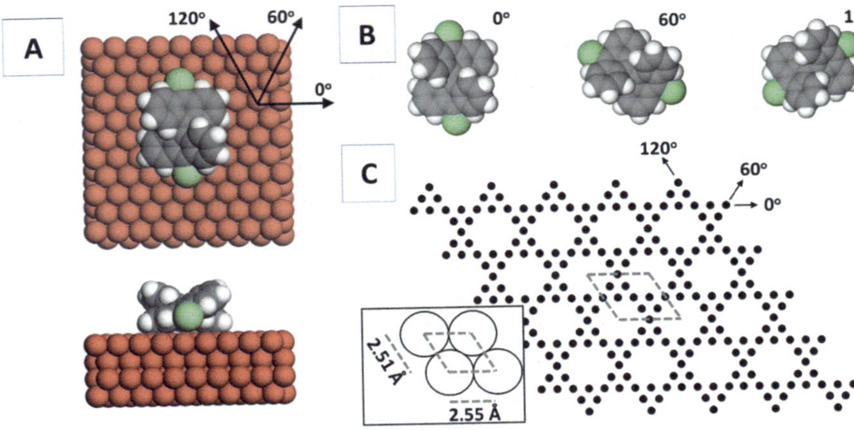

Fig. 2 (**a**) Conformation of a single Br_2BA precursor molecule adsorbed to Cu(111), as computed by density functional theory (DFT) calculations. Top image has the Cu(111) surface in the plane of the page. Bottom image has the plane of the surface perpendicular to the page. The arrows indicate the directions of the surface atomic planes. (**b**) Energetically preferred orientations for a Br_2BA molecule adsorbed to Cu(111), as identified by DFT calculations. Cu(111) surface (not explicitly shown) is in the plane of the page. (**c**) Energetically preferred adsorption sites for a Br_2BA molecule on Cu(111), as identified by DFT calculations. Rows of adsorption sites that lie in the direction of the arrow marked 0° (respectively 60°, 120°) allow the molecule to adopt the 0° (respectively 60°, 120°) orientation. The grey box indicates a single unit cell of the Cu(111) surface. The insert shows the locations of the Cu atoms in the unit cell. DFT calculations were performed in VASP, using a plane wave basis with 400 eV energy cut-off, PAW-PBE pseudopotentials, 2 × 2 × 1 Monkhorst–Pack k-points grid, and the rev-vdW-DF2 exchange correlation functional [16–19]

surface atoms ignored, around 2 weeks are required to generate a database containing around 2000 pairwise interactions and their interaction energies.

3.2 *Energy Prediction*

Let us now consider Step II in detail. For a given arrangement of n precursor molecules on a metal surface, the energy is predicted according to the equation

$$E = \sum_{k=1}^{n} u_k + \sum_{i,j,i<j}^{n} v_{ij}, \tag{1}$$

where u_k is the interaction energy between precursor molecule k and the surface, and v_{ij} is the interaction energy between precursor molecules i and j. The first summation is performed over all precursor molecules on the surface, and the second summation is performed over all pairs of precursor molecules on the surface. The first summation can be performed without additional effort. This is because the surface–

molecule interaction energies, for all adsorption sites and orientations for the precursor molecule, were calculated when compiling Fig. 2c. On the other hand, the second summation in Eq. (1) is unknown. This is because the number of possible precursor–precursor interactions is very large (in the order of 10^8) compared to the size of our database (which contains in the order of 10^3 precursor–precursor interactions). For this reason, we employed machine learning techniques [22] to calculate the second sum in Eq. (1).

The scheme for predicting the energy E for a given arrangement of precursor molecules is described in Fig. 3. For each precursor–precursor interaction, a *classifier* is first used to classify the interaction into one of two classes. Then, a *predictor* is used to predict the energy of the interaction, given its class. Once the energies are predicted for each precursor–precursor interaction on the surface, the second sum in Eq. (1) is calculated. The first sum is then calculated directly as described in the previous paragraph.

To construct the classifier and predictors it is necessary to define a *feature vector* for each precursor–precursor interaction in our sample. The feature vector is a vector of real-valued numbers which encode the physical features of the pairwise interaction. In the present study, we employed the *vectorized Coulomb matrix* for our feature vectors [23, 24]. Namely, for a given precursor–precursor interaction $\mathbf{x}$, we arbitrarily label one of the molecules as "molecule 1," the other molecule as "molecule 2," and define the feature vector as

$$\phi(\mathbf{x}) = \left(c_{11}, c_{12}, \ldots, c_{1m}, c_{21}, \ldots, c_{mm}\right) \tag{2}$$

Here m is the number of atoms in a single Br_2BA molecule,

$$c_{ij} = \frac{q_i q_j}{d_{ij}} \tag{3}$$

q_i is the atomic number of atom i (which resides in molecule 1) and q_j is the atomic number of atom j (which resides in molecule 2) and d_{ij} is the distance between atoms i and j. The components c_{ij} of the feature vectors are called *features*. Once the vectorized Coulomb matrices are calculated for all interactions in the database, the classifiers and predictors in the previous scheme can be constructed using machine learning techniques.

The specific machine learning techniques used to construct the classifier and predictors are called *support vector machines* and *kernel ridge regression*, respectively. In the support vector machines technique, the sample of precursor–precursor interactions are used to construct a so-called linear separating plane. Then, when the interaction energies are plotted as a function of their features, we find that the attractive interactions fall on one side of the separating plane, and the repulsive interactions will fall on the other side. This linear separating plane is then used by Classifier 1 to predict the class of the precursor–precursor interactions that are present on the surface (Fig. 3). In the case of this study, it is not possible to separate the sample data

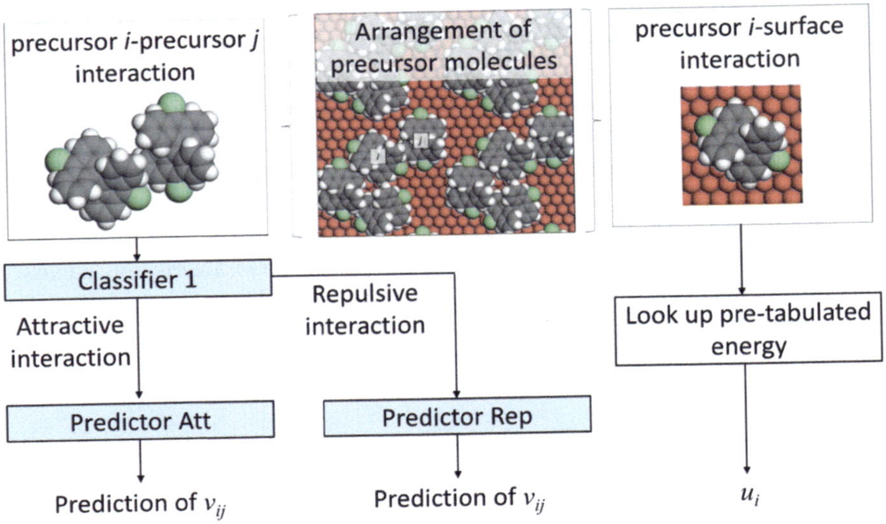

Fig. 3 Simplified scheme for calculating the energy E in Eq. (1) for a given arrangement of precursor molecules on a metal surface [15]. For every pair of precursor molecules on the surface, the following is performed (left-hand side): First, the pair of molecules passes through the component "Classifier 1," and is predicted to be either "attractive" or "repulsive." An interaction is considered "attractive" if its interaction energy is less than or equal to zero, and is considered "repulsive" otherwise. If the pair of molecules is predicted as being attractive, then it passes through the component "Predictor Att. Otherwise, the pair of molecules passes through "Predictor Rep." These predictors generate the terms v in Eq. (1). Classifier 1 and the two Predictors were constructed using machine learning techniques called support vector machines and kernelized ridge regression, respectively (see main text). Finally, for every precursor molecule on the surface, the surface–molecule interaction term is computed by identifying the adsorption site and orientation of the molecule, and then reading from a table of precomputed energies (right-hand side of the image)

with a simple linear separating plane. For this reason we employ the so-called kernel trick, which replaces the feature vectors in Eq. (2) with a new set of higher dimensional feature vectors for which linear separation is possible. In the kernel ridge regression technique, the feature vectors for the interactions are transformed (again using the kernel trick) in such a way that energy has a linear dependence on its features. A linear regression plane is fit to this data, which, in turn, is used by Predictor Att and Predictor Rep to predict the energies of the interactions on the surface. Support vector machines and kernel ridge regression are standard techniques in the machine learning field (see Ref. [22] for a detailed introduction to these methods). The accuracy of these methods in predicting test data is shown in Table 1 and Fig. 4.

Note that the choice of feature vector is very important to the performance of Classifier 1, Predictor Att, and Predictor Rep; if the feature vectors do a poor job of encoding the precursor–precursor interactions, then it will become difficult to construct classifiers and predictors with adequate performance. In the case of the vectorized Coulomb matrices, we are assuming that the classifications and predictions can be adequately performed on the basis of interatomic Coulombic interactions

Table 1 Performance of Classifier 1

Training set size	Test set size	Fail rate
2762 points	582 test points	1.70%

Training set size and test set size refer to the number of precursor–precursor interactions used to construct and test the classifier, respectively. Fail rate refers to the percentage of test points which were incorrectly classified by Classifier 1. Classifier 1 was constructed via the support vector machines method. Training and test data were obtained via DFT calculations. See main text and Fig. 2 caption for calculation details

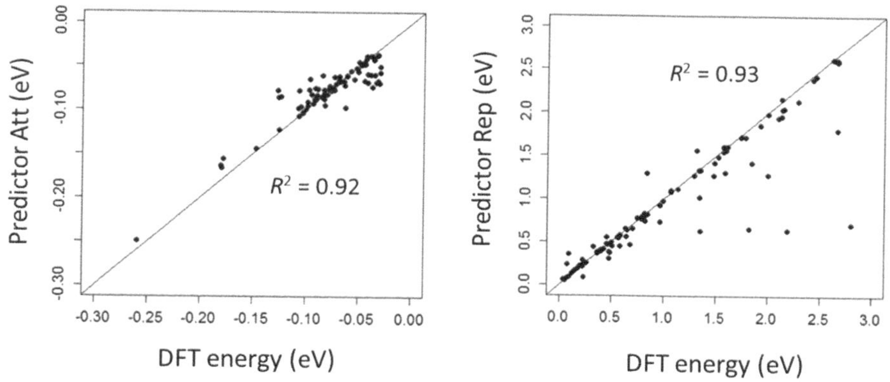

Fig. 4 Performance of predictor Att and predictor Rep on test data. The horizontal axis refers to the exact precursor–precursor interaction energy computed via DFT, and the vertical axis refers to the prediction. Predictor Att and Predictor Rep were constructed via kernelized ridge regression. Training set size/test set size: 2441/129 (Predictor Att), 1083/121 (Predictor Rep). See caption of Fig. 1 for DFT calculation details

alone. While the accuracy of our classifiers and predictors appears very good (Table 1, Fig. 4), it is possible that they could be improved further by incorporating information relating to exchange interactions and electron correlation into the feature vectors. This is clearly a direction for future research.

Using the scheme shown in Fig. 3, it is possible to predict the energy of an arrangement of precursor molecules on a metal surface, with account of the surface, within a few seconds of computational time. This is a very dramatic reduction in calculation time compared to a typical DFT calculation. While our scheme certainly solves problem (A), we now need to consider problems (B) and (C), which concern the actual prediction of the outcome of the molecular self-assembly process.

4 Equivalence Class Sampling for Model Predictions

Now we consider problems (B) and (C) described above. Given the energy calculation scheme developed in the previous section, we must predict the outcome of the molecular self-assembly process within a reasonable period of time.

Markov chain Monte Carlo (MCMC) is one of the most widely used approaches to carrying out such predictions [25]. To introduce this technique, let σ denote one specific way of arranging the precursor molecules on the surface. The probability $p(\sigma)$ that molecule arrangement σ results from the self-assembly process is given by the *Boltzmann distribution*, namely,

$$p(\sigma) = C\exp\left(-\varepsilon(\sigma)/k_B T\right) \tag{4}$$

where $\varepsilon(\sigma)$ is the energy of molecule arrangement σ (computed according to the scheme in Fig. 3), T is the temperature, k_B is Boltzmann's constant, C is a normalizer constant. Equation (4) assumes that the molecular self-assembly process proceeds until thermodynamic equilibrium is reached. The purpose of MCMC is to randomly generate a sample of precursor molecule arrangements, in such a way that the number of times that molecule arrangement σ appears in the sample is proportional to $p(\sigma)$. By analyzing the sample generated by MCMC, we can identify high probability molecule arrangements can calculate statistics such as average island size, average island shape, and so on. In this sense, MCMC solves problem (C) mentioned above.

Unfortunately, MCMC does not address problem (B) mentioned above. To explain why, let us consider in more detail how MCMC generates the sample of molecule arrangements (Fig. 5). Let Ω denote the set of all possible arrangements for the precursor molecules on the metal surface. Ω, which is referred to as the *configuration space*, can be thought of as a cloud of points, where each point in the cloud corresponds to one specific way of arranging the precursor molecules on the surface. In MCMC, we simulate a special random walk (known as the *Metropolis–Hastings chain*) which jumps between points in Ω. This random walk is simulated in such a way that the fraction of times that it visits point σ in the cloud becomes proportional to $p(\sigma)$ as the simulation time grows to infinity. More concretely, if we define the *frequency of visits to point* σ as

$$f_\sigma = \frac{n_\sigma(t)}{t} \tag{5}$$

where t is the total number of jumps the random walk has made since the beginning of its simulation, and $n_\sigma(t)$ is the number of times it has visited point σ during that time, then $f_\sigma \to p(\sigma)$ as $t \to \infty$. The difficulty with MCMC for the present case is that the number of possible ways of arranging the precursor molecules on the surface is incredibly large. This means that the number of points in Ω is enormous, and hence that impractically long simulation times are required for f_σ to converge to $p(\sigma)$.

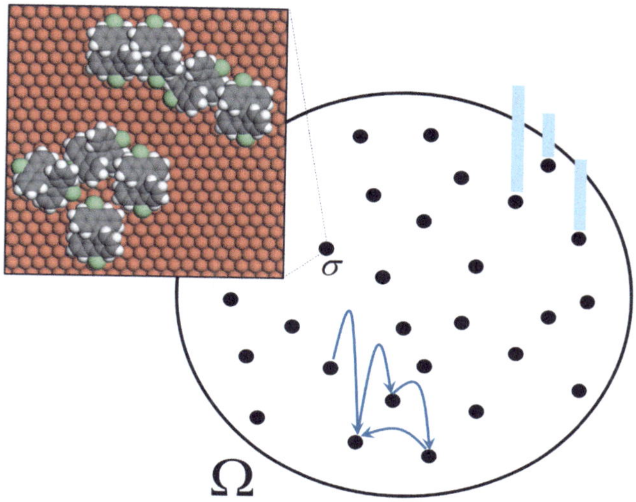

Fig. 5 Outline of Markov chain Monte Carlo (MCMC). The circle marked Ω is called the configuration space, and each point inside of Ω corresponds to one possible arrangement of precursor molecules on the surface (the number of precursor molecules and the surface size is identical for each point). One molecule arrangement (marked σ) is shown. The blue bars indicate the probability that the molecule arrangement will appear at thermodynamic equilibrium (see Eq. (4)). Probabilities for only some of the points are shown. The blue arrows indicate the path of a random walk. In MCMC, the random walk is designed so that the number of visits it makes to a point is proportional to the probability of the point. By this method, we can identify the arrangements of molecules which are likely to appear on the surface at thermodynamic equilibrium

4.1 Equivalence Class Sampling Idea

We created a variant of MCMC called *equivalence class sampling* (ECS), in which predictions relevant to molecular self-assembly can be achieved within a reasonable time period [26]. The basic idea behind ECS is to replace Ω with a much smaller configuration space, and then run the Metropolis–Hastings random walk on this configuration space to predict the outcome of the molecular self-assembly process.

In order to create the smaller configuration space, we first clarify exactly what it is that we want to predict. Consider the precursor molecule arrangement shown in Figs. 6a, and note that the molecules in this case can be separated into two subsets. These subsets are called *islands*, and correspond to groups of molecules that are close together (in a certain sense). Islands can be thought of as distinct, noninteracting structures that emerge from the molecular self-assembly process. Thus, our primary goal should be to predict the islands that result from the molecular self-assembly process. In other words, information such as the orientation or specific orientation of the islands on the surface is not necessary in order to make useful predictions for precursor molecule self-assembly.

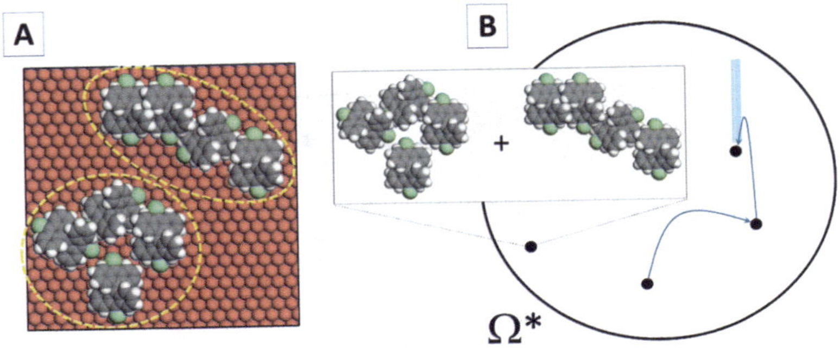

Fig. 6 Outline of equivalence class sampling (ECS). (**a**) One arrangement of precursor molecules on the surface. The yellow ovals indicate islands. (**b**) The reduced configuration space Ω*. Each point inside of Ω* corresponds to a possible combination of islands on the surface. Because we do not distinguish between islands that differ only in their position or orientation, the reduced configuration space is smaller than the space Ω* in Fig. 4. The MCMC procedure is performed on Ω* to identify combinations of islands that have a high probability of appearing at thermodynamic equilibrium

Having identified islands as the target for our predictions, we can form a *reduced configuration space* Ω*. The points in Ω* correspond to unique *combinations of islands* that could result from the molecular self-assembly process (Fig. 6b). The reduced configuration space Ω* is indeed much smaller than the original configuration space Ω. For each island combination, Ω must contain one point for each of the ways in which those islands can be arranged or oriented on the surface. The number of such points increases dramatically as the size of the surface and number of precursor molecules increases. A more subtle point is that number of points in Ω* does not increases as the size of the surface increases (providing that the number of precursor molecules is held fixed). It is therefore far easier to use the reduced configuration space for studying molecular self-assembly compared to Ω.

The major difficulty with ECS is setting up a Metropolis–Hastings random walk on the reduced configuration space Ω*. This is because the reduced configuration space is much more difficult to visualize than Ω, which makes its analysis less transparent. A successful construction of such a random walk, as well as a rigorous analysis of its properties is discussed in Ref. [26]. Note that this particular random walk is only accurate in the limit of low coverages of precursor molecules on the surface. While work to amend this situation is in progress, our implementation of ECS at present is limited to situations involving very low coverages of molecules on the surface.

4.2 Predictions for the Molecular Self-Assembly Process

A typical high-probability outcome for Br_2BA precursor self-assembly on Cu(111) is shown in Fig. 7a. These predictions were achieved using the ECS approach described above, where the energy $\varepsilon(\sigma)$ was computed according to the machine learning scheme discussed previously (Fig. 3). We find that our approach indeed predicts the formation of chain-shaped structures from the molecular self-assembly process, similar to what was seen in the STM images shown in Fig. 1. Figure 7b plots the probability of forming chain-shaped structures as a function of temperature. The probability of forming chain-shaped structures is predicted to increase to a maximum and then slowly decreases. This counterintuitive prediction is surprisingly correct, as confirmed by STM imaging shown in Fig. 7c. The reason for this effect is actually a consequence of entropy (see Ref. [15] for more detail). Note that the reason for the discrepancy between the predicted and experimental temperatures at which the peak occurs is due to the fact that the calculation assumed a much lower coverage of molecules on the surface than was used in the experiment. This is due to the limitation of ECS to cases involving low coverages of molecules on the surface.

As a further showcase of our method, we consider the case of $(CH_3)_2BA$ precursor self-assembly on Cu(111). $(CH_3)_2BA$ has an identical chemical structure to Br_2BA, except that the Br atoms are replaced by CH_3. In the case of $(CH_3)_2BA$,

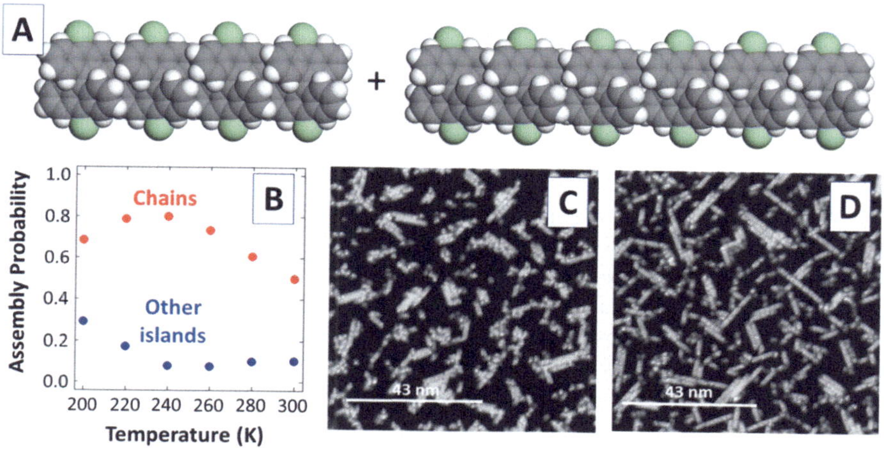

Fig. 7 Model predictions (Br_2BA self-assembly on Cu(111)). (**a**) A typical combination of islands predicted at room temperature. The surface (not explicitly shown) lies in the plane of the page. Both islands have a chain shape (**b**) Probability of assembly of chain-shaped islands as a function of temperature. (**c**, **d**) STM images of Cu(111) following Br_2BA deposition, following annealing at 360 °C and 400 °C, respectively. Model predictions in (**a**, **b**) were performed with equivalence class sampling, with energies calculated according to the scheme in Fig. 3 and ten precursor molecules, and a 50 unit cell × 50 unit cell Cu(111) surface with periodic boundary conditions. See Fig. 2 caption for STM imaging conditions

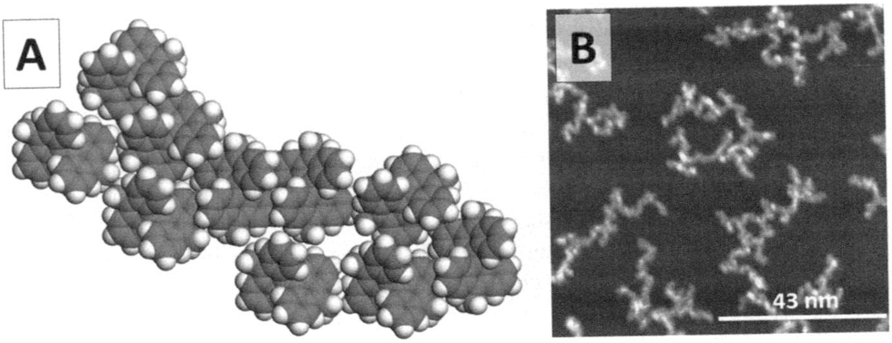

Fig. 8 Model predictions ($(CH_3)_2BA$ self-assembly on Cu(111)). (**a**) A typical island resulting from $(CH_3)_2BA$ self-assembly on a Cu(111) surface at 200 K. The surface (not explicitly shown) lies in the plane of the page. The probability for chain formation for this case is predicted to be less than 3%. (**b**) STM image of Cu(111) following $(CH_3)_2BA$ deposition, following annealing at 400 °C, respectively. Model predictions in (**a**) and (**b**) were performed with equivalence class sampling, with energies calculated according to the scheme in Fig. 3 and ten precursor molecules, and a 50 unit cell × 50 unit cell Cu(111) surface with periodic boundary conditions. An imaging temperature of 77 K was used to obtain the STM image. Other STM imaging conditions are shown in the Fig. 2 caption

chain-shaped islands almost never form (Fig. 8a). Instead, relatively formless islands consisting of mixtures of short chains and other structures emerge. This prediction also agrees with STM imaging (Fig. 8b), which confirms the absence of chain-shaped islands following self-assembly for this system. Unfortunately, the STM images are not of high enough resolution to examine the specific arrangement of molecules within these islands.

To summarize the situation so far, we have demonstrated that predictions on the outcome of the molecular self-assembly process for real chemical systems are possible via a combination of machine learning techniques and special Markov chain Monte Carlo sampling techniques. Predictions such as those shown here are not possible with the standard techniques from computational physics. Although some areas for technical improvement remain (such as the extension of equivalence class sampling to general coverage conditions), our approach can, in principle, be used to screen precursor molecules for those which form desired target structures upon self-assembly. Research in this direction has been reported in [28].

5 Another Approach: Bayesian Optimization for Model Predictions

Before concluding this chapter, we briefly mention how another machine learning method, called *Bayesian optimization*, shows immense potential for predicting the outcome of the molecular self-assembly process [29]. Unlike the methods described in Sect. 2, which were designed to predict numbers (energies), Bayesian optimization is designed to predict the "best object" from a large number of objects. Recent studies by other authors have shown that Bayesian optimization can considerably accelerate several time-consuming tasks in materials science, including as crystal database screening and structure optimizations [30–35]. Recently, Bayesian optimization was applied by Todorovic et al. to predict the adsorption conformation of single organic molecules on metal surfaces [36]. In the present case, the "objects" correspond to arrangements of molecules on the surface, and the "best object" is the one with the lowest energy. While Bayesian optimization does not overcome problems (A) or (C) mentioned above, it shows unprecedented performance when dealing with problem (B).

The basic concept behind Bayesian optimization is illustrated in Fig. 9. As in the previous section, we let Ω represent the configuration space of the system, where each point in Ω corresponds to one way of arranging the molecules on the surface. In Step I of Bayesian optimization, a small sample of data are acquired. In the present case, this sample corresponds to a sample of arrangements for the precursor molecules, as well as their energies. These energies are computed directly according to Eq. (1). In Step 2 of Bayesian optimization, a so-called prior probability density $f_\sigma(\varepsilon)$ is defined for each point σ in Ω. The prior probability density is defined so that the quantity

$$f_\sigma(\varepsilon)\Delta\varepsilon$$

measures our intuitive belief that the true energy of the system lies between ε and $\varepsilon + \Delta\varepsilon$ when the precursor molecules are arranged according to σ. In Step 3 of Bayesian optimization, the sample data is used to correct the prior probability den-

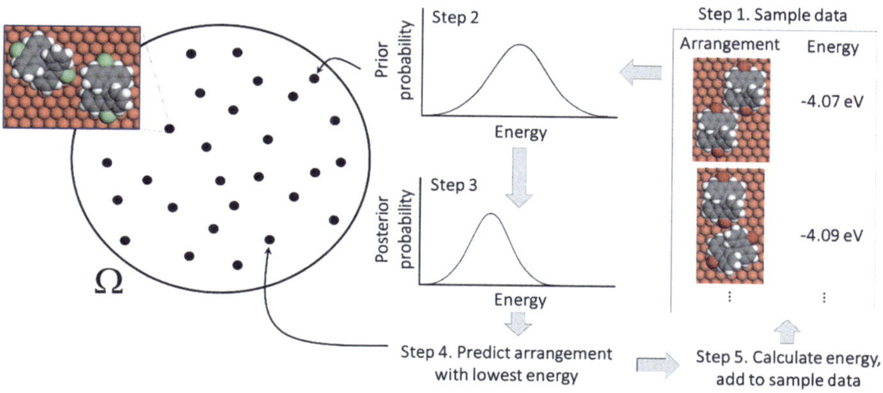

Fig. 9 Summary of the Bayesian optimization method. The case of two precursor molecules adsorbed to a metal surface is shown. Ω denotes the configuration space for the molecules. See main text for details

sity proposed in the first step. This step makes use of the famous Bayes's formula from probability theory,

$$g_\sigma\left(\varepsilon|S\right) \propto L\left(S|\varepsilon\right) f_\sigma\left(\varepsilon\right) \tag{6}$$

In Eq. (6), $L(S|\varepsilon)$ is referred to as the *likelihood function*, and is constructed on the basis of the sample data collected in the second step (S stands for sample data). The updated prior probability density, $g_\sigma(\varepsilon|S)$ is referred to as the *posterior distribution*. Finally, in Step 4 of Bayesian optimization, we use the posterior distribution to identify which point in Ω has the highest probability of being the energy-minimizing arrangement of precursor molecules. The energy for this precursor molecule arrangement is computed accurately from density functional theory (DFT), and the results are added to the original sample data from the first step. A single iteration of Bayesian optimization corresponds to Steps 2–4 above. In Bayesian optimization, the cycles are iterated until the minimum energy precursor arrangement in the configuration space is been found. A detailed introduction to Bayesian optimization specifically written for materials scientists is available in Ref. [37].

A major appeal of Bayesian optimization is that it allows for materials science expertize to be incorporated into the search procedure. Such materials science expertize can be incorporated into the prior probability density. In this sense, Bayesian optimization utilizes human knowledge and corrects it through the use of probability theory. If our expertize is not sufficient and our prior distribution turns out to be an inadequate predictor for the energy of the precursor molecule arrangements, then the Bayesian optimization method can still be applied. However, the number of iterations needed for the procedure to complete will generally be small when the prior probability distribution is well chosen.

We have applied Bayesian optimization to the simple case of two Br_2BA molecules adsorbed to Cu(111) [27]. In this case, we assumed that one of the molecules (molecule 1) was fixed at a specific point on the surface, and that the position of the other molecule (molecule 2) could be varied. To reduce the number of places where molecule 2 could reside, we assumed that it can only reside at adsorption sites and in the orientations described previously (see Fig. 2). Moreover, we excluded cases which were clearly unphysical (such as when the atoms of molecule 2 were far too close to those of molecule 1), or cases where the two molecules are far apart and not interacting. Under these assumptions, there are 294 distinct positions (adsorption site and orientation) for molecule 2. On our hardware, it takes between 20 and 30 h to accurately compute the energy of the system for any one of these 294 choices via density functional theory calculations (with full incorporation of the copper atoms). Thus, even for the simple case of two molecules adsorbed to Cu(111), tens of months would be needed to identify the most energetically stable arrangement of molecules via computational screening.

Figure 10 shows the result of applying Bayesian optimization to this system [27]. Starting with a random sample of ten positions for molecule 2 as well as their energies, we found that the most energetically stable position for molecule 2 could be

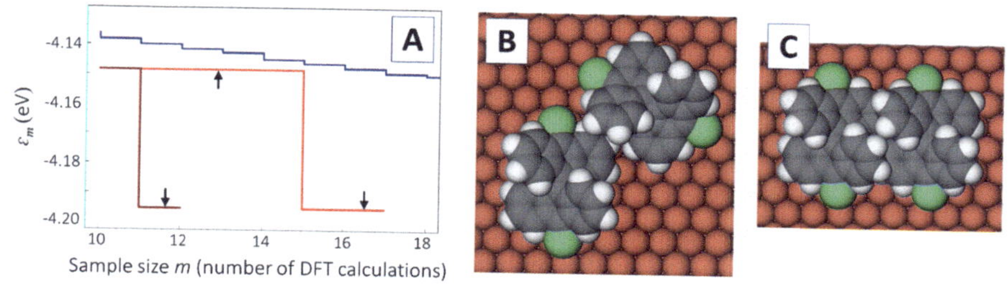

Fig. 10 Bayesian optimization applied to the case of two Br_2BA molecules on Cu(111). (**a**) The red and dark-red lines indicate two independent runs of the method. Both runs used initial samples of ten randomly chosen pairwise interactions. ε_m is the lowest energy in the sample, as calculated by Eq. (1). The blue line is the lower-bound to ε_m when uniform random sampling is used instead of Bayesian optimization. (**b**, **c**) The molecule arrangements corresponding to the stabilization energy are marked up-pointing arrow and down-pointing arrows in (**b**), respectively. The molecule arrangement in (**c**) is the lowest energy arrangement for the two molecules. Note that for both runs, both samples happened to contain molecule arrangement in (**b**). See Ref. [27] for more details

identified within only one cycle of Bayesian optimization. To confirm this remarkable result, we repeated the calculation using another, independently generated, random sample of ten positions, and found that the optimal positon could be identified within only five cycles of Bayesian optimization. In total, both cycles required only 11 and 15 energy calculations, respectively, which amounts to around 2.5 weeks of computation on our hardware. Bayesian optimization. Note that by coincidence, both trials contained the same low-energy case in their initial random sample, which is why both curves start at a relatively low energy close to −4.15 eV. While our trial had a "lucky" initial sample, the results, nonetheless, show that Bayesian optimization is significantly more efficient than random sampling (blue curve). The result is particularly remarkable considering that only around 5% of the 294 possible positions for molecule 2 were examined during the procedure.

While Bayesian optimization certainly shows great promise for studying molecular self-assembly, it is important to note three serious shortcomings. The first shortcoming is that the Bayesian optimization algorithm requires specification of several parameters (called *hyperparameters* in this context) in order to be used. The performance of Bayesian optimization depends critically on the value of these parameters, and so they need to be carefully chosen in order to achieve results similar to those in Fig. 10 In our study, we were able to specify the value of these parameters via a physically motivated procedure (see Ref. [27]), however for many systems such a procedure may not be possible. The second shortcoming is that, in order to identify the next object for the calculation (see Step 4 in Fig. 10), it is necessary that the entire space of objects can be generated and stored in the computer's memory. Moreover, the value of the posterior distribution for each of these objects must be calculated. For the case of molecular self-assembly involving tens or hundreds of molecules, the number of objects (ways of arranging the precursor molecules on the surface) becomes exceeding large, and Bayesian optimization becomes very difficult to implement. The final shortcoming is that, while Bayesian optimization so far has been applied to identify the minimum energy arrangement of molecules on the surface, further work is necessary in order to predict the minimum *free energy arrangement* of molecules on the surface. Such predictions are more relevant to the actual outcome of the molecular self-assembly process, as they incorporate the effects of thermodynamics. For this reason, MCMC methods are currently preferred over Bayesian optimization for studying precursor self-assembly. On the other hand, Bayesian optimization has a considerably shorter history than MCMC, and rigorous mathematical studies on this technique have only just begun. We therefore believe that methods for overcoming these shortcomings will be developed, and that Bayesian optimization can improve ability to predict the outcome of the molecular self-assembly process.

6 Conclusions

In order to predict the structures formed from the molecular self-assembly, novel theoretical methodologies which can handle the enormous space and time scales involved in this process are necessary. While there is room to develop new methodologies based on judicious physical approximations [3–5], modern techniques from applied mathematics, such as machine learning and Markov chain Monte Carlo, appear very promising in this area. The application of such techniques to molecular self-assembly is in its infancy, and at present it is difficult to gauge the limitations of these approaches. Nonetheless, with interest in such techniques rapidly increasing, we expect that these approaches will become properly evaluated as more researchers join this field.

As our work has shown, machine learning can be used to construct predictive models, and Markov chain Monte Carlo using reduced configuration spaces can be used to quickly solve these models. We have also demonstrated the use of machine learning in accelerating structure searches involving density functional theory calculations. While we have applied these techniques in the limited context of on-surface organic precursor molecule self-assembly, the general concepts can be extended to other types of systems, including to self-assembly processes that occur in biological environments.

Acknowledgments The work reported here has been supported by the following grants: Japan Science and Technology Agency PRESTO "Collaborative Mathematics for Real-World Issues" Grant No. 100167050008, JSPS Kakenhi Shingakujyutsu "Exploration of Nanostrucure-Property Relationships for Materials Innovation" Grant No. 836167050004, JSPS Kakenhi Wakate Kenkyu Grant No. 18K14126, and the World Premier Research Institute Initiative promoted by the Ministry of Education, Culture, Sports, Science and Technology of Japan (MEXT) for the Institute for Integrated Cell-Material Sciences (iCeMS), Kyoto University, and the Advanced Institute for Materials Research (AIMR), Tohoku University. Collaboration with Patrick Han (Tohoku University) and Taro Hitosugi (Tokyo Institute of Technology) is kindly acknowledged.

References

1. Whitesides GM, Grzybowski B (2002) Self-assembly at all scales. Science 295:2418
2. Stanchak J (2017) What's the most important unsolved problem in chemistry? ACS Axial Blog (18 Jan 2017). http://axial.acs.org/2017/01/18/most-important-problem-chemistry/
3. Roussel TJ et al (2014) Predicting supramolecular self-assembly on reconstructed metal surfaces. Nanoscale 6:7991
4. Copie G et al (2015) Surface-induced optimal packing of two-dimensional molecular networks. Phys Rev Lett 114:066101
5. Perkett MR, Hagan MF (2014) Using Markov state models to study self-assembly. J Chem Phys 140:214101
6. Wakayama Y (2016) On-surface molecular nanoarchitectonics: from self-assembly to directly assembly. Jpn J Appl Phys 55:1102AA
7. Wee A et al (2016) An update from Flatland. ACS Nano 10:8121
8. Han P et al (2014) Bottom-up graphene-nanoribbon fabrication reveals chiral edges and enantioselectivity. ACS Nano 8:9181
9. Han P et al (2015) Self-assembly strategy for fabricating connected graphene nanoribbons. ACS Nano 9:12035
10. Xu W, Lee T-W (2016) Recent progress in fabrication techniques of graphene nanoribbons. Mater Horiz 3:186–207
11. Mullen K (2014) Recent progress in fabrication techniques of graphene nanoribbons. ACS Nano 8:6531
12. Giustino F (2014) Materials modelling using density functional theory. Oxford University Press, Oxford
13. Li X, Packwood DM (2018) Substrate-molecule decoupling induced by self-assembly – implications for graphene nanoribbon fabrication. AIP Adv 8:045117
14. Gao DZ et al (2015) Efficient parametrization of complex molecule-surface force fields. J Comp Chem 36:1187
15. Packwood D et al (2017) Chemical and entropic control on the molecular self-assembly process. Nat Commun 8:14463
16. Kresse G, Furthmuller J (1996) Efficient iterative schemes for ab initio total-energy calculations using a plane-wave basis set. Phys Rev B 54:11169
17. Klimes J et al (2011) Van der Waals density functionals applied to solids. Phys Rev B 83:195131
18. Klimes J et al (2010) Chemical accuracy for the van der Waals density functional. J Phys Condens Matter 22:022201
19. van der Hamada I (2014) Waals density functional made accurate. Phys Rev B 89:121103
20. Obersteiner V et al (2017) Structure prediction for surface-induced phases of organic monolayers: overcoming the combinatorial bottleneck. Nano Lett 17:4453
21. Wasio N et al (2017) Correlated rotational switching in two-dimensional self-assembled molecular rotor arrays. Nat Commun 8:160577
22. Murphy K (2012) Machine learning: a probabilistic perspective. MIT Press, Cambridge, MA
23. Rupp M et al (2012) Fast and accurate modeling of molecular atomization energies with machine learning. Phys Rev Lett 108:058301
24. Hansen K et al (2013) Assessment and validation of machine learning methods for predicting molecular atomization energies. J Chem Theory Comput 9:3404
25. Robert CL, Casella G (2004) Monte Carlo statistical methods. Springer, New York, NY
26. Packwood DM, Han P, Hitosugi T (2016) State-space reduction and equivalence class sampling for a molecular self-assembly model. R Soc Open Sci 3:150681
27. Packwood D, Hitosugi T (2017) Rapid prediction of molecule arrangements on metal surfaces via Bayesian optimization. Appl Phys Express 10:065502
28. Packwood DM, Hitosugi T (2018) Materials informatics for self-assembly of functionalized organic precursors on metal surfaces. Nat Commun 9:2469

29. Snoek J, Larochelle H, Adams RP (2012) Advances in neural information processing systems. NIPS Conf 25:2951
30. Seko A et al (2015) Prediction of low-thermal-conductivity compounds with first-principles anharmonic lattice-dynamics calculations and Bayesian optimization. Phys Rev Lett 115:205901
31. Seko A et al (2014) Machine learning with systematic density functional theory calculations: application to melting temperatures of single- and binary-component solids. Phys Rev B 89:054303
32. Kiyohara S et al (2016) Acceleration of stable interface structure searching using a kriging approach. Jpn J Appl Phys 55:045502
33. Ueno T et al (2016) COMBO: an efficient Bayesian optimization library for materials science. Mater Discov 4:18
34. Ju S et al (2017) Designing nanostructures for phonon transport via Bayesian optimization. Phys Rev X 7:021024
35. Packwood DM (2020) Exploring the configuration spaces of surface materials using time-dependent diffraction patterns and unsupervised learning. Sci Rep 10:5868
36. Todorovic M et al (2019) Bayesian inference of atomistic structure in functional materials. NPJ Comput Mater 5:35
37. Packwood DM (2017) Bayesian optimization for materials science. Springer series in the mathematics of materials. Springer, New York, NY

DNA Nanotechnology to Disclose Molecular Events at the Nanoscale and Mesoscale Levels

Masayuki Endo

1 Introduction

DNA is the blueprint of life, which carries genetic information. At the same time, from a chemistry viewpoint, DNA is an excellent supramolecule characterized by the formation of sequence-specific molecular assemblies and defined double helical structure. Various methods have been developed to design and create nanoscale structures using the programmability and characteristic structural periodicity of DNA molecules [1, 2]. In particular, an innovative DNA self-assembly system called DNA origami has been developed for the design and construction of various two-dimensional (2D) [3] and three-dimensional (3D) nanostructures [4] based on well-established DNA nanotechnologies. Using DNA origami technology, researchers in various fields have conducted basic research in computer science and nanoscience with applications for material and biological sciences such as photonics, plasmonics, electronics, mechanics, diagnosis, and therapeutics [1, 2].

Visualization of target biomolecules is relatively a straightforward approach to study physical properties of molecules involved in various phenomena in living systems. For example, single-molecule imaging using a scanning probe microscope is a practical approach for investigating the motions of biomolecules during reactions. Atomic force microscopy (AFM) enables the direct observation of biomolecules, their various functional states, and chemical reactions under physiological conditions at nanoscale spatial resolution. To facilitate the observation of single biomolecules, a versatile DNA origami scaffold is often needed for the precise

M. Endo (✉)
Graduate School of Science and Institute for Integrated Cell-Material Sciences, Kyoto University, Kyoto, Japan
e-mail: endo@kuchem.kyoto-u.ac.jp

D. O. Wang, D. Packwood (eds.), *Cell-Inspired Materials and Engineering*, Fundamental Biomedical Technologies, https://doi.org/10.1007/978-3-030-55924-3_4

analysis of interactions and reactions [2, 5, 6]. Using this method, the detailed functional dynamics of the molecules can be visualized.

In the past decade, visualization of molecular movements during biological reactions at the subsecond timescale has been achieved using high-speed AFM (HS-AFM) [7–10]. The HS-AFM records images at 5–40 frames per second rate (depending on the scanning dimension and the number of data points), which enables imaging of the dynamic movements of biomolecules in real-time at molecular resolution [11, 12]. Combining the DNA origami system and HS-AFM imaging, dynamic movement of molecules during enzymatic reactions, DNA structural changes, DNA photoreactions, DNA catalytic reactions, and RNA interactions can be imaged at the single-molecule level [5, 13]. Soft surfaces such as lipid bilayers are used to directly observe the dynamic formation of DNA origami assemblies using HS-AFM. In addition, optical tweezers are employed for the single-molecule detection and to determine the physical properties of the biomolecules in the confined space of the DNA origami scaffold. The target-oriented design of DNA origami nanostructures and improvements to both HS-AFM imaging and optical tweezers techniques have allowed these imaging and detection systems to be extensively used in elucidating the physical properties of individual molecules, assemblies, and structures involved in both biological and nonbiological events.

2 DNA Origami for Nanostructure Construction

DNA origami is a designable 2D DNA nanostructure formed by the programmed self-assembly of DNA molecules to assume a defined size and shape (approximately 100 nm), originally developed by Paul Rothemund at the California Institute of Technology in 2006 [3]. In this method, a long single-stranded DNA (M13mp18, 7249 bases) and multiple predesigned short complementary DNA strands (called "staple strands" whose sequences are designed according to the intended structures) were mixed and heated at 85 °C, then allowed to gradually cool down to room temperature (annealing) for about 1 h. This procedure results in the formation of a new DNA structure in the predesigned shape that can be observed by AFM (Fig. 1). During the annealing process, staple DNA strands hybridize complementarily to predefined positions along the long single-stranded DNA and induce folding into its predesigned 2D structure. Staple DNA strands were also used to conjugate additional DNA strands, hairpin DNAs, functional molecules, and nanoparticles at designated positions to the DNA origami structures [14, 15]. Therefore, specified positions within DNA origami can be targeted based on their sequence information (addressability). The DNA origami method has been a major scientific breakthrough to design nanoscaled structures reliably and arrange molecules in a programmed fashion.

We have pioneered a DNA origami-based method to both directly visualize the motions of biomolecules and synthetic molecules in a defined DNA origami nanostructure and nanospace using HS-AFM, and to regulate their reactions in the

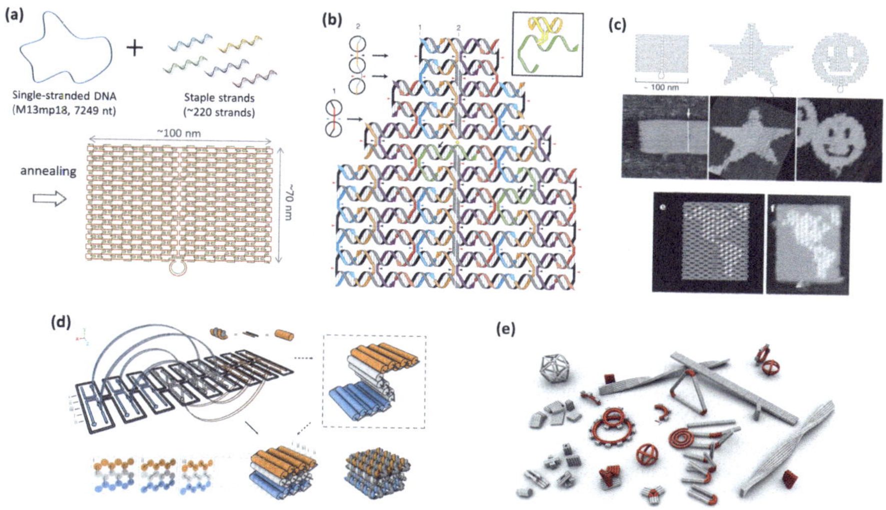

Fig. 1 Design and preparation of DNA origami structures. (**a**) Assembly of the DNA origami structure by annealing long single-stranded DNA and short complementary DNA strands (staple strands). (**b**) Geometric diagram of the two-dimensional DNA origami structure. The black strand represents the scaffold strand and the colored strands represent the staple strands. To maintain flatness, the separation of adjacent crossovers are 32 bp for three helical turns. Inset: staple DNA with a dumbbell hairpin to display a dot on the DNA origami. (**c**) The designed DNA origami structures and their AFM images. Hairpin DNAs are introduced onto the DNA origami surface to display a figure with white dots as pixels. (**d**) Design and geometry of the three-dimensional (3D) DNA origami structure. The pleated sheet and crossovers were used for connection in the vertical direction and these are folded into layered and honeycomb structures. (**e**) The designed 3D DNA origami structures

nanostructures [16]. In addition, physical properties of a target molecule can be elucidated by single-molecule manipulation in the designed DNA origami nanospace. The assembling process of DNA nanostructures has also been visualized on a lipid bilayer surface using HS-AFM. These developed methods can be utilized to determine the dynamic events of individual molecules in real-time at the molecule resolution.

3 Direct Observation and Regulation of Enzyme Reactions in the DNA Nanostructures

Several studies have been conducted using HS-AFM to observe the dynamic motions of the enzymes interacting with a dsDNA substrate [17–19]. These studies show the difficulty in obtaining a homogeneous population of dsDNA substrates, because dsDNA forms various shapes. To overcome this problem, we created an observation scaffold called a "DNA frame" using DNA origami in which substrate

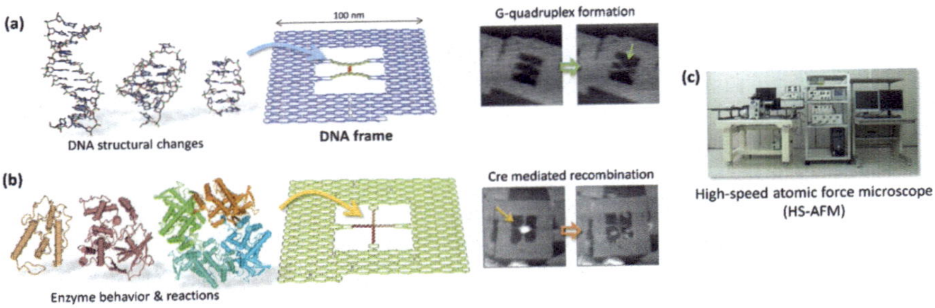

Fig. 2 Single molecule observation of the DNA structural changes and enzyme reactions in the DNA origami frame using high-speed atomic force microscope (HS-AFM). (**a**) A target DNA structure of interest is attached in the DNA origami scaffold called "DNA frame" and the structural change of the target DNA is directly observed using HS-AFM. (**b**) A DNA substrate of the target enzyme and protein is attached in the DNA frame, and the interaction with the target enzyme and protein and the behavior and reaction are directly observed using HS-AFM. (**c**) HS-AFM used in the experiment (RIBM Nano Live Vision)

dsDNAs are incorporated. This robust DNA frame could accommodate two different dsDNA fragments in its cavity (approximately 40 nm × 40 nm) and control their physical properties such as the tension, rotation, helicity, and topology (Fig. 2) [13, 16]. We used the DNA frame system for visualization of the dynamic movement and reaction and elucidation of the reactions including DNA methylase [13, 20], repair enzymes, 10–11 translocation (TET) enzyme [21], DNA recombinase [22], Holliday junction resolvase [23, 24], and Cas9 nuclease [25].

3.1 DNA Methylation

DNA-modifying enzyme often requires bending of the dsDNA substrates at a specific position to facilitate the reaction. DNA methylation enzyme, *Eco*RI methyltransferase (M.*Eco*RI) bends dsDNA by 55–59°, enabling the methyl-transfer reaction to proceed [26]. To visually examine the DNA bending effect on methylation with M.*Eco*RI and the consequence of methylation on DNA cleavage, two dsDNA fragments of different lengths, 64 bp (tense strand) and 74 bp (relaxed strand) dsDNA, were incorporated into the DNA frame structure (Fig. 3a) [13]. Sixty-four-bp-dsDNA fits the length across within the DNA frame thus its conformational flexibility is suppressed, while the 74 bp-dsDNA has ten extra base-pairs allowing the flexibility for local dsDNA bending. The dynamic movements of the dsDNAs and the formation of M.*Eco*RI complexes with dsDNAs were observed using HS-AFM (Fig. 3b). After this step, we applied *Eco*RI restriction enzyme to the same framework to visualize DNA cleavage, a process that DNA methylation inhibits. AFM analysis revealed that the 74 bp-dsDNA was less effectively cleaved compared to the 64 bp-dsDNA. This result indicated that the methylation occurred

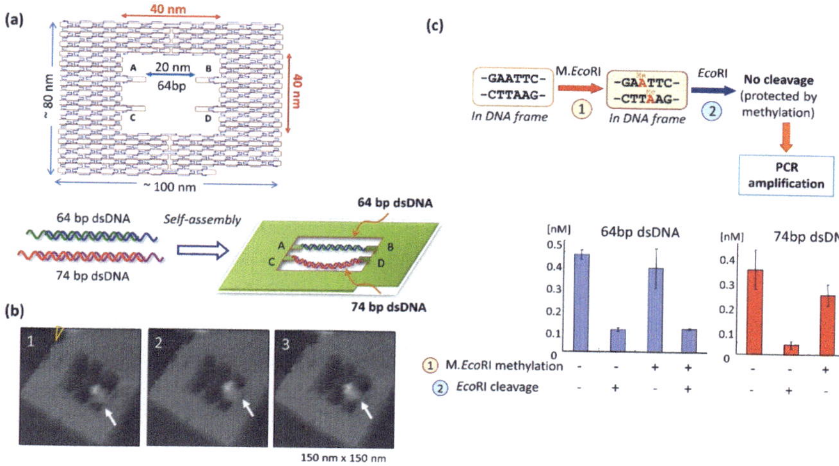

Fig. 3 Regulation of the methylation with M. *Eco*RI in the DNA origami frame and HS-AFM images of M. *Eco*RI movement. (**a**) DNA frame structure designed to incorporate two different dsDNAs, tense 64 bp-dsDNA and relaxed 74 bp-dsDNA, which contain a specific sequence for M. *Eco*RI at the center. (**b**) HS-AFM images of M. *Eco*RI moving on the 64 bp-dsDNA in the DNA frame. Successive AFM images were taken at a scanning rate of 1.0 frame/s. (**c**) Quantification of DNA methylation in the DNA frame. If methylation occurs, the target sequence is protected from the subsequent cleavage by restriction enzyme *Eco*RI, resulting in amplification of the target sequence by PCR. The graphs show the quantification of methylation of the target site in 64 bp- and 74 bp-dsDNA by quantitative PCR amplification

preferentially in the relaxed 74 bp-dsDNA. This result was supported by quantitative polymerase chain reaction (qPCR), which showed more efficient amplification of the 74 bp-dsDNA substrate (Fig. 3c). These results demonstrate the importance of structural flexibility for efficient methyl transfer with M.*Eco*RI and the potential use of DNA frame to apply tension to dsDNA to regulate their methylation potential.

3.2 DNA Base-Excision Repair

DNA repair is an indispensable biological function to preserve genetic information from mutations such as transversion (a single purine is changed for a pyrimidine, or vice versa) during replication [27]. The DNA base-excision repair enzymes, 8-oxoguanine glycosylase (hOgg1) and T4 pyrimidine dimer glycosylase (PDG) are the enzymes responsible for initiating the base excision repair pathway by recognizing the damaged target base and catalyzing the breakage of the base-sugar glycosyl bond. These enzymes often require local structural changes in the target DNA strands such as DNA bending; hOgg1 bends dsDNA about 70° to flip out the 8-oxoguanine base for the reaction to proceed [28]. PDG also bends the double helix by 60° to flip out the 3′-side of adenine in the opposite strand of the thymine dimer [29–32]. We employed similar DNA frames we designed to regulate M.*Eco*RI activity for the analysis of repair reactions in the DNA frame [20]. Various dsDNAs containing a damaged base were incorporated into a DNA frame as dsDNA cassettes, and the enzyme movement, structural effect of glycosylase/AP lyase activity including cleavage of the DNA strand and the trapping of reaction intermediates, were visualized on the 64 bp- and 74 bp-dsDNA in the DNA frame. The enzymes favored the relaxed 74 bp-dsDNA over tense 64 bp-dsDNA and generated an intermediate, termed the Schiff base, that when reduced by $NaBH_4$, formed an irreversible covalent complex between the DNA and enzymes visible under AFM. The DNA frame system has detailed the repair process by allowing direct observation of the events involved in DNA repair reactions such as binding, sliding, the catalytic reaction, and dissociation.

3.3 DNA Recombination

DNA recombination plays an important role in triggering genetic diversity and mediating DNA integration into the host genome [33]. The Cre/*loxP* recombination is catalyzed by Cre recombinase that recognizes a 34 bp *loxP* sequence. During recombination, four Cre recombinases and two *loxP* sites form a synapsed structure in which the DNA strands resemble models of four-way Holliday junction (HJ), followed by the exchange sets of strands to resolve the intermediate into recombinant products (Fig. 4a) [34, 35].

To visualize Cre-*lox*P recombination, substrate dsDNAs containing the *lox*P sequences were incorporated into the DNA frame in either antiparallel or parallel arrangement and incubated with Cre recombinase (Fig. 4b) [22]. The Cre–DNA complex and recombinant products were clearly observed in the DNA frame, demonstrating the occurrence of recombination in the nanospace (Fig. 4c). Time-lapse imaging showed that the Cre–DNA complex formed first, followed by the recombinant product (Fig. 4d). During observation of the Cre–DNA complex with HS-AFM, the tetrameric Cre in the synaptic complex dissociated into four monomers, and the recombinant product simultaneously appeared (Fig. 4e). However, Cre-tetramer bound to parallel DNA strands dissociated without giving recombinant products.

The DNA frame can be used to control the geometry of the Holliday junction (HJ) such as the angle relating the intersecting DNA strands and impose structural stress. We prepared DNA frames where the HJ intermediates crossed at either 90° or 60°, and found that HJ intermediates crossing at 90° were resolved into typical expected products [36, 37]. On the other hand, the reaction with HJ intermediates crossing at 60° changed the proportion of the resolution products that formed. Therefore, the structural stress imposed on the HJ intermediates in the DNA frame can regulate the direction of recombination. The desired geometric arrangements of substrate dsDNAs using DNA frames are valuable for studying recombination events, which are controlled by the orientation of substrate dsDNAs and the angle of HJ intermediates.

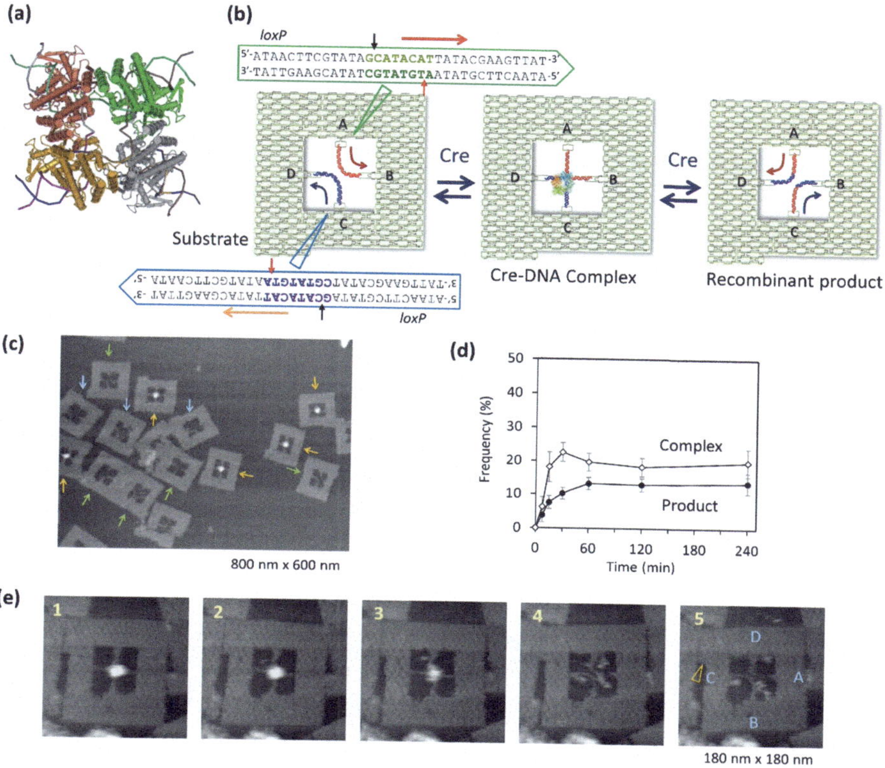

Fig. 4 Regulation and single-molecule observation of Cre-mediated DNA recombination in the DNA frame. (**a**) Crystal structure of the Cre tetramer bound to the Holliday junction intermediate (PDB: 3CRX). (**b**) Substrate dsDNAs having a *loxP* sequence in the antiparallel orientation were placed onto the DNA frame. In the DNA recombination, Cre tetramer forms a synaptic complex and subsequently generates a recombinant product, which is easily identified by the topological arrangement of two dsDNAs as different loop structures and cross-shape. (**c**) AFM image of the substrate in the DNA frame after incubation with Cre. The Cre–DNA complex, starting substrate, and recombinant product are indicated by orange, green, and blue arrows, respectively. (**d**) Time-course analysis of the formation of the Cre–DNA complex and recombinant product. (**e**) HS-AFM images of the dissociation of the Cre tetramer from the dsDNAs into four Cre monomers and the appearance of a recombinant product. Successive AFM images were taken at a scanning rate of 1.0 frame/s

The HJ-containing DNA frame was used for the observing the resolution of HJs by Rec U resolvase [24]. The resolved products were observed in the DNA frame, indicating that the HJ in the DNA frame can also be a substrate for the resolvase. Furthermore, an enzyme with unknown function was also characterized using the HJ-containing DNA frame. We visualized the binding preference and the endonuclease activity of monokaryotic chloroplast 1 (MOC1 in *Arabidopsis thaliana*) by using HS-AFM [23]. The interaction of MOC1 with the center of the HJ and symmetric cleavage of the HJ structure were observed in the DNA frame. Using the

DNA frame for observing DNA recombination and resolution, we first identified the MOC1 functions as an HJ resolvase in chloroplasts.

3.4 Cas9-Induced DNA Cleavage

The CRISPR-cas9 system in genome editing has been widely investigated because of its potential therapeutic applications [38]. Cas9 interacts with a short guide RNA (sgRNA) to hybridize to and cleave a target dsDNA (Fig. 5a) [39]. To examine the effect of torsional constraints imposed on dsDNA substrates on Cas9 cleavage, we prepared "locked" and "rotatable" DNA substrates using a DNA origami frame (Fig. 5b) [25]. After the addition of Cas9/sgRNA to the substrate in the DNA frame, complex formation and dsDNA cleavage were visualized using AFM (Fig. 5c). The cleavage of DNA substrates by Cas9 was quantified using quantitative PCR (qPCR) (Fig. 5d). The results revealed that when the nontarget DNA strand was under torsional constraint, cleavage efficiency of Cas9 of the target strand was significantly reduced, whereas torsional constraints on the target strand had little effect on

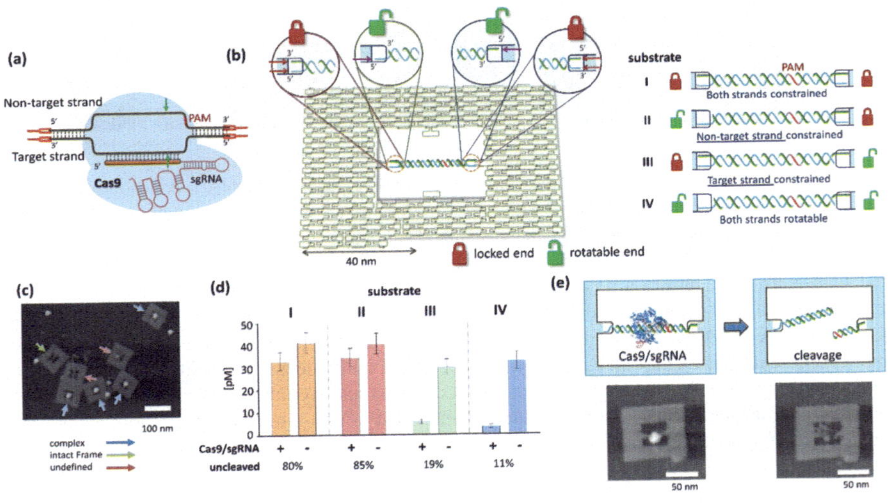

Fig. 5 Control of Cas9 cleavage reaction by introducing constrains on the substrate dsDNA using the DNA frame. (**a**) Cas9/sgRNA complexed with dsDNA by hybridization of sgRNA to the target DNA strand. Four terminals of dsNDA strands were construed. PAM and cleavage site (green arrow) are indicated. (**b**) DNA frame and the constrained substrate dsDNAs incorporated into the DNA frame. Substrate I: both strands fully constrained, substrate II: nontarget strand constrained, substrate III: target strand constrained, substrate IV, both strands rotatable. (**c**) AFM image of the reaction of Cas9 with substrate dsDNA in the DNA frame. (**d**) Quantification of the Cas9 cleavage reaction with four substrates with constraints by quantitative PCR. (**e**) HS-AFM images of the complex formation of Cas9 with substrate dsDNA in the DNA frame (left) and after the reaction (right)

cleavage efficiency. This study suggests that highly ordered and constrained DNA structures could impede Cas9 genome-editing efficiency. Moreover, we monitored the binding of Cas9/sgRNA complex to the DNA substrates and characterized the dissociation of the complex and dsDNA cleavage using HS-AFM (Fig. 5e).

4 DNA Binding Proteins and RNA Polymerase

4.1 *Zαβ Protein Binding to Z-Form DNA*

The DNA-binding domain of double-stranded RNA adenosine deaminase (ADAR1), called the Zα domain, specifically binds to the Z-form of CG repeat-containing dsDNA in the equilibrium state of the B–Z transition under physiological conditions [40]. Crystal structure studies revealed that two Zα domains form a stable complex with a Z-forming CG repeat sequence. For control of the Zα protein binding, we designed and constructed structurally constrained dsDNAs by introducing them into a DNA frame nanostructure, in which the rotation of dsDNAs were suppressed. The Zα protein should bind preferentially to the rotatable double helix rather than the constrained dsDNA. Using the same strategy used for the Cas9 study (Fig. 5), the dsDNA containing the CG repeat was attached to the two connection sites of the DNA frame using four terminals of DNA strands to prevent free rotation of the dsDNA substrate. Binding of the Zαβ protein to the constrained substrate was minimal (2.7%) compared to the binding observed with the rotatable dsDNA substrate (29%). While both substrates have same target sequence, Zαβ binding can be controlled by just regulating the rotation of dsDNA substrates using the DNA frame. The results showed that the rotational freedom of the double helix is an important factor for protein binding and reactions. The method described here can be used to investigate the effects of double-helical rotation on protein binding by constraining dsDNAs in the designed DNA scaffold.

4.2 *Cooperative Binding of Sox2-Pax6*

In embryonic stem cells, Sox2, a core transcription factor that maintains pluripotency, works along with Oct4 by binding in tandem to DNA elements in the numerous stemness-related genes [41, 42]. We examined Sox2–Pax6 complex formation on regulatory element DNA at the single-molecule level using AFM [43]. The crystal structure of the high mobility group (HMG) domain of Sox2 [Sox2(HMG)] in complex with dsDNA suggests that Sox2 binds to the minor groove of dsDNA and bends the DNA toward the major groove [44]. Using a DNA origami scaffold containing two dsDNA strands with different levels of tensile force, preferential Sox2 binding to the DNA element was examined. The AFM observations indicated that

Sox2 preferentially bound to more relaxed dsDNA substrate. It was revealed that DNA bending is necessary for Sox2 binding. When Sox2 was bound to the dsDNA, Pax6 bound adjacent to Sox2, whereas Pax6 did not bind to dsDNA alone. In addition, it was revealed that the two transcription factors bind cooperatively by observing increased occupancy of Sox2–Pax6 on the DNA element compared to that of Sox2 alone. These results indicate that Sox2 first binds to the more relaxed dsDNA and bends the target site, then Pax6 binds cooperatively and stabilizes the Sox2-Pax6 heterodimer complexed with dsDNA. By imposing tension to the dsDNA substrate, the binding ability and interaction of two different transcription factors could be investigated, and the cooperative interaction between these proteins could be characterized using this method.

4.3 Movement of Photoresponsive Transcription Factor GAL4

Transcription factors function as molecular switches to regulate transcription; this dynamic behavior involving DNA needs further elucidation in order to understand the switching mechanism. DNA binding proteins, including transcription factors, play an important role in gene expression and control of cellular functions. Transcription factors function as molecular switches to regulate transcription; this dynamic behavior involving DNA needs further elucidation in order to understand the switching mechanism. Direct observation of the transcription factors and enzymes has been carried out and several searching mechanisms for target sites, including sliding, hopping, intersegmental transfer, and three dimensional diffusion, have been proposed [45–47]. However, the detailed observation of these proteins at nanoscale resolution has not been achieved. We demonstrated the direct visualization of the dynamic behavior of transcription factor GAL4 with photoswitching functions (GAL4-VVD) [48] in the DNA origami structure (Fig. 6) [49]. Using high-speed atomic force microscopy (HS-AFM), we observed photo-induced complex formation between GAL4-VVD and substrate DNAs. Dynamic behaviors of GAL4-VVD, such as binding, sliding, stalling, and dissociation with two substrate DNA strands containing specific GAL4 binding sites, were observed. We also observed interstrand hopping on two double-stranded (ds) DNAs. On a long substrate DNA strand that contained five binding sites, a series of GAL4-VVD/DNA interactions, including binding, sliding, stalling, and dissociation, could be identified (Fig. 6b). We also found a clear difference in the movement of GAL4-VVD while sliding or stalling in the AFM images. Detailed analysis of kymograph revealed that GAL4-VVD randomly moved on the dsDNA using sliding and hopping for rapidly search of specific binding sites, and then stalled to the specific sites for the stable complex formation (Fig. 6c). These results suggest the existence of different conformational mode of the protein for sliding and stalling. This single-molecule imaging system with the nanoscale resolution provides an insight into the searching mechanism used by the DNA binding proteins.

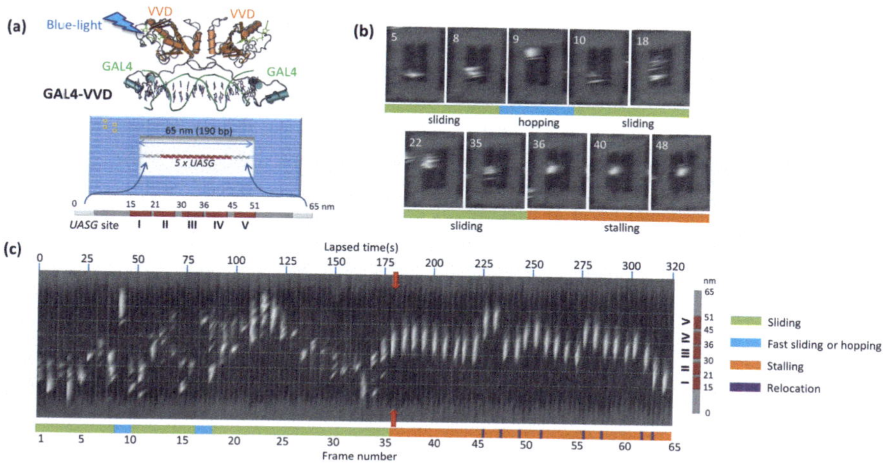

Fig. 6 Direct observation of GAL4-VVD on the dsDNA bridge in the DNA frame. (**a**) GAL4-VVD complexed with dsDNA. Binding is achieved by dimerization VVD part with blue-light irradiation. Illustration of DNA frame with dsDNA containing five *UASG* sites (158 bp) and linkers at both ends to connect to the DNA frame. (**b**) Time-lapsed HS-AFM images of the single GAL4-VVD movement on the substrate dsDNA in the DNA frame at various times. Frame number is indicated in the images. Scanning speed 0.2 frame/s. (**c**) Kymograph of the movement of GAL4-VVD on the dsDNA with 5 × *UASG* sites. Vertical and horizontal direction show the position on the dsDNA and lapsed time/frame number, respectively. The movements were categorized into sliding, hopping, stalling, and relocation

4.4 Transcription with RNA Polymerase

Transcription, a process of RNA synthesis from DNA template by RNA polymerase (RNAP), is the first step of central dogma of molecular biology. Transcription process involves a series of RNAP-mediated events, including binding to dsDNA, sliding along the dsDNA, synthesizing RNA copies, and dissociation from the dsDNA. We created a direct observation system using T7 polymerase and a dsDNA template (1.0 kb) containing the T7 promoter with its two ends attached to the designed DNA frame (Fig. 7a) [50].

Successive images of RNAP sliding on the dsDNA template were obtained. By analyzing the shape of the template dsDNA and the position of RNAP, it was determined that RNAP moved in the area where the dsDNA template was also mobile (Fig. 7b). The one-dimensional diffusion coefficient ($D = \langle \Delta l^2 \rangle / 2\,t$) obtained from the analysis of the AFM images was $5.1 \pm 0.7 \times 10^{-12}$ cm^2/s. Next, direct observation of transcription including RNA synthesis was examined on the dsDNA template-attached scaffold. In the presence of nucleotide triphosphates (NTPs), the transcript was synthesized from the dsDNA template-attached scaffold. RNAP movement from the promoter region and dissociation from the dsDNA template were observed on the HS-AFM images (Fig. 6c). In a series of reactions, single RNAP appeared around the promoter region (time = 2 s). The RNAP started to move downstream on

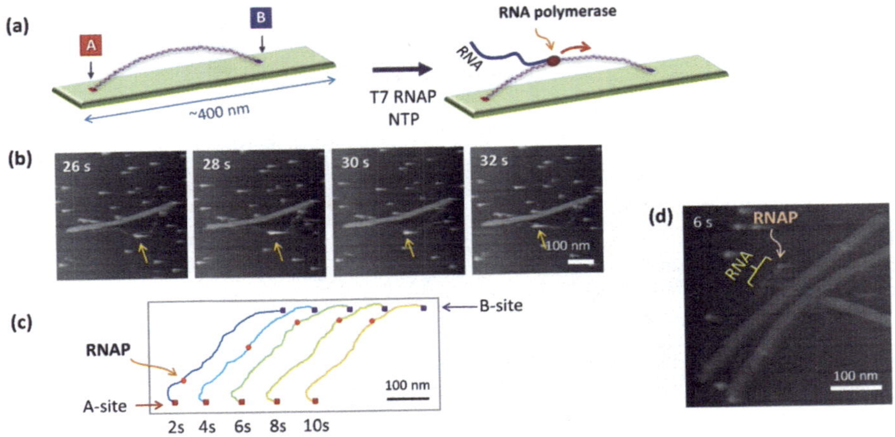

Fig. 7 AFM observation of the movement and reaction of RNA polymerase using a tape-like DNA scaffold. (**a**) Structure of a template dsDNA-attached scaffold and transcription by addition of RNAP and NTP. (**b**) Successive AFM images of RNAP sliding along the template dsDNA attached to the scaffold. (**c**) The traces of RNAP movement on the template dsDNA during transcription. (**d**) HS-AFM image of RNAP and RNA transcript production in transcription

the dsDNA template, with some unclear AFM scanning traces (time = 4–10 s), and continued to move to the middle of the dsDNA template. The RNAP finally dissociated from the dsDNA. In the expanded image (time = 6 s), unclear AFM scanning trace would show the RNA transcript synthesized by RNAP (Fig. 6d). The speed of the RNAP movement during transcription was not constant during the time-scale of AFM measurements due to contact between dsDNA template and the mica surface.

Transcription including RNAP sliding and RNA synthesis was successfully visualized using a dsDNA template-attached scaffold and HS-AFM. These observation systems can be used to characterize wide variety of events with DNA-binding proteins and enzymes that move along dsDNA during reactions. The combination of the DNA origami scaffold and HS-AFM is an innovative system for observing enzyme reactions and associated events including complex formation, catalysis, and dissociation at the single-molecule resolution.

5 Direct Observation and Regulation of DNA Structural Changes in the DNA Nanostructure

Structural variations and conformational changes in the DNA of living systems are closely linked to the regulation of their biological functions such as gene expression [51, 52]. The DNA frame system allows the introduction of various dsDNAs with unrestricted sequences for observing reactions, and DNA frames can also control the physical properties of dsDNAs such as their tensions and rotational freedom.

Here, the DNA frame system is used to the visualize DNA structural changes: (1) formation/disruption of G-quadruplexes, (2) G-quadruplex formation assembled from four strands, (3) formation/dissociation of dsDNA into GQ/i-motif structures, (4) triple helix formation, and (5) helical rotation in the B–Z transition.

5.1 G-Quadruplex Formation and Disruption

In the fields of structural and molecular biology, G-quadruplexes (GQ), four-stranded helix structures, are of great interest due to their variation in structure and biological function [53]. GQs are thought to be closely involved in biological functions such as the regulation of gene expression and cell fate control [54]. To detect GQ structure formation and disruption, we employed the DNA origami system to monitor the shapes of the dsDNA chains [55]. To place the G-rich sequences, two dsDNA chains containing single-stranded triple guanine (GGG) overhangs at their centers were prepared for interstrand GQ formation [56]. Three G-tracts were placed in the upper dsDNA chain, whereas the lower dsDNA chain had a single G-tract (Fig. 8a). Initially, the two dsDNA chains were not contact with each other. In the presence of K^+, the two dsDNA chains in the DNA frame clearly showed an X-shaped structure with 44% yield, demonstrating formation of an Interstrand GQ. The dynamic formation of the GQ was further examined in real-time by HS-AFM. During scanning of the sample in the presence of K^+, the two dsDNA

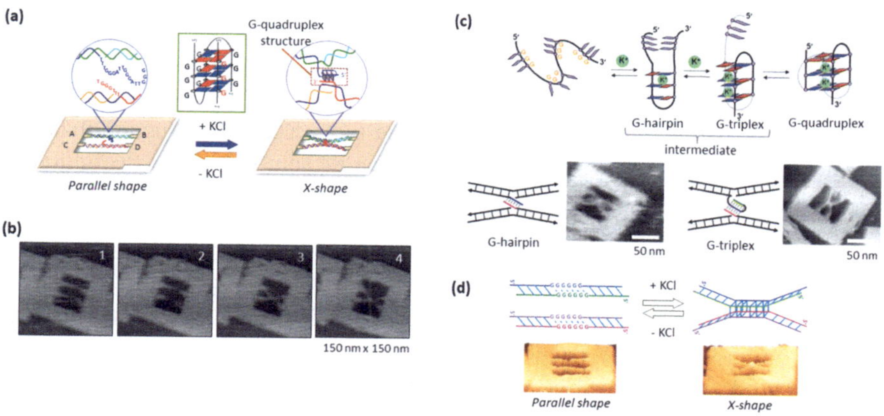

Fig. 8 Direct observation of G-quadruplex (GQ) formation by employing structural change of two supporting dsDNA chains placed in the DNA frame. (**a**) In the presence of KCl, the separated state of two supporting dsDNA chains changes to the X-shape by connecting at the center of two dsDNA chains via GQ formation. (**b**) HS-AFM images of the formation of an X-shape structure by the association of two dsDNA chains via GQ formation. Successive HS-AFM images were taken at a scanning rate of 0.2 frames/s. (**c**) Formation of G-hairpin and G-triplex intermediates in the DNA frame and their AFM images. (**d**) G-quadruplex formation from four G-tracts in the DNA frame. Antiparallel four G_6 strands were assembled to form GQ in the presence of KCl

chains with G-tracks remained separated for a given period, and then suddenly formed the X-shaped structure (Fig. 8b). In a similar fashion, the GQ disruption was directly observed in buffer without K^+. The X-shape remained unchanged for a while, and then reverted to the separated state during AFM scanning.

In addition, intermediate state such as G-triplex was observed using the DNA frame [57]. Using a DNA origami frame as a scaffold for structural control, these unprecedented solution-state structures of tetra-molecular antiparallel and (3 + 1)-type GQ intermediates such as G-hairpin and G-triplex were directly visualized at the single-molecule level with nanometer resolution (Fig. 8c). These intermediates were stable and formed with relatively high yield. A possible model of GQ folding was proposed.

Thus, single-molecule observation of the dynamic formation and disruption of GQ was successfully achieved by monitoring global structural changes of the two incorporated dsDNA chains in the DNA frame using HS-AFM.

5.2 G-Quadruplex Formation Using Four-Strand DNA Assembly

This observation system was used to visualize the formation of a synaptic GQ assembled from four DNA strands containing various G-tracts in the DNA frame [58]. By incorporating two duplex DNAs with G–G mismatch repeats in the center, we examined the conformational changes inside the DNA origami frame by monitoring the topology changes of the strands (Fig. 8d). In the absence of K^+, incorporated duplexes did not interact and laid parallel to each other. Addition of K^+ induced the formation of a GQ structure by stably binding the duplexes to each other in their centers. GQ formation allowed for DNA synapsis without disturbing the duplex regions of the participating sequences, resulting in an X-shaped structure that was monitored by AFM. Further, the GQ formation in K^+ solution and its disruption in K^+-free buffer were analyzed in real-time. The orientation of the GQ is often difficult to control and investigate using traditional biochemical methods. This method using DNA origami could successfully control the strand orientations, topology and stoichiometry of the GQ.

Next, we examined the formation of a four-stranded GQ induced by a GQ-binding ligand, bisquinolinium pyridine dicarboxamide [59]. Substrate dsDNAs containing 3–6 G–G mismatches in their centers were incorporated into a DNA origami frame. In the absence of ligand, the dsDNAs of interest did not interact, as visualized by their parallel-shape in AFM images. The presence of the ligand induced the GQ formation characterized by an X-shape. A biotin molecule was connected to the ligand via a linker. Therefore, addition of streptavidin to the ligand-induced GQ caused streptavidin to localize in the center of the X-shape, indicating that the ligand was bound to the GQ. A sequence of real-time images of the ligand-induced GQ formation and its reverse conformational switching by removing the ligand was captured by HS-AFM. Unprecedented intermediate-like states were recorded in the real-time analysis.

HIV-1 nucleocapsid proteins (NCps) facilitate remodeling of nucleic acids to fold thermodynamically stable conformations, and are thus called nucleic acid chaperones. To date, little is known regarding their stoichiometry, NCp–NCp interactions, and chaperone activity on GQ formation. Here, we carried out the direct and real-time analysis on the properties of proteolytic intermediate NCp15 and mature NCp7 using DNA origami scaffold [60]. The protein particles were found to predominantly exist in a monomeric form, while dimeric and multimeric forms were observed both in free solution and bound to the GQ structure. The formation and the dissociation events of the GQ were well documented in real-time and the intermediate-like states were visualized.

5.3 *Topological Control of G-Quadruplex and I-Motif Formation*

Tandem G-rich repeat sequences, which form GQ structures, are often observed in the promoter regions [61]. Promoter sequences such as a *c-myc* promoter contain a G-rich sequence, which forms a GQ structure that is involved in transcriptional regulation of the gene [54]. In the dsDNA in this region, a complementary C-rich repeat sequence forms an i-motif structure [62]. A number of studies have reported on GQ and/or i-motif formation in promoter regions containing GQ- and i-motif-forming sequences [63–66]. The i-motif structure is physically induced in acidic conditions, because i-motif formation requires a hemiprotonated cytosine dimer formation [62]. The i-motif formation at neutral pH has also been observed in dsDNA with negative superhelicity [65] and in a molecular crowding environment [67].

We have demonstrated visualization of the formation and resolution of dsDNA containing a GQ forming and its counterpart i-motif forming sequence in a DNA nanostructure at the single-molecule level (Fig. 9) [68]. Formation and dissociation of the GQ/i-motif complementary sequence in the insulin-linked polymorphic region (ILPR) core sequence were manipulated in the DNA frame structure. The ILPR promoter region is thought to separate into the GQ and i-motif structure during the regulation of transcription [64]. GQ/i-motif dsDNA was introduced into the DNA frame to observe GQ and i-motif formation under various conditions (Fig. 9a). In this system, topologically controlled dsDNA was prepared using the sequential manipulation of DNA strands in a series of programmed operations. Using strand displacement and the addition and removal of K^+, the topologically controlled GQ/i-motif dsDNA in the DNA frame was obtained in high yield. The resolution of dsDNA containing the GQ/i-motif sequence was performed by controlling the pH and K^+ conditions, which were altered either individually or together during formation of the GQ and i-motif structure (Fig. 9b). Compared to the initial

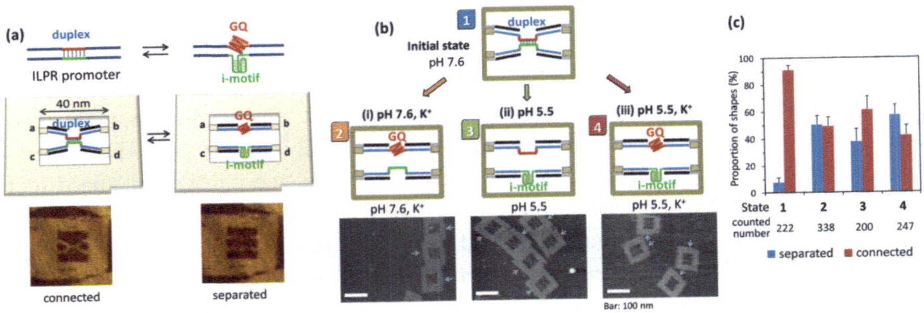

Fig. 9 Direct observation of G-quadruplex (GQ) and i-motif formation from a topologically controlled dsDNA in the DNA frame. (**a**) dsDNA in the ILPR promoter separates into the GQ and i-motif. The separation of topologically controlled dsDNA into the GQ and i-motif in the DNA frame was observed and the AFM images were captured. (**b**) Manipulation of the dsDNA separation in the DNA frame under acidic conditions and with the addition of K^+. State 1: dsDNA at pH 7.6, state 2: GQ formation using (1) pH 7.6 and K^+, state 3: i-motif formation using (2) pH 5.5, state 4: GQ and i-motif formation using (3) pH 5.5 and K^+. AFM images of states 2, 3, and 4. (**c**) Summary of the observed separated and connected DNA strands in the DNA frame for states 1–4

topology-controlled dsDNA (~90% connected), three of the conditions resulted in 40–60% dissociation of the GQ/i-motif dsDNA (Fig. 9c). The dsDNA was easily separated using GQ and i-motif forming condition, specifically by addition of K^+ and acidic conditions. Furthermore, dissociation of the dsDNA under GQ and i-motif forming conditions was directly observed by HS-AFM. These results indicate that the dsDNA containing GQ- and i-motif sequences can be effectively dissolved when the DNA duplex is helically loosened in the DNA frame.

5.4 *Triple Helix Formation*

Triple helices are a well-known DNA structural variant in which the third DNA strand binds to one of the Watson–Crick base pair strands via Hoogsteen base pairing in the major groove [69]. Single-molecule imaging of triple-helix formation was carried out using DNA nanostructures [70]. The binding of the third strand to dsDNA in a DNA origami frame was examined using two different types of triplet base pairs including a parallel and an antiparallel third strand [69, 71]. The target single-stranded DNA and the third strand were incorporated into the DNA frame, and the third strand binding was controlled with the addition of the complementary strand and formation of Watson–Crick base pairing. Triple-helix formation was monitored by observing the conformational changes in the incorporated DNA strands placed in the DNA frame. In the case of the parallel triplex formation, 24% of incorporated strand formed a triple helix after the addition of the target strand (one of the strands for Watson–Crick base pairing). Similarly, in the case of the antiparallel triplex formation, 31% of incorporated strand formed a triple helix.

These results indicate that the Watson–Crick base pairings induced and controlled binding of the third strand and triple helix formation. Triple-helix formation was also examined using a photocaged third strand. The third-strand binding was directly observed using HS-AFM combined with photoirradiation. Binding of the third strand could also be controlled by the duplex formation and photocaged strands in the designed nanospace.

5.5 B–Z Transition in the Equilibrium State

Right-handed B-form dsDNA containing a CG repeat sequence is known to transition to the left-handed Z-form structure by increasing the salt concentration [72]. A nanomechanical device employing the rotation of the B–Z transition was created on a DNA nanostructure, and the rotation of the nanostructure was investigated by FRET [73]. We directly visualized rotary motion of a B–Z conformational transition of DNA double helix in the DNA frame structure [74]. To visualize the B–Z transition, dsDNA containing a 5-methyl-CG six-repeat sequence $(mCG)_6$ and a flag marker containing three-helix-bundled DNA connected by crossovers were introduced to the DNA frame (Fig. 10). The $(mCG)_6$ repeat can promote formation of the Z-form even at low salt concentrations [75]. One dsDNA with the $(mCG)_6$ sequence and a flag marker was introduced to the top as B–Z transition system, while the other dsDNA containing a random sequence and a flag marker was introduced to the bottom as a control. To allow for rotation during the B–Z transition, four connectors

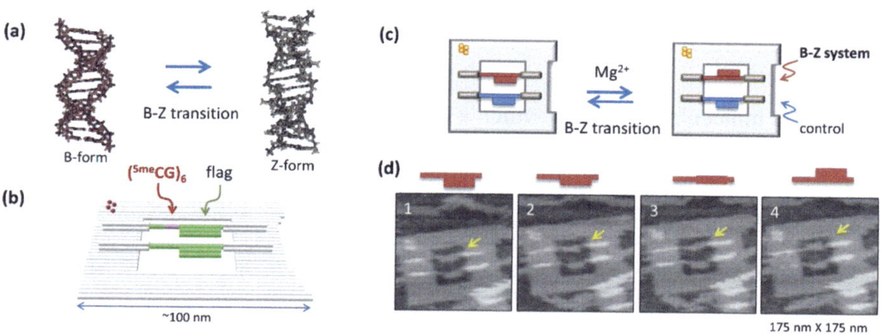

Fig. 10 Direct observation of the B–Z transition in the DNA frame. (**a**) B-form and Z-form dsDNA structure and B–Z transition. (**b**) Single-molecule observation system for the B–Z transition. Two dsDNAs having a $(^{5me}CG)_6$ sequence (B–Z system) and a random sequence (control) were introduced to the top and bottom site in the DNA frame, respectively. To observe dsDNA rotation, both ssDNA linkers in the left terminal of the dsDNA were fixed to the connector, while one ssDNA linker in the right terminal was attached to the connector. (**c**) The dsDNAs have a flag marker to observe the rotation of the dsDNA shaft during the B–Z transition. (**d**) HS-AFM images of the flipping motion of the flag marker at the top site (yellow arrow). Four successive images are presented at a scanning rate of 0.2 frames/s

were introduced in the DNA frame to lift both ends of dsDNAs from the scaffold surface (Fig. 10a). In addition, to observe rotation of the double helix, both of the single-stranded DNA (ssDNA) linkers in the left terminal of the dsDNA were fixed to the connector, while one ssDNA linker in the right terminal was attached to the connector. During the B–Z transition, the flag marker was expected to rotate around the dsDNA shaft, and the rotary motion could be observed by monitoring the position of the flag marker. By increasing the concentration of Mg^{2+} ions, 70% of the flag marker within the B–Z system rotated to the upper side, whereas 76% of the flag in the control remained unchanged (Fig. 10b). Furthermore, by controlling the concentration of Mg^{2+} ions, the rotation of the flag marker within the B–Z system was examined by HS-AFM under equilibrium conditions for the B–Z transition. Flag movement of the B–Z system was observed during HS-AFM scanning (Fig. 10c). The successive images also showed the height change of the flag marker, indicating that the rotation of the flag marker occurred around the dsDNA shaft of the B–Z transition system.

Using a DNA origami scaffold and HS-AFM system, important DNA conformational changes including G-quadruplex formation, triple helix formation, G-quadruplex/i-motif formation from the dsDNA, and B–Z transition were successfully imaged. The observation system used in these experiments can be used as a general strategy for investigating various DNA structural changes and molecular switches working at the single-molecule level. This method could also be applied to the single-molecule imaging of chemical reactions such as bond formation and cleavage.

6 Direct Observation of Artificial Molecular Systems Using DNA Nanostructures

Controlled molecular system in which a reaction is initiated by a stimulus, such as photoirradiation and metal ions, is an interesting target for single-molecule observation. Here, observation systems for (1) photo-induced hybridization and dehybridization of photoswitching DNA strands, (2) the metal ion-induced base pair formation, (3) the Zn^{2+}-induced DNAzyme reaction, and (4) the formation of ATP-induced kissing RNA complex were constructed and visualized by HS-AFM. The stepwise binding of proteins to RNA to form RNA nanostructures was observed by HS-AFM. In addition, these RNA nanostructures were newly designed and constructed using the DNA origami method.

6.1 Light-Induced DNA Strand Interaction

The direct observation of hybridization and dissociation of dsDNA at molecular resolution is quite challenging. AFM-based single-molecule imaging can visualize whole nanostructures by directly monitoring the shape of DNA strands. Photoswitching DNAs containing azobenzene molecules were employed to observe the hybridization and dissociation of dsDNA [76, 77]. Photoswitchable DNA strands can hybridize in the *trans*-form of the azobenzene moiety and dissociate in the *cis*-form by photoisomerization using UV irradiation. The dissociated DNA strands in the *cis*-form hybridize again upon visible light (Vis) irradiation. We incorporated a pair of photoswitchable DNA strands connected to individual supporting dsDNA chains in the cavity of the DNA frame to visualized the photo-induced hybridization and dissociation at the single molecular level (Fig. 11a) [78]. The hybridized photoswitching dsDNA in the center of the supporting dsDNA chains in the DNA frame was clearly visualized (Fig. 11b). Using this origami system, hybridization and dissociation was identified by the global structural change of the two supporting dsDNA chains as either an X-shape or a separated shape in the DNA frame, respectively. Hybridization and dissociation were observed directly using HS-AFM. Dissociation of the two dsDNA chains that were in contact at the center (X-shape) was imaged during UV irradiation. The contact of the two separated

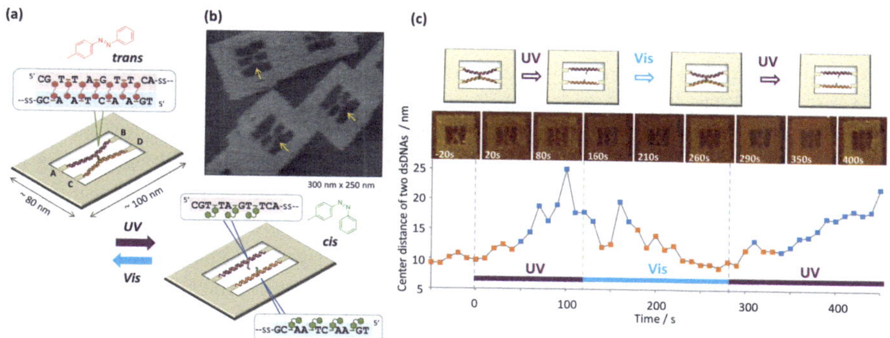

Fig. 11 Direct observation of the hybridization and dissociation of a pair of photoswitching DNA strands. (**a**) Single-molecule observation system. In the *trans*-form of an azobenzene moiety, two photo-responsive domains hybridize to form a duplex. In the *cis*-form induced by UV irradiation, the two domains dissociate. Two dsDNA chains containing photoswitching DNA strands were placed in the DNA frame structure to observe the dissociation and hybridization after UV and Vis irradiation, respectively. Two different dsDNA chains containing different photoswitching DNA strands were connected between the specific sites in the DNA frame via the corresponding overhanging ssDNAs. (**b**) AFM image of the photoswitching DNA duplex supported by two dsDNA chains in the DNA frame (orange arrow). (**c**) Single-molecule switching activity of photoswitching DNAs in the DNA frame. The repeating dissociation and hybridization were visualized by HS-AFM during alternating UV/Vis photoirradiation. The distance between the centers of two dsDNA chains was plotted. The appearance of the X-shape and separated shape is shown as an orange and blue rectangle in the graph, respectively

dsDNA chains in the center was imaged again during Vis irradiation (Fig. 11c). Alternating dissociation and hybridization of the photoswitchable DNA strands were visualized at the single-molecule level in the DNA frame.

Next, we incorporated two switching motifs, a pair of photoswitchable DNAs and another pair of G-telomeric repeats for GQ formation, into the DNA frame using six connectors [79]. These functional DNA strands with different response systems were incorporated into three parallel dsDNAs. Transformation from the photo-induced duplex dissociation to GQ formation and reverse process from the GQ dissociation to photo-induced duplex formation were achieved by UV irradiation with addition of K^+ and by Vis irradiation with removal of K^+, respectively. In addition, a series of configuration conversions with reversible dual switching in the DNA frame was successfully observed by HS-AFM. Photoirradiation and K^+ were used as different stimuli to switch the interactions between the three functional DNA strands in a logical manner.

6.2 *Metal Ion-Induced Base Pair Formation*

The expansion of the combination of base pairing systems and functionalization of bases has been widely investigated, and metal ion-mediated base pairing systems have attracted particular attention [80, 81]. The most commonly used metal ion-mediated mismatch base pairing systems are cytosine-Ag(I)-cytosine (C-Ag-C) and thymine-Hg(II)-thymine (T-Hg-T) [82, 83]. Because of the possibility of intrastrand base pair formation induced by metal ions, long dsDNA containing consecutive C-Ag-C and T-Hg-T base pairs have not been characterized. We demonstrated single-molecule imaging of metal-ion induced double-stranded DNA formation in DNA nanostructures [84]. The formation of the metal ion-mediated base pairing in a DNA origami frame was examined using C-Ag-C and T-Hg-T metallo-base pairs. The target DNA strands containing 20-mer consecutive C or T were incorporated into the DNA frame, and the binding was controlled by the addition of metal ions. Double-stranded DNA formation was monitored by observing the structural changes in the incorporated DNA strands using high-speed AFM. Using the T-Hg-T base pair, we observed the dynamic formation of unique dsDNA and its dissociation. The formation of an unusual shape of dsDNA with consecutive T-Hg-T base pairs was visualized in the designed nanoscale structure. This system can be used for the nanoscale construction of nanowires and devices, which can assemble on the surface of DNA nanostructures.

6.3 Zn^{2+}-Dependent DNA Cleavage by DNAzyme

Catalytic nucleic acids (DNAzymes) are widely used for biochemical purposes such as detection of target molecules, amplification, and switching. Metal-ion-dependent DNAzymes are one specific class of DNAzymes that consist of various DNA sequences and recognize a variety of substrates. These metal-ion-dependent DNAzymes have been used for amplifying labels for sensing, functional components for logic gates, and stimuli-responsive DNA switches [85, 86]. Here, we performed the single-molecule imaging of the catalytic reaction of the Zn^{2+}-dependent DNAzyme [87] in a DNA frame structure [88]. The DNAzyme and substrate strands attached to two supporting dsDNAs were assembled into the DNA frame in two different configurations. The reaction was monitored by observing configuration changes of the two incorporated DNA strands in the DNA frame, such as an H-shape to a parallel configuration and an X-shape to a double-loop configuration. These configuration changes were clearly observed in accordance with the progress of the reaction. Separation of the strands after the cleavage reaction by Zn^{2+}-DNAzyme was also dependent on the lengths of the dissociation domains of DNAzyme and substrate strands. Separation of the supporting dsDNAs, induced by substrate cleavage by the DNAzyme, was directly visualized in two configurations by HS-AFM. This nanostructure-based AFM imaging is applicable for monitoring various chemical and biochemical catalytic reactions at the single-molecule level.

6.4 Riboswitch and Kissing Complexes of RNA

The structural diversity of RNA molecules is an important property, that mediates their unique functions such as specific complex formation and catalysis. In addition, RNA molecules uniquely form complexes through specific hairpin loops, called kissing complexes, which enables assembly of complementary RNA loops via Watson-Crick base pairing [89]. Kissing complexes are widely investigated and used to construct various RNA architectures such as polygonal structures and three-dimensional assemblies [90, 91]. Molecular switches have also been created by combining a kissing loop with a ligand-binding aptamer to control interactions between RNA molecules [92]. Two kinds of RNA aptamers were incorporated into a DNA origami frame and used AFM to observe their ligand-responsive interactions at the single-molecule level [93]. A designed RNA aptamer called a GTP-switch has a guanosine triphosphate (GTP)-responsive domain that can bind to a specific target RNA hairpin called Aptakiss in the presence of GTP [92]. These RNA aptamers were attached to supporting dsDNA strands to incorporate them into the DNA frame individually. In the absence and presence of GTP, shape changes of the supporting dsDNA strands with the RNA aptamers were observed in the DNA origami frame, which were induced by the GTP-switch. In addition, the switching function within the nanospace was improved using a cover strand over the kissing loop of the

GTP-switch and by deleting one base from this kissing loop. These newly designed ligand-responsive aptamers can be used for the controlled assembly of various DNA and RNA nanostructures.

7 Direct Observation of a Mobile DNA Nanomachine on the DNA Origami Surface

Considering the programmability of DNA sequences and manipulation of the association and dissociation of duplexes, DNA has been used as various dynamic molecular machines [94]. DNA nanomachines such as DNA tweezers and various walking devices have been developed by manipulating duplex formation using strand displacement [94]. DNA motor systems powered by enzyme cleavage and photocleavage were designed and created on the DNA origami surface and their walking motions were directly observed by HS-AFM.

7.1 A DNA Motor System Created on a DNA Origami Scaffold

A DNA transportation system was constructed using a mobile DNA nanomachine that could move along a designed track on the DNA origami surface (Fig. 12). The track on the DNA scaffold was constructed to observe the multistep movement of a mobile DNA strand (DNA motor) [95]. Multiple ssDNAs (stators) were introduced onto the rectangular DNA origami tile as a motor track to hybridize with a complementary mobile DNA motor strand (Fig. 12a). When the DNA motor strand hybridizes to the specific stator strand, the stator/motor duplex is subsequently cleaved by nicking enzyme Nt.BbvCI, which removes a short ssDNA from the stator (Fig. 12b) [96]. The motor strand then binds to the neighboring intact stator by branch migration and finally moves one step forward. The DNA tile carrying the motor strand at the initial position was incubated with Nt.BbvCI to examine the migration of the motor strand along the DNA motor track. The motor strand was imaged as a single spot of the duplex on the DNA origami scaffold, which was easily distinguished from the invisible single-stranded stators. The time-dependent movement of the motor strand along the motor track was observed (Fig. 12c). Furthermore, the movement of the motor strand was directly visualized by HS-AFM. During HS-AFM scanning, the stator/motor duplex spot showed back and forth movement along the motor track, followed by a period of decreased movement, and finally the spot moved forward (Fig. 12d). From the kymograph analysis, the distance of the motor-strand movement corresponded to the distance between adjacent stators, indicating that movement occurred stepwise on the track. In addition, the intermediate state of the branch migration was identified by HS-AFM imaging, which was observed in real-time as suppressed movement during AFM scanning.

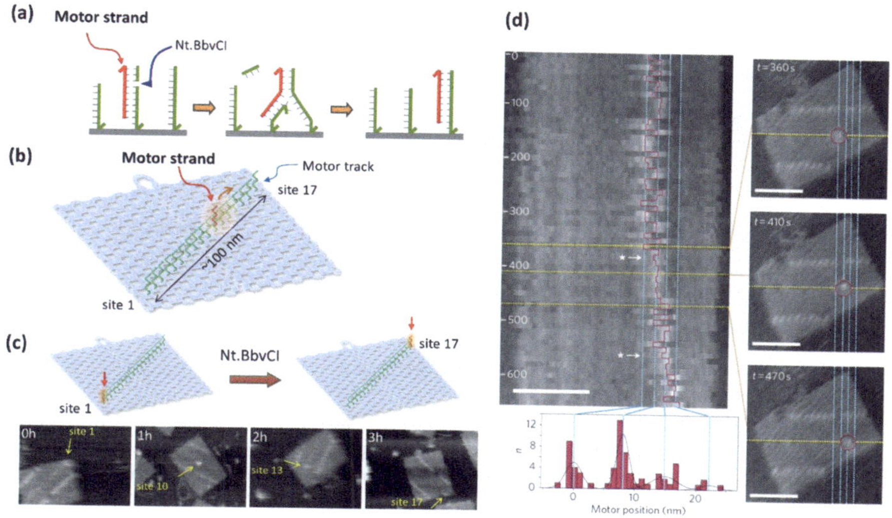

Fig. 12 Direct observation of DNA motor movement on the DNA origami surface. (**a**) Mechanism of the DNA motor movement. When a stator strand (green ssDNA) forming a duplex with a DNA motor strand (red ssDNA) is cleaved by nicking enzyme Nt.BbvCI, the motor strand moves to the adjacent stator strand by branch migration. (**b**) A DNA otor track consisting of 17 stators (green ssDNAs) was constructed on the DNA origami scaffold, and the movement of the DNA motor (red ssDNA) was examined. (**c**) AFM images of time-dependent movement of DNA motor. A DNA duplex containing a DNA motor strand was visualized as a white dot. (**d**) Stepwise movement of the DNA motor observed by HS-AFM, and kymograph and distribution of the motor positions. Scale bar 50 nm

7.2 *Single-Molecule Operation of DNA Motor Using the Programmed Instruction*

The method was further applied to regulate the transportation of a DNA motor on a branched motor track. In this system, the direction of the DNA motor movement was precisely controlled by blocking and releasing strands with predefined instructions (Fig. 13) [97]. This system used a branched DNA motor track with four destinations and three junctions. The direction of the motor strand was controlled by open or closed gates located at each junction (Fig. 13a). In the initial operation, the gates of the two junctions opened by removing the block strands using specific releasing strands (Fig. 13b). Then the motor strand moved along the branched motor track based on the open and closed states of the gates. After gate operation and motor movement, the motor strand followed the instruction and accumulated at the target destination, which was observed by AFM (Fig. 13c). In addition, the movement of the motor strand having a quencher was monitored in a bulk solution using fluorescence quenching method, in which the quencher was attached to the motor strand and four different fluorescence dyes were placed close to the final destinations. Selective fluorescence quenching at the target destination was observed by

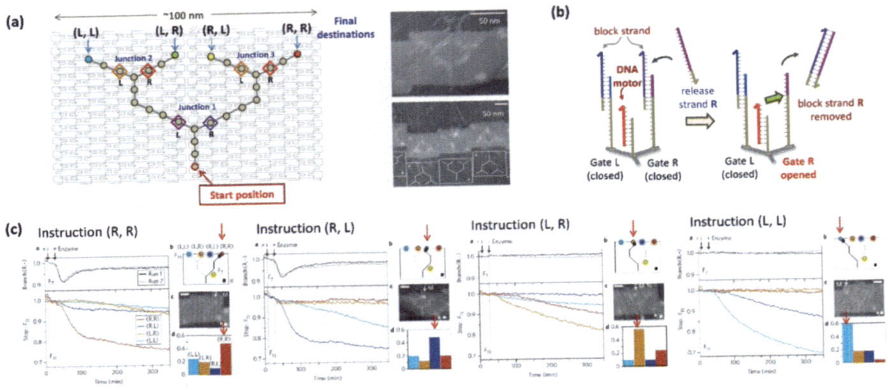

Fig. 13 Molecular transportation using a branched DNA motor track constructed on the DNA origami scaffold. (**a**) A branched DNA motor track with three controllable gates at junctions and four final destinations. AFM images of the motor track on the DNA scaffold. (**b**) Mechanism of controlled close/open of the gate at the junction. In the initial state, both left and right stator have a blocked strand to stop the motor strand moving forward. By adding the specific release strand to remove the block strand, the motor strand can move to the unblocked stator. (**c**) Motor movement to the four final destinations. For example, Instruction (R, R) means that the right (R) side at Junction 1 and right side at Junction 3 are opened by adding the corresponding release strands. The positions of the motor strands were monitored by fluorescence quenching and AFM imaging. Different fluorescence dyes were incorporated close to the four destinations, and the fluorescence intensities of four dyes were monitored during the movement of the quencher-attached DNA motor. Distribution ratio of four destinations was obtained from the AFM images

following the programmed instructions. Using this controllable DNA motor system constructed on the DNA origami tile, single motor movements can be regulated with nanoscale precision using operation instructions such as selective gate opening.

A programmed system was constructed on a DNA origami scaffold using a predesigned DNA track. Using HS-AFM, the detailed motion of the DNA motor was directly observed and analyzed. Related works, such as a DNA spider [98] and a programmed assembly line [99], also used mobile DNA nanomachines and DNA origami scaffolds for the construction of nanoscale transportation systems.

7.3 *Photo-Controlled DNA Motor System Constructed on the DNA Origami*

Photochemical reactions are the most commonly used and effective method for regulating molecular systems in a noninvasive fashion. Here, we created a nonenzymatic DNA motor system using a pyrene-attached DNA strand and disulfide-containing stator strands on the DNA origami tile (Fig. 14a) [100]. A single-stranded DNA carrying two pyrene molecules was employed as a photo-controllable DNA motor that was assembled on the DNA tile carrying a linear track composed of four

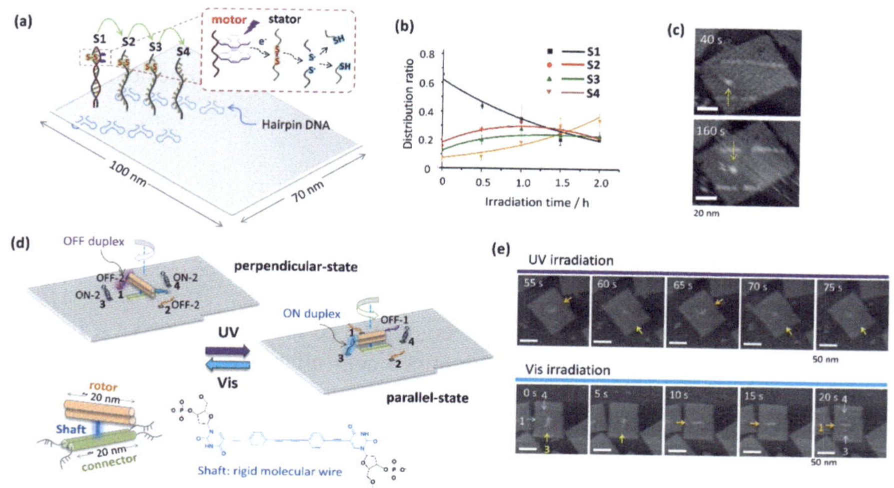

Fig. 14 Photo-controlled DNA nanomachine constructed on the DNA origami scaffold. Dynamic movements of the photo-controlled DNA motor and rotor were observed. (**a**) Pyrene-attached DNA motor strand and four stator strands are introduced to the DNA scaffold. The reaction mechanism of the photo-induced electron transfer from pyrene to a disulfide bond in the stator strand to reduce it to be cleaved. After the cleavage of the disulfide bond, the motor strand moves to the next stator. (**b**) Distribution of the DNA motor at four positions after photoirradiation at 0, 0.5, 1, 1.5, and 2 h. The data were fitted to calculate the stepping rates at three individual steps. (**c**) HS-AFM images of the one-step movement of the DNA motor during photoirradiation and AFM scanning. (**d**) DNA rotator system constructed on the DNA origami scaffold. A double dsDNA bundled (double crossover) structure with two different photoswitches at the ends are connected to the DNA origami scaffold by a phenylene ethynylene molecular wire. The DNA scaffold has two off-switches and two on-switches to trap the rotor to the vertical- and perpendicular-state, respectively. (**e**) HS-AFM images (top) of the rotor movement from the perpendicular-state to the parallel-state during UV irradiation. HS-AFM images (bottom) of the rotor movement from the parallel-state to the perpendicular-state during Vis irradiation

stator strands. The excited pyrene molecules ($\lambda_{ex} = 350$ nm) of the motor reduced the disulfide bond in the stator strands via a photo-induced electron transfer, followed by the cleavage of the disulfide bond [101]. During UV irradiation, the DNA motor migrated from the cleaved stator to the next intact one on the DNA tile continuously until reaching the final stator. The entire walking process of the motor was determined by characterizing the distribution ratios of motor/stator duplexes at four anchorage sites on the tile under different irradiation times (Fig. 14b). The observed stepping rate constants for the photoresponsive DNA motor were calculated from these data. Finally, the photo-induced movement of the motor was observed using HS-AFM. The one-step movement of the photoresponsive DNA motor during UV irradiation was observed in real-time by HS-AFM (Fig. 14c).

The stepwise movements of the photoresponsive motor on the DNA origami scaffold were fueled by an optical source, and the autonomous unidirectional motions were regulated remotely. The dynamic mechanical behaviors of DNA in a limited nanospace were successfully characterized at the single molecule level. The photonic modification of DNA-based nanomaterials shows promise for biological applications such as cargo transport and manual configuration changes of biomolecules in mesoscopic systems.

7.4 *Photo-Controlled DNA Rotator System Constructed on the DNA Origami*

Various DNA-based nanodevices have been developed on the nanometer scale that are regulated using light However, programmed controllability is still a major challenge for these artificial nanodevices. We constructed a rotary DNA nanomachine on a nanostructure, in which the rotations were driven and also reversibly regulated by different wavelengths of light (Fig. 14d) [102]. A DNA rotor consisting of a stiff bar-shaped double-crossover molecule supported by a rigid phenylene ethynylene molecular wire was placed on top of a rectangular DNA tile [103]. Two types of photoresponsive DNA switches were combined together at each end of the rotor to trigger rotary motions in the DNA nanostructure. By employing two pairs of switching motifs, the rotor could be regulated to rotate at a specified angle to be locked in two reconfigurable states by switching photo-irradiations between UV and visible light. Two reconfigurable states (perpendicular and parallel) of the rotor were obtained, in which the angular changes were characterized by AFM and fluorescence quenching assays. In addition, the real-time rotational motions of the rotor were directly visualized by HS-AFM. Reversible rotary motions and switching between the two states (from perpendicular to parallel and from parallel to perpendicular states) were directly visualized on the DNA tile surface with nanometer scale precision during UV and visible light irradiation, respectively (Fig. 14e). This rotary system needed rotation of the 20-nm bar-shaped rotor to transition between the two states (the perpendicular and parallel states), which had a much larger mobile range compared to single molecular switches. The system can be utilized to construct switchable functional systems that work on the solid surface. This rotary nanostructure represents a unique prototype for DNA-based nanodevices.

7.5 *Transcription Regulation System Integrating RNA Polymerase and Genes on the DNA Origami*

In synthetic biology, the control of gene expression requires a multistep processing of biological signals. The key steps are sensing the environment, computing information and outputting products [104, 105]. The complexity of genetic circuits remains low because it is difficult to completely avoid cross talk between the circuits. We created an orthogonal self-contained nanodevice by integrating a reactor and sensing system onto a DNA origami-based nanochip that contains T7 RNA polymerase (RNAP) and multiple target-gene substrates (Fig. 15) [106]. This gene nanochip orthogonally transcribes its own genes, and the exact positioning of molecules onto a DNA origami allows to rationally design gene expression levels by controlling the intermolecular distances between the RNAP and the target genes (Fig. 15a). To proof our concept, we first examined the relationship between the positioning and transcription activity. We changed the distance between the RNAP

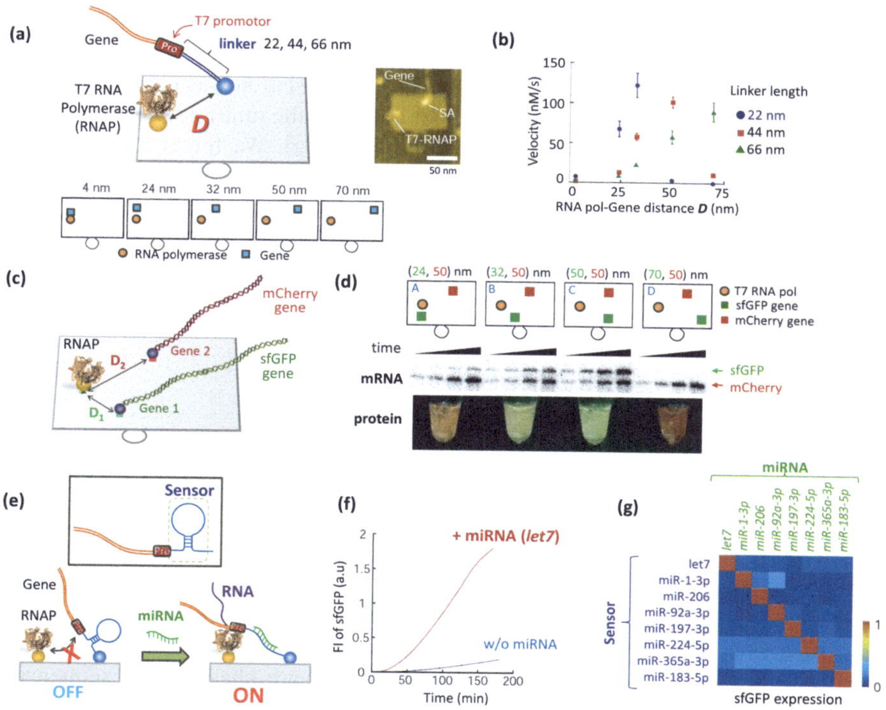

Fig. 15 DNA origami gene nanochip. (**a**) Transcription regulation system controlled by distance between T7 RNA polymerase (RNAP) and gene and linker length to T7 promoter. The distance between RNAP and gene was controlled on the DNA origami. AFM image of the assembled DNA origami gene nanochip. (**b**) Effect of distance between RNAP and gene with three different linker lengths on transcription activity. (**c**) Dual gene expression system. RNAP and two different genes (sfGFP and mCherry) were placed on the DNA origami nanochip. (**d**) Expression of RNA and protein using dual gene expression system. The distances between RNA polymerase and two genes are regulated. mCherry gene/RNAP was separated by 50 nm and the sfGFP/RNA distance was changed to 24, 32, 50, and 70 nm. For the in vitro translation, PURE system was used and the system was incubated at 37 °C for 5, 10, 30, and 60 min using ca.1 nM nanochip. (**e**) Length-responsive switching system for transcription activation. By addition of the specific miRNA, length of the sensor part is extended and consequently transcription starts. (**f**) Transcription/translation activity of let-7 sensor nanochip in the presence or absence of let-7 miRNA. (**g**) Orthogonality of the sensor and miRNA. High specificity of the sensor/miRNA combination and orthogonality were observed by just replacing the sensor part of nanochip

and gene and used three different linker length to T7 promotor. We observed the dependence of the distance between RNAP and gene on the activity and found the appropriate distance (Fig. 15b). Next, we introduced two difference genes onto the DNA origami to control the dual gene expression level (Fig. 15c). Using two fluorescence protein genes [sfGFP (green) and mCherry (red)], the distance between RNAP-mCherry gene was fixed and the distance between RNAP-sfGFP gene was changed. As the result, the expression of both RNA and protein expression were rationally controlled (Fig. 15d). We also incorporate switching part to respond to specific small RNA (miRNA) to initiate the transcription (Fig. 15e). Stem-loop structure was employed to make a miRNA-responsive switch. Binding of miRNA opens the stem-loop and extends the length of the linker part to initiate the transcription. Using this switching system, the transcription/translation was initiated only in the presence of miRNA (Fig. 15f). We further designed the multiple sensor/miRNA pairs to examine the response of these pairs. (Fig. 15g). We found these sensor/miRNA pairs worked robustly and orthogonally. The results show that the nanochips compute their miRNA profiles by just replacing the sensor part. By combining multiple sensors, the nanochips can function as a logic-tip, which response to the input miRNAs and give correct output transcription signal. We developed the new strategy to integrate molecular sensors and transcription output to create a gene nanochips. These systems can provide a basis for large-scale, integrated genetic circuits.

8 Direct Observation of Assembly of DNA Origami Structures

Selective self-assembly of the nanomaterials is one of the fundamental issues in chemical science, physics, and nanotechnology. A unique property of DNA origami can be used to create a programmed assembly system [3]. The π-stacking interaction between the edges of DNA origami tiles and shape- and sequence-complementarity were used for the programmed assembly of the multiple DNA origami units. Photoswitching DNA strands have been employed for the selective assembly and disassembly of the DNA origami units, which was directly observed on a soft surface such as a lipid bilayer. In addition, large-sized assembly of DNA origami units was realized on the lipid bilayer surface and the assembly process was visualized by HS-AFM.

8.1 Programmed Assembly System Using DNA Origami

DNA origami structures have been employed for selective positioning of functional molecules and nanoparticles [1, 2]. For further development of the programmed assembly of multiple functionalized DNA origami structures, novel methods should be explored. We developed a novel method for assembling multiple DNA origami structures using designed DNA origami rectangles named "DNA jigsaw pieces" possessing sequence-programmed connectors (Fig. 16a) [107, 108]. We introduced shape- and sequence-complementarity into the concavity and convex connectors in the rectangular tiles for selective connection through nonselective π-stacking interactions between the side-edges of the DNA tiles. Single DNA tile units were assembled into unidirectional nanostructures with the correct alignment and uniform

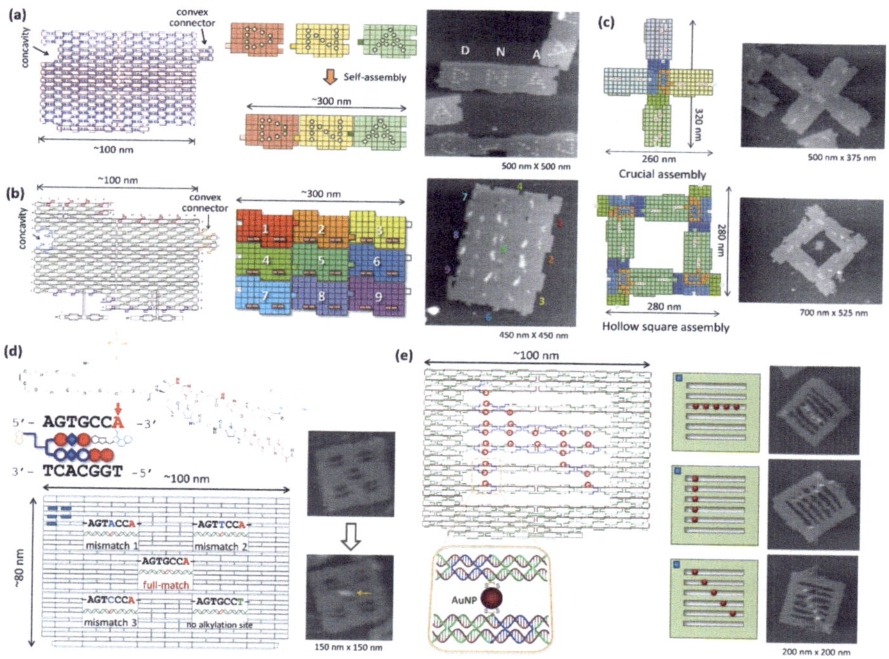

Fig. 16 Programmed assembly system using designed DNA origami. (**a**) 1D assembly system using different DNA jigsaw pieces with shape- and sequence-complementarity. AFM image of the three-piece assembly showing the letter D N A with hairpin DNAs. (**b**) 2D assembly system using nine different DNA jigsaw pieces with shape- and sequence-complementarity. AFM image of the 3 × 3 assembly. (**c**) 2D assembly using a four-way connector. Crucial and hollow square assembly. (**d**) Site-selective modification of a five well frame DNA origami scaffold with five different dsDNAs using a PI polyamide-alkylating agent conjugate. Center cavity has a full-matched sequence for selective PI polyamide binding. AFM images of before (top) and after the reaction and addition of streptavidin (bottom). (**e**) DNA origami scaffold with six slits. Slit cavities were modified with bis-thiol groups at the designated sites (red circle) to trap AuNP. After the addition of 5 nm AuNP, AuNPs were attached to the expected positions

orientation. Three to five different DNA tiles were assembled into predesigned and ordered nanostructures in a programmed manner. In addition, three-, four-, and five-letter words have been displayed using this programmed DNA jigsaw piece system.

This programmed assembly system can be expanded into two-dimensional (2D) assembly using specially designed DNA origami tiles [109]. The same 1D assembly system used previously was employed and the shape- and sequence-complementarity were added to selectively connect to the vertical direction to the DNA helical axis (Fig. 16b). Nine DNA jigsaw pieces were designed and prepared to form a 3 × 3 2D-assembly. For assembling into the vertical direction, three DNA jigsaw pieces could be connected to form a vertical trimer assembly. Next, nine different jigsaw pieces were directly assembled for the formation of a 3 × 3 2D-assembly, but the yield was low (~10%). By performing the assembly in four different ways, the step-wise self-assembly from the three vertical trimer assemblies resulted in a 35% yield of the target 2D-assembly. The surfaces of the jigsaw pieces were decorated with hairpin DNAs to display the letters of the alphabet, and the self-assembled structure displayed the word in nanoscale.

The direction of the DNA origami assembly was expanded in both horizontal and vertical directions. The designed dual-directional DNA origami connector, which can allow connection at all four edges, has sequence-programmed connection sites to both horizontal and vertical edges (Fig. 16c) [110]. A four-way connector that oriented the helical-axis element of all four edges outside was designed. In addition, concavity and convex connectors were introduced to allow the selective connection of different DNA origami components via these connection sites. Using this connector, 2D assemblies such as cruciate and closed hollow square assemblies were constructed. These programmed assembly systems represent the first successful preparation of selective multiple DNA origami assemblies.

8.2 *Site-Selective Modification of DNA Origami Scaffold*

The selective modification of a DNA origami scaffold was performed using a sequence-selective DNA binder pyrrole-imidazole (PI) polyamide. We observed sequence-selective alkylation of target double-stranded DNA (dsDNA) using a PI polyamide in a designed DNA origami scaffold having five cavities (Fig. 16d) [111]. Doubly functionalized PI polyamide was designed by the introduction of the alkylating agent 1-(chloromethyl)-5-hydroxy-1,2-dihydro-3H-benz[e]indole (*seco*-CBI) and biotin for sequence-selective alkylation at the target sequence and subsequent streptavidin labeling, respectively [112]. A newly designed DNA origami scaffold, termed as a "five-well DNA frame" carrying five different dsDNA sequences in its cavities, was used for the detailed analysis of the sequence-selectivity and alkylation. The 64-mer dsDNAs were introduced to five individual wells, in which the target sequence AGTXCCA/TGGYACT (XY = AT, TA, GC, CG) was employed as fully matched (X = G) and one-base mismatched (X = A,T,C) sequences. The fully matched sequence was alkylated as 88% selectivity over other mismatched sequences. In addition, the PI polyamide failed to attach to a target sequence lacking the alkylation site after washing and streptavidin treatment. Therefore, the PI polyamide discriminated the one mismatched nucleotide at the single-molecule level, and alkylation anchored the PI polyamide to the target dsDNA. Zn-finger proteins were also used for the sequence-selective binding to the target dsDNAs, which were incorporated into the five-well frame scaffold [113]. Two well-characterized Zn-finger proteins, zif268 and AZP4, were chemically modified and fused to fluorescent proteins, which maintained selective binding to the target sites at the single-molecular level.

Site-selective modification of a DNA scaffold by nanomaterials such as gold nanoparticle (AuNP) is important for the construction of plasmonic and optical nanodevices [114]. We designed a novel DNA scaffold, DNA slit, for the programmed positioning of AuNPs [115]. Various patterns of thiolated DNA slits were constructed by programing the positions of thiol groups (Fig. 16e). After the addition of AuNP, AuNPs were correctly placed at the predesigned positions in the thiolated DNA slits, indicating that the thiolated staples and the slit cavities guide the correct assembly of AuNPs.

Using sequence-selective synthetic molecules and proteins and site-selective modification of the scaffold, target molecules and nanoparticles can be selectively incorporated into the DNA origami scaffold. These functionalized origami structures can be utilized as functional units for further programmed assembly.

8.3 Photo-Controlled Assembly and Disassembly of DNA Origami

The advantage of using a photoreaction is that the target reactions can be initiated and controlled by irradiation of a specific wavelength of light. Single-molecule dissociation and hybridization of photoswitching DNA strands containing azobenzene molecules have been observed in the DNA origami scaffold by using HS-AFM [78]. The method is also applied for the assembly and disassembly of the DNA origami nanostructures. Using the hexagonal-shaped DNA origami carrying four photoswitching DNA strands, the hexagonal dimer in the initial state (*trans*-azobenzene) was dissociated into monomers (*cis*-azobenzene) by UV irradiation, and then reassembled to form a dimer by visible light (Vis) irradiation (Fig. 17a) [116]. The assembly and disassembly are reversibly controlled by UV/Vis irradiation in solution. The dimer formation and dissociation in solution were examined by gel electrophoresis, and the dynamic formation and dissociation in solution were monitored by fluorescence quenching (Fig. 17b). Repeated assembly and disassembly of the hexagons were observed in high-yield by alternating UV/Vis irradiation. Various shapes such as linear and ring-shaped assemblies were prepared by programming the positions of the photoswitching DNA strands (Fig. 17c).

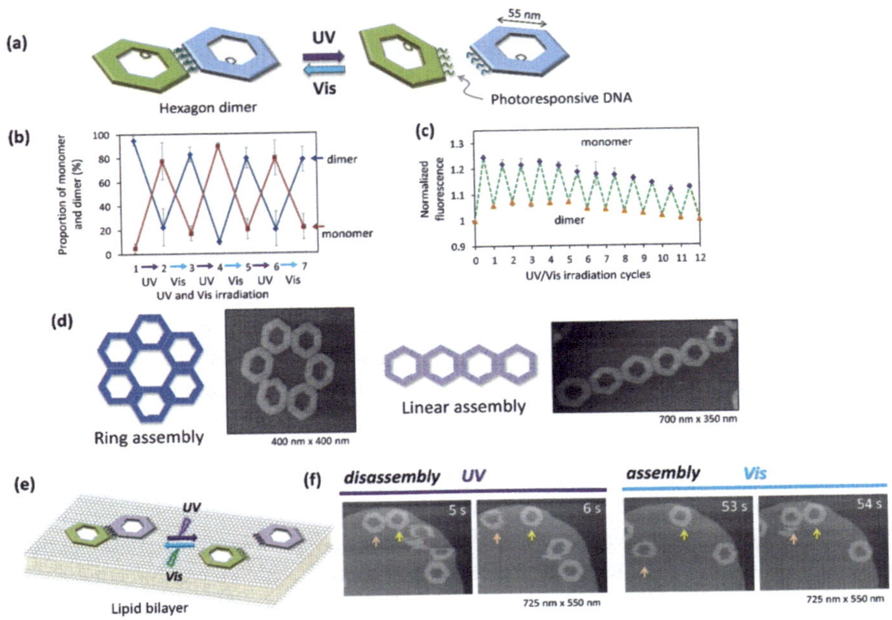

Fig. 17 Direct observation of the assembly and disassembly of photoresponsive hexagonal origami units. (**a**) Assembly and disassembly of hexagonal origami units possessing four photo-switching DNA strands by UV/Vis irradiation. (**b**) Dimer formation and dissociation of photoresponsive hexagons by alternating UV/Vis irradiation, which was measured by gel electrophoresis. (**c**) Dynamic dimer formation and dissociation of photoresponsive hexagons in solution by alternating UV/Vis irradiation, which was characterized by fluorescence quenching. (**d**) Formation of ring-shape and linear assemblies and AFM images. (**e**) Observation of dynamic assembly and disassembly on the lipid bilayer by UV/Vis irradiation. (**f**) Direct HS-AFM observation of dissociation of dimer following UV irradiation (left) and assembly by Vis irradiation (right)

8.4 Direct Observation of Assembly and Disassembly of Hexagonal DNA Origami on the Lipid Bilayer

AFM observation is basically carried out on a mica surface to fix the structure for clear imaging. The dynamic movement of the molecules can be imaged using a fluidic lipid-bilayer surface. Here, we observed dynamic assembly and disassembly of photoresponsive hexagonal origami structures on the lipid bilayer surface using HS-AFM (Fig. 17d) [117]. The fluidic property of lipid bilayer can be adjusted according to the target molecules and structures [118]. Cholesterol moieties were introduced in the hexagon for interaction with the lipid. The cholesterol-modified photoresponsive hexagon dimers were loaded onto the lipid bilayer and their diffusion was observed by HS-AFM. When these photoresponsive hexagon dimers were exposed to UV light during AFM scanning, the dimer immediately dissociated into monomers (Fig. 17e). The monomers were then irradiated with visible light; as a result, the monomers attached to form a correct dimer after the diffusion and collision. The formed dimer was dissociated again by a second round of UV irradiation. The reversible assembly and disassembly of hexagonal origami was directly imaged on the lipid bilayer surface. These photo-controlled manipulations of the interactions of DNA nanostructures can also be expanded to the construction of various switching nanodevices.

8.5 *Large-Sized Assembly of DNA Origami and Visualization on a Lipid Bilayer*

The lipid bilayer was further used to assemble DNA origami structures into large-sized assemblies. Here a lipid-bilayer-assisted assembly was performed for assembling various DNA origami monomers into two-dimensional lattices (Fig. 18) [119]. DNA origami structures were electrostatically adsorbed onto the lipid bilayer surface in the presence of divalent cation. The origami structures were mobile on the lipid bilayer surface and assembled into large 2D lattices in the range of micrometers (Fig. 18a). A cross-shaped DNA origami monomer was employed to form a

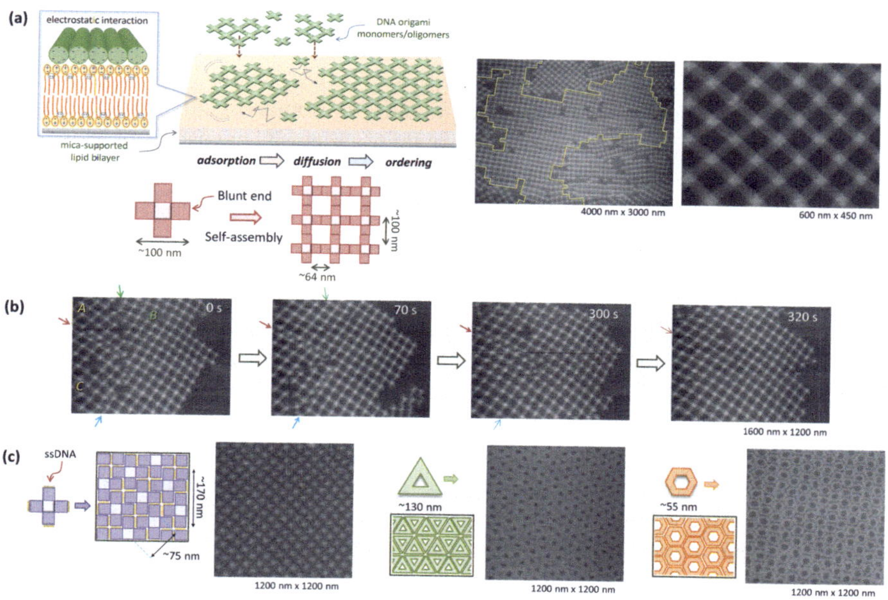

Fig. 18 Direct observation of assembly of DNA origami monomer into micrometer-sized structures on the lipid bilayer surface. (**a**) Assembly of cross-shaped origami monomers into lattice structures via π-π interaction of blunt ends. AFM image of lattice structures. (**b**) HS-AFM images of ordered lattice formation from the clacked lattice. (**c**) Packed assembly of cross-shaped origami units with ssDNA linkers at the ends (left), triangular origami units (middle), and hexagonal origami units in a shape-fitted manner

lattice by π-π interaction of the blunt ends between the monomers [120]. The dynamic processes including attachment and detachment of monomers and reorganization of lattices were visualized using HS-AFM. In addition, the biotinylated lattices were modified using streptavidin, and the binding of streptavidin to the lattice was observed by HS-AFM. Using the cross-shaped origami monomer with ssDNA linkers at the ends to prevent the interaction, the monomers were packed into 2D assemblies on the lipid bilayer surface in a shape-fitted fashion (Fig. 18b). Other monomers including the triangular and hexagonal monomers were also assembled into packed micrometer-sized assemblies.

DNA origami can be assembled into predesigned micrometer-scale 2D structures using the lipid bilayer for the surface of assembly. The method developed here can be used for the precise designing and preparation of micrometer-scale DNA architectures with various functions. Using the lattice prepared on the lipid bilayer, a square-shaped DNA origami was assembled into the lattice space via size-fitting and hybridization. Space-filling and diagonal assembly patterns in the lattice were constructed on the lipid bilayer. In addition, dynamic assembly of the square DNA origami into the lattice was observed using HS-AFM.

8.6 Programmed and Hierarchical Self-Assembly of DNA Origami into DNA Origami Frameworks

Ordered DNA origami arrays have the potential to compartmentalize space into distinct periodic domains that can incorporate a variety of nanoscale objects. We used the cavities of a preassembled 2D DNA origami framework to incorporate square-shaped DNA origami structures (SQ origami) (Fig. 19) [121]. The lattice was self-assembled on a lipid bilayer membrane from cross-shaped DNA origami structures and subsequently exposed to the SQ-origami. High-speed AFM revealed the dynamic adsorption/desorption behavior of the SQ-origamis, which resulted in continuous changing of their arrangements in the lattice (Fig. 19a). These dynamic SQ-origamis were trapped in the lattice cavities by increasing the Mg^{2+} concentration. Next, single-stranded DNAs (ssDNA) were introduced to SQ (A_8-ssDNA) and lattice (T_8-ssDNA) for hybridization to stabilize the assembled the SQ inside the lattice cavities (Fig. 19b). Using the hybridization, most of the cavities were filled with SQ and the movement of the SQ was almost suppressed. Furthermore, two cross-shaped origami components (cross-A and cross-B origami) which can form lattice via specific hybridization were used to form AB-lattice (Fig. 19c). Adhesive ssDNAs were only incorporated to cross-A, which enable the controlled assembly of SQ-ssDNA into the AB-lattice. After the assembly of SQ into the AB-lattice, we obtained a checkerboard-like pattern of SQ in the AB-lattice. The lattice type and hybridization patterns can be designed; therefore, this strategy offers a platform to create hierarchically assembled nanostructures or nanosystems consisting of multiple DNA origami components.

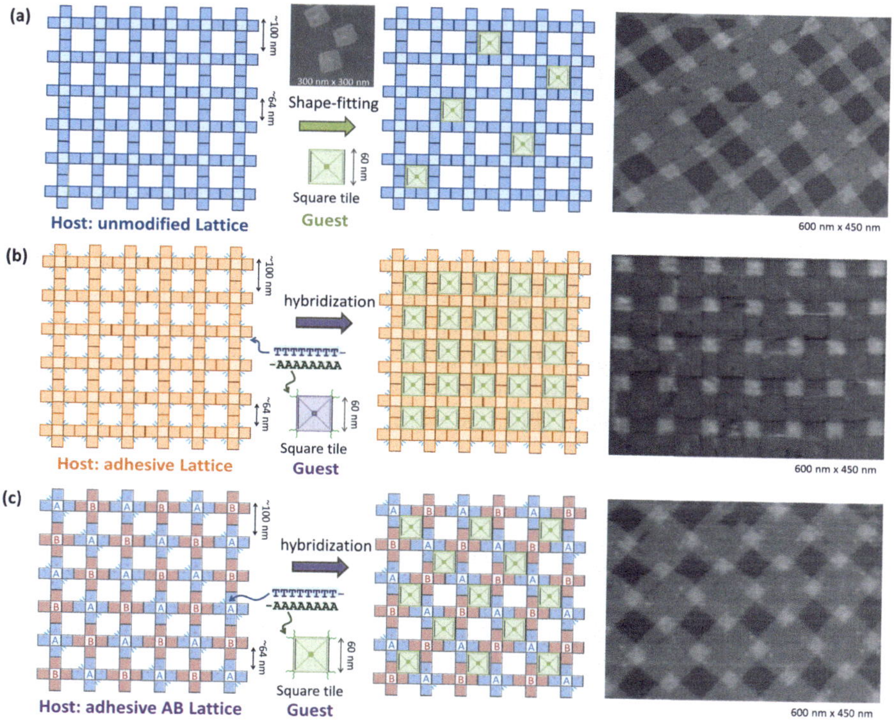

Fig. 19 Hierarchical incorporation of square tile (SQ: guest) to the assembled lattice (host) on the lipid bilayer surface. (**a**) Assembly of SQ into the unmodified lattice via shape fitting. AFM image of SQ incorporated to spaces of the lattice. (**b**) Assembly of SQ with A_8-ssDNAs into the T8-ssDNA-modified lattice via hybridization. AFM image of SQ with A_8-ssDNAs incorporated to spaces of the T_8-ssDNA-modified lattice. (**c**) Assembly of SQ with A_8-ssDNAs into the T8-ssDNA-modified AB lattice via hybridization. The AB lattice was prepared by selective hybridization of cross A and cross B at four ends. AFM image of SQ with A_8-ssDNAs incorporated to spaces of the T_8-ssDNA-modified lattice

8.7 Extended DNA Origami Using RNA as Constructing Materials

Structural changes of RNA induced by protein binding were directly observed by HS-AFM. The ribosomal protein L7Ae binds to specific positions of three kink-turns in double-stranded RNA (dsRNA) to form a triangular structure [122]. Using HS-AFM, the stepwise binding of three L7Ae proteins to the circular dsRNA to form the triangular RNA-protein structure was clearly observed [123]. By changing the number of base pairs of dsRNA between kink turns to regulate the geometry of the complex, superhelical RNA-protein structures could be produced with different numbers of triangle helices. Cell fate control was achieved by conjugating the L7Ae with caspase 8, which can induce apoptosis. Similar to the structural changes and reactions on DNA molecules, detailed RNA-RNA and RNA-protein interactions and reactions can be visualized in real-time at nanoscale resolution using the nucleic acid scaffolds and HS-AFM system.

In RNA nanotechnology, designed RNA molecules are widely explored because of their usefulness originating from their structural and functional diversity [124]. We constructed DNA-RNA hybrid nanostructures such as a seven-helix-bundled rectangle and a six-helix-bundled tube from a single-stranded RNA template and DNA staple strands [125]. The chemically modified RNA–DNA hybrid origami structures were also prepared by using RNA templates containing modified uracils. In addition, an RNA scaffold and staple RNA strands were used to form RNA origami nanostructures [126]. After annealing of the mixtures, a seven-helix bundled RNA tile and a six-helix bundled RNA tube were observed as predesigned shapes. These nanostructures were easily functionalized by introducing chemical modifications into the RNA scaffolds. Thus, the design and preparation method of DNA origami was extended and utilized to construct RNA nanostructures.

9 Three-Dimensional DNA Origami

The mechanical control of structural changes in DNA nanostructures has been widely investigated in the field of DNA nanotechnology. These mechanical 3D structural changes include conformational change such as switching and isomerization. DNA strand displacement is one of the most popular methods, in which a DNA strand having a toehold sequence is removed from the initial duplex by hybridization of a fully matched complementary strand [94]. The structural change in the nanostructure can also be controlled by the strand displacement. This method can be applied for the regulation of biological reactions, such as the activation of transcription induced by structural changes in 3D structures.

9.1 Observation of Structural Changes in 3D DNA Origami

Observation and analysis of structural change in the 3D DNA origami is important for designing dynamic DNA origami system. Physical and structural properties of the 3D origami structures were examined using the designed 3D structures and HS-AFM. We designed and constructed three different prism structures (triangular, square, and hexagonal prism) by folding the successively connected three, four, and six rectangles with connecting staple strands (Fig. 20a) [127]. The 3D structures formed were analyzed by AFM and cryo-electron microscopy. The breaking of the prism structures was imaged and characterized by HS-AFM (Fig. 20b). During HS-AFM scanning, one of the edges between the folded rectangles was found to be broken, yielding an opened 2D structure. From this analysis, the formation of the prism structure from multiple rectangles was confirmed. In addition, a cubic

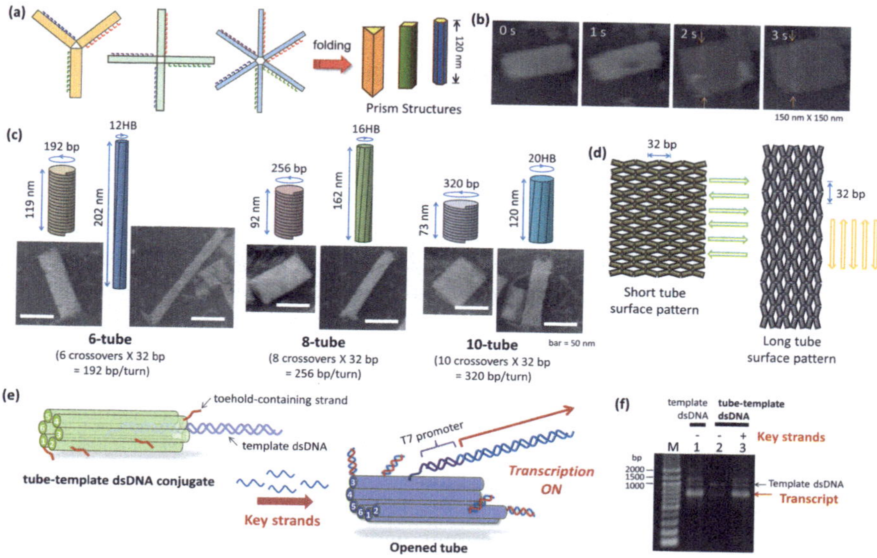

Fig. 20 Configuration changes of the 3D origami structures. (**a**) DNA prism structures folded from the multiple rectangles. (**b**) HS-AFM images of opening of triangular prism. (**c**) Helical tubular structures constructed in three different sizes. Expected short tubes and unexpected long tubes were formed. AFM images of tubes. (**d**) Geometry of helical direction of short (expected design) and long (unexpected isomer) tubes. From this analysis, the short and long tube are isomer with different geometry of crossover connection. (**e**) Transcription activation system. Closed tube-template dsDNA conjugate is opened by addition of full-complementary strands to toehold-containing strands (key strands). Transcription was activated by exposure of the promoter site after the tube opening. (**f**) Analysis of the transcripts without and with addition of the key strands (1% agarose gel electrophoresis)

structure was designed and constructed by folding six connected square structures [128]. The size of the 3D structure was characterized by the dynamic light scattering (DLS) method. The configurational changes from the cubic box to the opened form were directly observed using HS-AFM. This type of design can be applied in switchable dynamic structural changes for wrapping the molecules and nanomaterials and initiating the reactions using strand displacement in a programmed manner.

9.2 Dynamic Conformational Change in Helical DNA Nanotubes

A new type of tubular structures differing from the conventional DNA tubes was created. We designed and constructed novel helical tubular structures with various sizes using DNA origami [129]. The size-controlled tubular structures, which contained 192, 256, and 320 bp in one round of the tube, were designed, and designated as 6-tube, 8-tube, and 10-tube, respectively. After the preparation of the designed helical tubes, expected short tubes and unexpected long ones were observed (Fig. 20c). Detailed analyses of the surface patterns of the tubes by AFM and electron microscope revealed that the short tubes had mainly a left-handed helical structure. The unexpected long tubes mainly formed a right-handed helical structure and extended to the directions of the double-helical axes. Therefore, the short tube and long one are the structural isomer assembling from the same starting point to different structures. These two structures can be formed by the isomerization of different forms of Holliday junction crossovers (Fig. 20d). The folding pathways of the helical tubes were estimated by analyzing the proportions of short and long tubes with different annealing conditions. Depending on the number of base pairs involved in one round of the tube, the population of left-/right-handed and short/long tubes was altered. Using the smallest diameter 6-tube, the short tube was found to be a major product, while in the case of the largest diameter 10-tube, the long predesigned tube was a major product. These results indicate that the bending stress caused by the stiffness of double helices and nonnatural helical pitch determine the structural variation of the tubes. These helical tubes were further used for the isomerization of Holliday junction crossovers between long and short isomers by stretching using optical tweezers.

9.3 Transcription Activation Using the Structural Manipulation of DNA Origami

A transcription regulation system initiated by changes in the 3D nanostructure was designed and constructed by using configurational change of a 3D closed structure to a 2D opened one. Using the toehold-system, specific DNA strands induced the opening of the tubular structure. We designed a tubular DNA nanostructure and incorporated a template dsDNA inside at the promoter region to control the binding of RNA polymerase (Fig. 20e) [130]. A tube opening system using specific DNA strands was preinstalled in this tubular nanostructure. For the construction of the transcription activation system, the PCR-amplified template dsDNA was incorporated into the tubular structure. The tubular structure attached to the template dsDNA was opened using the specific DNA strands complementary to the toehold-containing strands (key strands). In vitro transcription from the tube-attached template dsDNA was examined using T7 RNA polymerase. The transcription product from the purified tube-template dsDNA was observed by addition of the specific DNA strands for tube opening and subsequent exposure of the promotor region, while the transcription was suppressed without adding the specific strands (Fig. 20f). The specific DNA strands induced the 3D structural change and subsequent transcription. Therefore, transcription can be controlled by the mechanical operation of 3D DNA nanostructures. This method can be applied to mechanical switches for the expression of various biological reactions [131].

10 Photo-Controlled Devices for Delivery and Molecular Switch

One of the targets in DNA nanotechnology is the creation of molecular machines, molecular robots, and mechanical devices. The photo-induced initiation of the reaction is one of the most effective methods because the reaction is noninvasive and spatially and temporally controllable. The combination of the photoreaction with DNA origami structures is attractive for controlling the dynamic conformational changes in DNA origami structures. These photoresponsive 3D origami structures can be used as carriers for the controlled molecular delivery to the cell and dynamic switching devices for morphological changes of nanomaterials.

10.1 Photo-Controlled DNA Origami Nanocapsule

The development of molecular delivery systems to incorporate target molecules into the cell is an important application in DNA nanotechnology. We designed a photo-functionalized octahedral-shaped DNA nanocapsule (NC) for the preparation of a carrier for nanomaterials [132]. A photo-controllable open/closed system using hybridization and dissociation of photoswitching DNA strands was incorporated between two square pyramids of the NC (Fig. 21a). The opening and closing of the two square pyramids of the NC was controlled by UV and visible light irradiation. The reversible opening and closing were examined by gel electrophoresis and AFM, and the open behavior was directly observed by HS-AFM (Fig. 21b). In addition,

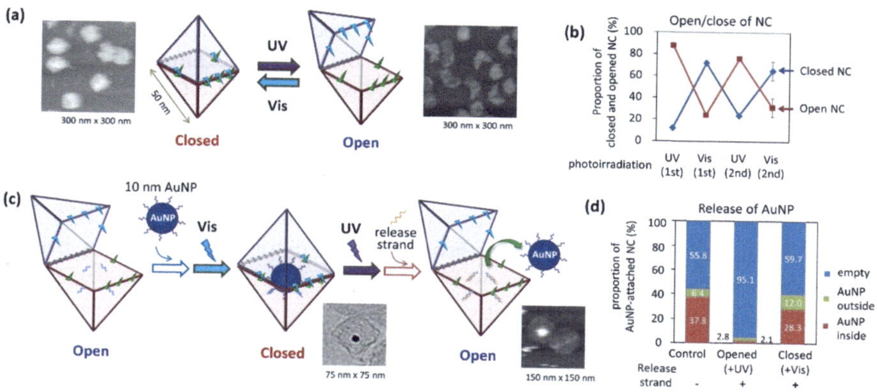

Fig. 21 DNA nanocapsule for molecular delivery. (**a**) Photoresponsive NC to control opening and closing by UV and visible light irradiation, respectively. AFM images of closed and open NC. (**b**) Reversible control of opening and closing by UV/vis irradiation. (**c**) Inclusion of gold nanoparticle (AuNP) inside the NC (cryo-EM image) and release of the AuNP from the NC with open/close/open operation by UV/Vis/UV irradiation. AFM image captured after opening of the NC with AuNP. (**d**) Release of AuNP from the NC from the closed and opened NC

the toehold system was introduced into the NC for the inclusion and release of a gold nanoparticle (AuNP) by a combination of photoirradiation and strand displacement (Fig. 21c). The encapsulation of the DNA-modified AuNP into the NC was carried out via hybridization of a specific DNA strand (capture strand). The DNA-modified AuNP was incorporated into the opened NC, and following the incubation for hybridization to trap the AuNP, the NC was closed by Vis irradiation. Inclusion of the AuNP was observed by cryo-electron microscopy. AuNP incorporated inside the NC was in 49% yield. Next, the release of AuNP was examined by the addition of toehold-containing complementary DNA (release strand). After UV irradiation and subsequent strand displacement, the AuNP was completely released from the NC by opening of the NC, while the AuNP release was suppressed without UV irradiation (Fig. 21d). The inclusion and release of the AuNP were successfully achieved using the open/close system controlled by photoirradiation. These nano-sized photoresponsive NCs can be applied as an intelligent carrier for the delivery of nanomaterials to cells similar to a virus capsid.

10.2 Manipulation of Photo-Controlled DNA Nanocapsule in the Cell

Development of a nano-sized carrier for cellular incorporation is one of the important research topics in nanobiotechnology. Because of their potential applications for imaging, diagnosis, and therapeutics, nano-sized signal- and stimuli-responsive carriers are attracting attention. We employed a DNA capsule (NC) with a photoinducible mechanical unlocking system for creation of a carrier for delivery system to the cells [133]. A photocage system was introduced into the NC for control of opening of the NC with photoirradiation (Fig. 22). The opening of the NC was observed by AFM (Fig. 22a), and the dynamic opening of the NC was examined by fluorescence recovery from the quenching (Fig. 22b). Next, the caged-NC was introduced to the cell and observed in the cytoplasm without toxicity. Photoirradiation was then performed after the caged-NC was incorporated into the cell to examine the unlocking of the caged-NC (Fig. 22c). After photoirradiation, the fluorescence increased in the cell, showing that the quenched dye was recovered by opening the NC. The increased fluorescence was observed throughout the cytoplasm and was partially localized in some organelles (arrows in Fig. 22c right). Furthermore, we performed single-cell photoirradiation to selectively open the caged-NC in the target cells (Fig. 22d). The caged-NC was labeled with Alexa 488 dye, and the opening of the NC was monitored by quenching and recovery of the fluorescence of the Alexa 647 dye/BHQ-3 quencher pair attached to the front corner of the NC. The single-cell photoirradiation was performed using a laser with a 405 nm wavelength (Fig. 22d-e). Three cells that had taken up the caged-NC were selected and individually irradiated (yellow circled cells 1–3). Before laser irradiation (left), the cells showed a

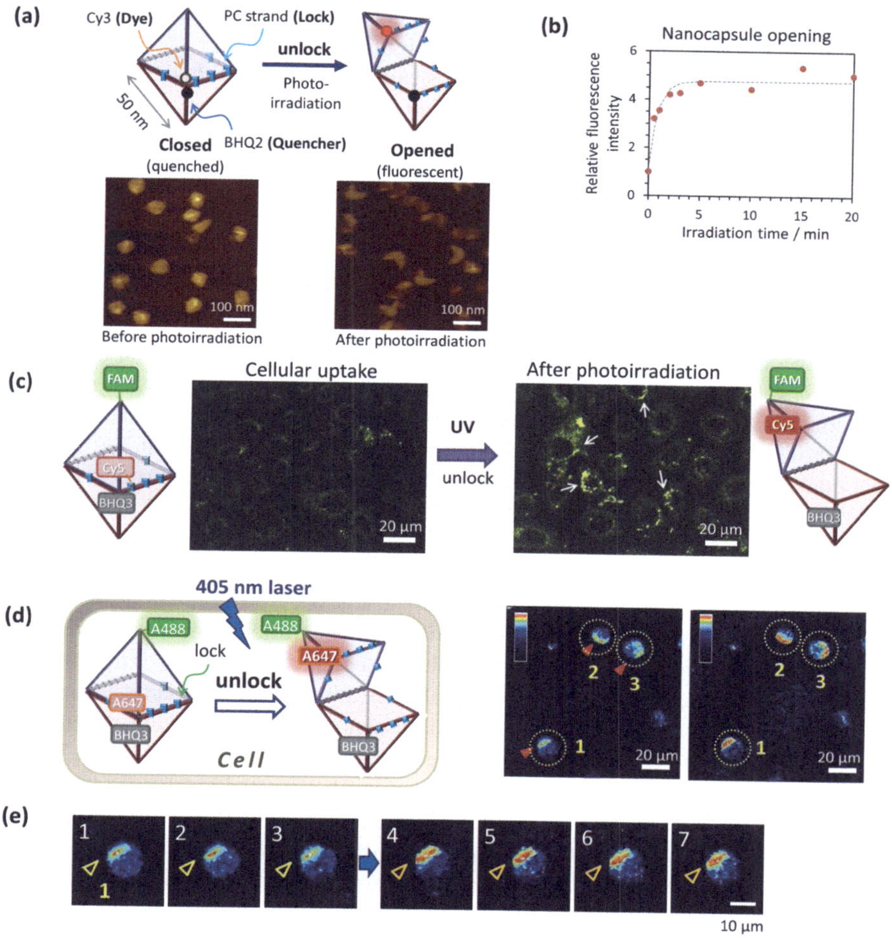

Fig. 22 Photocaged DNA nanocapsule as a delivery carrier to the cell and its manipulation. (**a**) Photocaged NC to control opening by UV light irradiation. AFM images of caged and uncaged NC. (**b**) Fluorescent dye and quencher are attached to the front corner of the caged-NC. Time-dependent uncaging of NC with UV irradiation. The fluorescence recovery of the caged-NC labeled with Cy3/BHQ-2 after photoirradiation was monitored. (**c**) Confocal laser scanning microscope (CLMS) images of the cells after incubation with FAM-labeled caged-NC with Cy3 dye/BHQ-2 quencher without (left) and with photoirradiation (right). (**d**) Single particle unlocking of the caged-NC with photoirradiation to single cell. Unlocking of caged-NC with 405 nm laser irradiation. Cells images before (left) and after (right) irradiation are shown. Three individual cells (yellow circles 1–3) were irradiated sequentially. The caged-NC was labeled with Alexa 488, and the opening of the NC was monitored by the recovery of fluorescence quenching of the Alexa 647/BHQ-3 quencher during laser irradiation. Fluorescence microscope images were reconstituted using ratio images of the fluorescence intensity of Alexa 647/Alexa 488. (**e**) Change of fluorescence ratio images of single cell before (images 1–3) and after (images 4–5) laser irradiation

modest red color (Alexa 647 intensity). After irradiation, a significant increase of the red-colored area was visible in the cells (right).

Using this NC system, gold nanoparticles can be included inside the large cavity as a cargo, which does not change the photoresponsive properties. Therefore, the caged-NC system could be applied for an intelligent delivery system for relatively large nanomaterials and proteins to the cells similar to the native virus system.

10.3 Photo-Controlled DNA Origami Nanoscissors

One of the key goals of nanotechnology is to construct molecular robots, mechanical devices, and nanomachines. To create such objects, stimuli-responsive molecular actuators are required. We directly observed the dynamic opening and closing of photo-controllable DNA origami nanoscissors by HS-AFM [134]. First, the conformational change between the open and closed states controlled by adjustment of the surrounding salt concentration could be directly observed during AFM scanning [135]. Then, light-responsive DNA strands were incorporated into the nanoscissors to control these conformational changes by photoirradiation (Fig. 23a). Using photoswitching DNA strands, photoresponsive nanoscissors were created, in which the open and closed conformations could be distinguished after respective irradiation with UV and visible light, via gel electrophoresis and AFM imaging. Additionally, these reversible changes in shape during photoirradiation were directly visualized by HS-AFM. Under UV irradiation, the closed nanoscissors opened during fluctuation, and then the opened nanoscissors closed under Vis irradiation (Fig. 23b). These results indicate that the opening and closing can be reversibly observed by just switching the light. Moreover, four photoswitchable nanoscissors were assembled into a tetramer. A scissor-actuator-like higher-order object whose configurations could be controlled by the open and close switching as induced by UV and Vis light irradiation. As seen here, the capability of switching is preserved even while the structures are attached on surfaces, which may offer interesting possibilities for constructing light-responsive electronic or photonic devices using DNA origami switches [136].

10.4 Photo-Controlled Plasmonic Switching Device

One of the targets of the controllable DNA nanostructures are optical devices. Synthetic molecular machines typically operate at the nanometer scale or below. Using the 10- to 100-nm-sized DNA nanostructure and molecular machines, controlled operation of individual molecular machines to a larger dimension should be achieved with many practical applications. Here, we created a light-driven plasmonic nanosystem, which has azobenzene-modified molecular switches attached to the host nanostructure and expresses reversible chiroptical function with large amplitude modulation [137]. Two bar-shaped DNA nanostructures connected at the center as a pivot was functionalized with photoswitching DNA strands to control the locked and relaxed states by visible and UV light irradiation, respectively (Fig. 23c). Then, two gold nanorods (AuNRs) were assembled onto a reconfigurable DNA origami scaffold to create the plasmonic nanostructure (Fig. 23d). In the locked state, expected peaks in the circular dichroism (CD) spectra were observed, while no peak was observed in the relaxed state because of the random positioning of the two AuNRs (Fig. 23e). The reversible switching of the relaxed and locked states of the plasmonic nanostructure was observed by alternating UV and visible light irradiation, respectively (Fig. 23f). In this system, the two states can be clearly identified and light is utilized as both an energy source and an information probe. An all-optically controlled plasmonic nanosystem was constructed on a designed DNA nanostructure. This system can amplify the subnanometer conformation changes of azobenzene through the active host nanostructure and consequently translate the light-induced molecular motion of azobenzene into reversible plasmonic chiroptical response, which can be in situ read out by optical spectroscopy. Light can reversely "write" and "erase" the conformation states of the nanostructure through photoi-

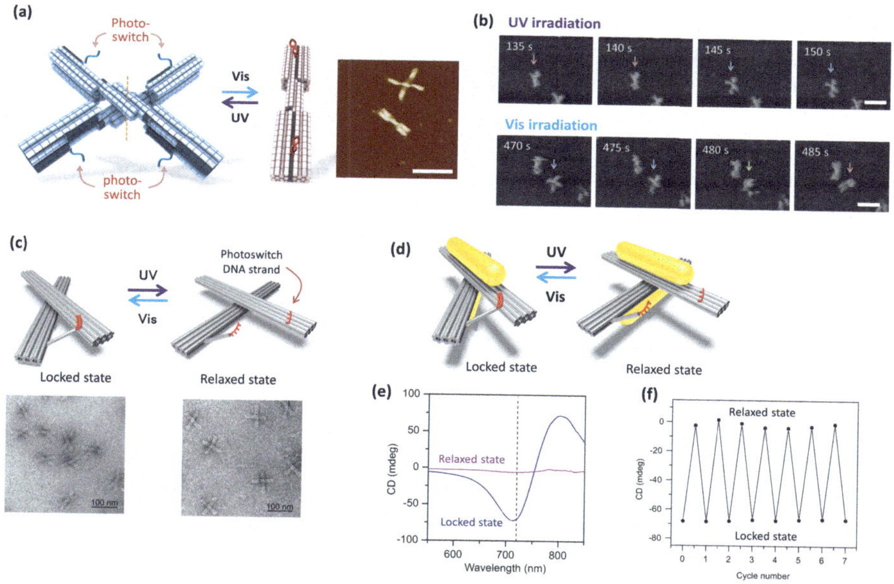

Fig. 23 DNA origami-based dynamic nanodevices. Photoswitchable DNA nanoscissors and photoswitchable plasmonic nanodevice. (**a**) Photo-controlled DNA nanoscissors possessing photoswitching DNA strands. AFM image of closed and opened conformation. (**b**) HS-AFM images of conformational changes of DNA nanoscissors under UV and Vis irradiation. (**c**) Photoswitchable dynamic DNA nanostructure for locked and relaxed state controlled by photoswitching DNA strands. TEM images of locked and relaxed states. (**d**) Plasmonic nanostructure with two gold nanorods. Locked and relaxed state can be controlled by UV/Vis irradiation. (**e**) CD spectra of the chiral plasmonic state of the locked (CD band) and relaxed (suppressed band) states. (**f**) Reversible plasmonic response induced by alternating UV/Vis irradiation

somerization at a localized region. The plasmonic nanosystem bears unique features of optical addressability, reversibility, and modulability, which are crucial for developing all-optical molecular devices with desired functionalities.

This reversible open and closed system is used to initialize the configuration of the nanostructures and change into the specific shapes repeatedly by just irradiating with the specific wavelength of light. The photonic modification of DNA based nanomachines can be used for biological applications such as cargo transport and manual configuration change of biomolecules in mesoscopic systems. Further, these nano-sized photoswitchable molecular devices could be applied for shape-changeable nanomaterials and may potentially become useful tools to regulate biochemical reactions.

11 Single-Molecule Sensing and Manipulation by Optical Tweezers

Mechanochemistry is an emerging field that investigates the coupling between mechanical and chemical processes. Under mechanical stress, the stability of covalent or noncovalent interactions changes, which either strengthens or weakens molecular structures. In mechanochemical sensing, various mechanical signals can be monitored. Compared to fluorescence signals that are subject to numerous background noises, mechanical signals, such as tension in a molecule, experience little environmental interference. The high signal-to-noise ratio is an advantage of mechanochemical sensing. The mechanochemical sensing and force measurements are employed for the experiment by combination with DNA origami scaffolds.

11.1 Single-Molecule Detection by Force Sensing DNA Origami Device

Single-molecular biosensing offers ultimate detection limit. However, its throughput is often compromised due to restricted platforms, in which sensing events are carried out once at a time in most cases [138]. To solve this problem, DNA origami nanostructures were introduced as expanded platforms in a new sensing strategy that exploits mechanochemical principles. As a proof of concept, six sensing probes were incorporated at different locations of a seven-tile DNA origami template (Fig. 24). Binding of a target molecule to any of these probes induces rearrangement of the 2D or 3D origami nanostructure, which is monitored in real time using optical tweezers. A DNA aptamer for platelet derived growth factor (PDGF) was used to connect DNA tiles by hybridization with a complementary strand (Fig. 24a). Using this system, the duplex dissociation and expansion of the length after the addition of 25 nM PDGF was clearly detected by force sensing at the single-molecule level

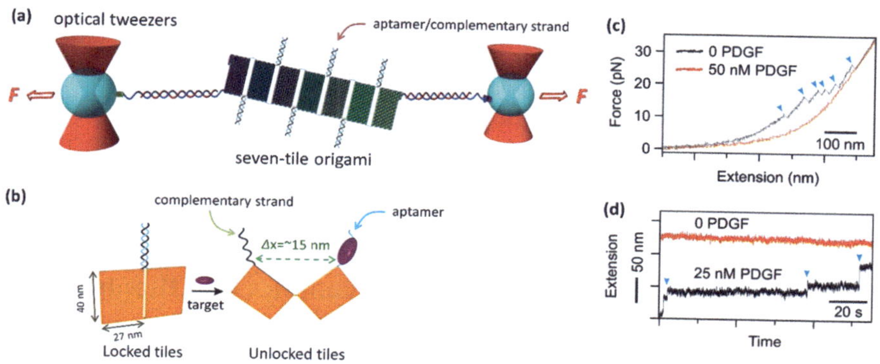

Fig. 24 DNA origami sensing device for detection of a single biomolecule by force-sensing using optical tweezers. (**a**) Seven DNA origami tiles are connected by six aptamer-containing dsDNAs. (**b**) Aptamer-containing dsDNA strand dissociates by binding of a target molecule to the aptamer, and the length change can be detected subsequently by optical tweezers. (**c**) Force-extension curve for detection of PDGF. Six length changes due to binding of PDGF were detected. (**d**) Real-time detection of PDGF binding by monitoring the change in length

(Fig. 24b-d). Furthermore, detection of two different molecules was performed using one platform by introducing different aptamers to different types of connections between tiles. After the addition of PDGF and specific DNA strand, bindings of two molecules were detected and identified in different stretching lengths at the single-molecule level. This platform is able to detect 10 pM PDGF within 10 min, while differentiating the PDGF and the DNA target in a multiplexing fashion. Using the versatile DNA origami assembly, this mechanochemical platform offers the solution for high-throughput single-molecule sensing.

11.2 Dynamic Configurational Change of Helical DNA Nanotubes

Isomerization of Holliday junction is one of the basic topics in DNA nanotechnology. In DNA origami, a single-stranded DNA template is shaped into a desired nanostructure using DNA staples that form Holliday junction crossovers with the template. Novel helical tubular structures were designed and constructed using the DNA origami [129]. These helical DNA origami tubes have geometrical isomers called short and long tubes, in which the direction of the Holliday junction crossovers are different from each other. If these helical tube isomers are stretched mechanically, the cooperative isomerization of the helical tubes can be examined. Mechanical properties of isomerization of helical tubes were further characterized using optical tweezers [139]. By stretching the short tube to vertical direction using optical tweezers, the mechanical isomerization was observed between two conformations of DNA nanotubes at 10–35 pN. The collective actions of individual

Holliday junctions are only possible in DNA origami with rotational symmetric arrangements of Holliday junctions in the DNA nanotubes. The results indicate that Holliday junctions control mechanical behaviors of DNA assemblies. The mechanical properties observed here provide insights for designing better DNA nanostructures. In addition, the unprecedented mechanical isomerization process offers new strategies for the development of nanosensors and actuators.

12 Observation of Single-Molecule Dynamics and Biophysics in the DNA Nanocages

The folding and unfolding of macromolecules inside cells often occur in a confined space. For example, these occur in chaperone or proteasome machinery, and entry/exit channels of ribosomes and polymerases [140, 141]. These observations are considered as the confined volume effect [142, 143], in which the entropy of unfolded biopolymer decreases compared to that of the folded conformation, leading to a more stabilized folded state. Molecular simulations have suggested that confined space increases the stability of a folded structure due to entropic effects [144].

12.1 G-Quadruplex in the DNA Nanocages

Using DNA origami nanocages, we investigated the effect of confined space on the property of individual human telomeric DNA G-quadruplex (GQ) [145]. A GQ-forming DNA strand was placed inside DNA origami nanocages of three different sizes (6, 9, and 15 nm) (Fig. 25). The repulsion due to negative charges from the DNA origami and the telomeric sequence prevents the direct interaction between them, allowing clear-cut evaluation of the spatial confinement. To target the folded structure inside the cage, we used optical tweezers to investigate the mechanical and thermodynamic stability and transition kinetics of the GQ with respect to those in a crowded or a diluted buffer [146, 147]. By targeted mechanical unfolding of the GQ while leaving the nanocage intact, the formation of GQ structure was observed inside the three different nanocages (6 × 6 nm, 9 × 9 nm, and 15 × 15 nm inside). The formed GQ in the 6 × 6 nm nanocage appeared as an unusual basket-type structure (6.9 nm stretch), while the GQ in the medium and large cages formed usual hybrid-type structure (8.2 nm stretch), indicating that the confined space affects the shape of the GQ structure. Mechanical and thermodynamic stabilities of the GQ inside the nanocage significantly increased compared to those of GQ without the cage (Table 1). In addition, the stability increased with decreasing the size of the nanocage. For the folding kinetics, folding rate of the GQ in the confined space increased in two-order of magnitude compared to that of the GQ without nanocage.

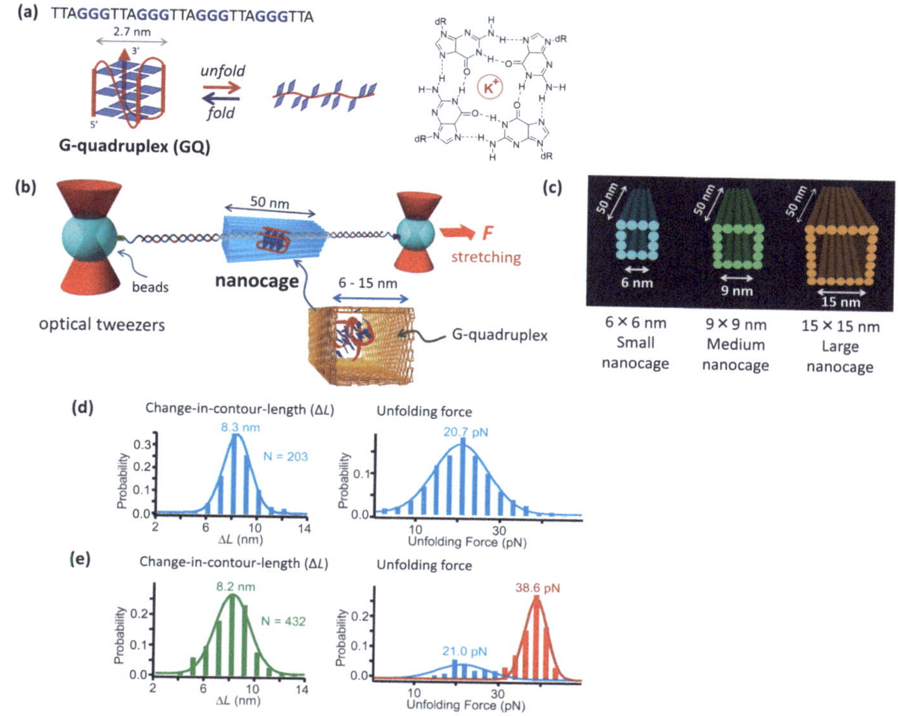

Fig. 25 Investigation of the physical properties of biomolecules in the confined space of a DNA origami nanocage. (**a**) G-quadruplex (GQ) structure and the folding and unfolding of the GQ structure. (**b**) GQ in a DNA nanocage and investigation of the physical property of unfolding/folding of GQ in the nanocage using optical tweezers. (**c**) Three different nanocages used in the experiment. Nanocage with 6, 9, and 15 nm space, respectively. (**d**) Histograms of contour-length changes for unfolding (left) and mechanical force for unfolding of GQ without nanocage (right). (**e**) Histograms of contour-length changes for unfolding (left) and mechanical force for unfolding of GQ in the 9 × 9 nm nanocage (right)

Table 1 Change-in-contour-length (ΔL) and force of GQ unfolding, and ΔG_{unfold} in the nanocages

Nanocage	ΔL (nm)	Unfolding force (pN)	ΔG_{unfold} (kcal/mol)
6 × 6 nm	6.9 ± 0.1	35.9 ± 0.2	13.4 ± 1.4
9 × 9 nm	8.2 ± 0.2	38.6 ± 0.2	14.2 ± 1.0
15 × 15 nm	8.2 ± 0.2	27.9 ± 2.0	8.2 ± 0.4
No nanocage	8.3 ± 0.1	20.7 ± 1.4	7.1 ± 0.2

Telomeric GQ was mechanically unfolded inside DNA origami nanocages, mimicking the exit channels in telomerase or polymerases. In such a confined space, GQ structures display fast folding kinetics with increased mechanical and thermodynamic stabilities. These effects in a confined space scale inversely with nanocage size, and are stronger than those in molecularly crowded conditions. These results suggest the possibility of simultaneous folding of GQ inside telomerase or other motor proteins during their respective enzymatic processes.

12.2 Estimation of the Environment Inside the DNA Nanocages

Due to the small size of a nanoconfinement, property of water contained inside is rather challenging to probe. We measured the amount of water molecules released during the folding of individual G-quadruplex (GQ) and i-motif (iM) structures, from which water activities are estimated in the DNA nanocages (cross sections: 9 × 9 nm to 15 × 15 nm) [148]. We first examined the properties of i-motif inside the nanocages by folding and unfolding using optical tweezers (Fig. 26a). i-Motif forms

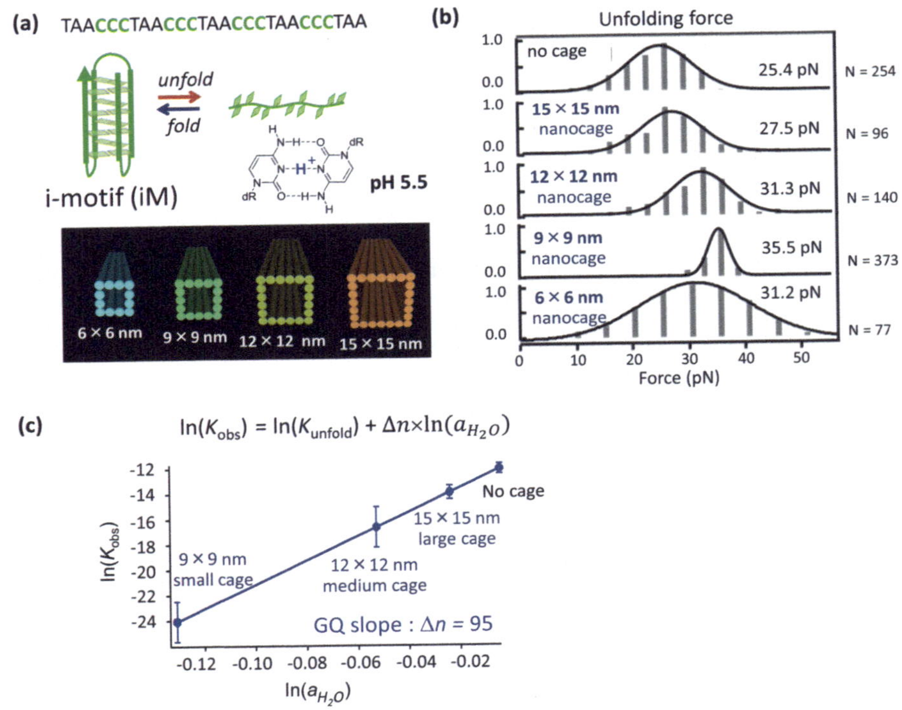

Fig. 26 Investigation of the physical properties of i-motif in the DNA origami nanocages and the environment inside the nanocages. (**a**) i-Motif (iM) structure and the folding and unfolding of the iM structure under the acidic condition (pH 5.5). Four different nanocages used in the experiment; nanocages with 6, 9, 12, and 15 nm space, respectively. (**b**) Histograms of mechanical force for unfolding of iM in the four different nanocages. (**c**) Estimation of water activity inside the nanocages. The obtained K_{obs} values were potted on the following equation $\ln(K_{obs}) = \ln(K_{unfold}) + \Delta n \times \ln\left(a_{H_2O}\right)$ $(\Delta n = 95)$

Table 2 Free energy change ΔG_{unfold} for i-motif unfolding in the nanocages

Nanocage	ΔG_{unfold} (kcal/mol)
No cage	7.10 ± 0.48
15 × 15 nm	7.78 ± 0.80
12 × 12 nm	8.50 ± 0.91
9 × 9 nm	11.3 ± 0.99

Table 3 Estimated water activity (a_{H_2O}) in the nanocages

Nanocage	a_{H_2O}
No cage	0.99
15 × 15 nm (large)	0.98
12 × 12 nm (medium)	0.95
9 × 9 nm (small)	0.88

a tetraplex using and consecutive C sequence and protonated-C/C pairing under acidic condition. We observed the formation of iM in the 9 × 9 nm nanocage at pH 5.5. Interestingly, iM structure formed even at near neutral pH such as 6.0 and 6.5 in the nanocage, while iM itself did not form under acidic near neutral conditions. These indicate the nanocage can stabilize the iM structure at higher pH. Next, we examined the size effect of nanocages on the iM stability (Fig. 26b). When the smallest 6 × 6 nm nanocage was used, iM structure did not form. Similar to the results of GQ cases, both mechanical and thermal stability of iM increased by decreasing the size of nanocages (Table 2). Finally, we estimated the water activity inside the nanocage using the obtained values of free energy change of GQ cases. Using the equation, $\ln(K_{obs}) = \ln(K_{unfold}) + \Delta n \times \ln\left(a_{H_2O}\right)$ and reported $\Delta n = 95$ [149] for plotting the obtained K_{obs} on the linear slope (Fig. 26c), we obtained the water activity inside each nanocage (Table 3). We found water activities decrease with reducing nanocage size. In the 9 × 9 nm nanocage, water activity was reduced beyond the reach of regular cosolutes such as polyethylene glycol (PEG). With this set of nanocages, we were able to retrieve the change in water molecules throughout the folding trajectory of GQ or iM. Therefore, the overall loss of available water molecules for unfolding drives the increase stabilization of GQ and iM in the nanocages with reduced water activity.

13 Conclusions and Perspectives

DNA origami technology opened up a new era in molecular science and technology. This innovative technology allows us to design and construct various nanostructures and precise placement of molecules, and to manipulate the target molecules and control molecular devices and machines. The author employed the designed nanospaces to image the molecules of interest and regulate their reactions. The single-molecule observation system has been created to elucidate the biophysical and biochemical properties of target molecules by HS-AFM. The method can be utilized to observe various reactions involved in essential biological events such as replication, transcription, and translation. Single-molecule imaging of chemical reactions such as complex formations, photoreactions, and bond formation/cleavage has been achieved using this system. In addition, direct imaging of the mobile molecular nanomachine on DNA origami structures has been demonstrated using the HS-AFM system. The detailed mechanism for movement of a DNA motor was analyzed at the single-molecule level. The method can be applied for development of molecular machines that transport specific molecules, which can realize more complicated movements controlled by light input. The programed assembly systems were created to control the arrangement of the DNA origami, and the lipid bilayer was used to expand the HS-AFM imaging for direct observation of the dynamic formation of micrometer-sized assemblies. Dynamic conformational changes of the 3D origami structures were investigated by HS-AFM, and nanodevices such as plasmonic and molecular delivery devices were created. For desired observations, target-oriented nanospaces were created, and the mechanical properties of the G-quadruplex and i-motif was investigated by force sensing with optical tweezers, which revealed that confined spaces stabilized these structures with decreased water activity. Using the designed DNA origami scaffold and improving the single-molecule measurement technique, single-molecule imaging and sensing have been realized to elucidate the physical properties of various molecules both in the biological and nonbiological environments. The molecular motor system [95, 97], transcription-based nanochip [106], and channel/receptor system [150] have been constructed on the DNA origami structures, which show the robust performance as predesigned.

It is obvious that DNA origami is the most reliable and robust molecular system in the self-assembly-based technology. Now that the basic properties of DNA origami have been elucidated, practical utilization of the nanodevices and the material and biological applications are the next stage. The author believes that DNA nanotechnology has great potential to revolutionize the framework of present science and technology.

Acknowledgments This research was financially supported by a Grant-in-Aid for Scientific Research on innovative areas "Molecular Robotics" of MEXT, and JSPS KAKENHI. This research was also supported by JST CREST. The author thanks Tokuyama Science and Technology Foundation, Sumitomo Foundation, Iketani Science and Technology Foundation, Nagase Science and Technology Foundation, Mitsubishi Foundation, Asahi Glass Foundation, Sekisui Chemical Foundation, Kurata Memorial Hitachi Foundation, Novartis Foundation, Naito Foundation, Uehara Memorial Foundation, and Nakatani Foundation for financial supports. The author thanks Prof. Hiroshi Sugiyama for supporting the research. Contributions of Dr. Yuki Suzuki, Dr. Arivazhagan Rajendran, Dr. Yangyang Yang, Dr. Yousuke Katsuda, Dr. Seigi Yamamoto, Ms. Kumi Hidaka, and Ms. Tomoko Emura for the research are greatly acknowledged. The author thanks all the students and collaborators.

References

1. Endo M, Sugiyama H (2009) Chemical approaches to DNA nanotechnology. Chembiochem 10:2420–2443
2. Endo M, Yang Y, Sugiyama H (2013b) DNA origami technology for biomaterials applications. Biomater Sci 1:347–360
3. Rothemund PW (2006) Folding DNA to create nanoscale shapes and patterns. Nature 440:297–302
4. Douglas SM, Dietz H, Liedl T, Hogberg B, Graf F, Shih WM (2009) Self-assembly of DNA into nanoscale three-dimensional shapes. Nature 459:414–418
5. Rajendran A, Endo M, Sugiyama H (2012) Single-molecule analysis using DNA origami. Angew Chem Int Ed 51:874–890
6. Torring T, Voigt NV, Nangreave J, Yan H, Gothelf KV (2011) DNA origami: a quantum leap for self-assembly of complex structures. Chem Soc Rev 40:5636–5646
7. Ando T, Kodera N (2012) Visualization of mobility by atomic force microscopy. Methods Mol Biol 896:57–69
8. Ando T, Kodera N, Takai E, Maruyama D, Saito K, Toda A (2001) A high-speed atomic force microscope for studying biological macromolecules. Proc Natl Acad Sci U S A 98:12468–12472
9. Rajendran A, Endo M, Sugiyama H (2014c) State-of-the-art high-speed atomic force microscopy for investigation of single-molecular dynamics of proteins. Chem Rev 114:1493–1520
10. Uchihashi T, Kodera N, Ando T (2012) Guide to video recording of structure dynamics and dynamic processes of proteins by high-speed atomic force microscopy. Nat Protoc 7:1193–1206
11. Schitter G, Astrom KJ, DeMartini BE, Thurner PJ, Turner KL, Hansma PK (2007) Design and modeling of a high-speed AFM-scanner. IEEE T Contr Syst T 15:906–915
12. Walters DA, Cleveland JP, Thomson NH, Hansma PK, Wendman MA, Gurley G, Elings V (1996) Short cantilevers for atomic force microscopy. Rev Sci Instrum 67:3583–3590
13. Endo M, Katsuda Y, Hidaka K, Sugiyama H (2010a) Regulation of DNA methylation using different tensions of double strands constructed in a defined DNA nanostructure. J Am Chem Soc 132:1592–1597
14. Sharma J, Chhabra R, Andersen CS, Gothelf KV, Yan H, Liu Y (2008) Toward reliable gold nanoparticle patterning on self-assembled DNA nanoscaffold. J Am Chem Soc 130:7820–7821
15. Voigt NV, Torring T, Rotaru A, Jacobsen MF, Ravnsbaek JB, Subramani R, Mamdouh W, Kjems J, Mokhir A, Besenbacher F et al (2010) Single-molecule chemical reactions on DNA origami. Nat Nanotechnol 5:200–203
16. Endo M, Sugiyama H (2014) Single-molecule imaging of dynamic motions of biomolecules in DNA origami nanostructures using high-speed atomic force microscopy. Acc Chem Res 47:1645–1653
17. Crampton N, Yokokawa M, Dryden DT, Edwardson JM, Rao DN, Takeyasu K, Yoshimura SH, Henderson RM (2007) Fast-scan atomic force microscopy reveals that the type III restriction enzyme EcoP15I is capable of DNA translocation and looping. Proc Natl Acad Sci U S A 104:12755–12760
18. Gilmore JL, Suzuki Y, Tamulaitis G, Siksnys V, Takeyasu K, Lyubchenko YL (2009) Single-molecule dynamics of the DNA-EcoRII protein complexes revealed with high-speed atomic force microscopy. Biochemistry 48:10492–10498
19. Suzuki Y, Gilmore JL, Yoshimura SH, Henderson RM, Lyubchenko YL, Takeyasu K (2011) Visual analysis of concerted cleavage by type IIF restriction enzyme SfiI in subsecond time region. Biophys J 101:2992–2998
20. Endo M, Katsuda Y, Hidaka K, Sugiyama H (2010b) A versatile DNA nanochip for direct analysis of DNA base-excision repair. Angew Chem Int Ed 49:9412–9416

21. Xing X, Sato S, Wong NK, Hidaka K, Sugiyama H, Endo M (2020) Direct observation and analysis of TET-mediated oxidation processes in a DNA origami nanochip. Nucleic Acids Res 48:4041–4051
22. Suzuki Y, Endo M, Katsuda Y, Ou K, Hidaka K, Sugiyama H (2014b) DNA origami based visualization system for studying site-specific recombination events. J Am Chem Soc 136:211–218
23. Kobayashi Y, Misumi O, Odahara M, Ishibashi K, Hirono M, Hidaka K, Endo M, Sugiyama H, Iwasaki H, Kuroiwa T et al (2017) Holliday junction resolvases mediate chloroplast nucleoid segregation. Science 356:631–634
24. Suzuki Y, Endo M, Canas C, Ayora S, Alonso JC, Sugiyama H, Takeyasu K (2014a) Direct analysis of Holliday junction resolving enzyme in a DNA origami nanostructure. Nucleic Acids Res 42:7421–7428
25. Raz MH, Hidaka K, Sturla SJ, Sugiyama H, Endo M (2016) Torsional constraints of DNA substrates impact Cas9 cleavage. J Am Chem Soc 138:13842–13845
26. Youngblood B, Reich NO (2006) Conformational transitions as determinants of specificity for the DNA methyltransferase EcoRI. J Biol Chem 281:26821–26831
27. David SS, O'Shea VL, Kundu S (2007) Base-excision repair of oxidative DNA damage. Nature 447:941–950
28. Bruner SD, Norman DP, Verdine GL (2000) Structural basis for recognition and repair of the endogenous mutagen 8-oxoguanine in DNA. Nature 403:859–866
29. Morikawa K, Matsumoto O, Tsujimoto M, Katayanagi K, Ariyoshi M, Doi T, Ikehara M, Inaoka T, Ohtsuka E (1992) X-ray structure of T4 endonuclease V: an excision repair enzyme specific for a pyrimidine dimer. Science 256:523–526
30. Nash HM, Bruner SD, Scharer OD, Kawate T, Addona TA, Spooner E, Lane WS, Verdine GL (1996) Cloning of a yeast 8-oxoguanine DNA glycosylase reveals the existence of a base-excision DNA-repair protein superfamily. Curr Biol 6:968–980
31. Vassylyev DG, Kashiwagi T, Mikami Y, Ariyoshi M, Iwai S, Ohtsuka E, Morikawa K (1995) Atomic model of a pyrimidine dimer excision repair enzyme complexed with a DNA substrate: structural basis for damaged DNA recognition. Cell 83:773–782
32. Vassylyev DG, Morikawa K (1997) DNA-repair enzymes. Curr Opin Struct Biol 7:103–109
33. Grindley ND, Whiteson KL, Rice PA (2006) Mechanisms of site-specific recombination. Annu Rev Biochem 75:567–605
34. Guo F, Gopaul DN, Van Duyne GD (1997) Structure of Cre recombinase complexed with DNA in a site-specific recombination synapse. Nature 389:40–46
35. Van Duyne GD (2001) A structural view of cre-loxp site-specific recombination. Annu Rev Biophys Biomol Struct 30:87–104
36. Gopaul DN, Guo F, Van Duyne GD (1998) Structure of the Holliday junction intermediate in Cre-loxP site-specific recombination. EMBO J 17:4175–4187
37. Guo F, Gopaul DN, Van Duyne GD (1999) Asymmetric DNA bending in the Cre-loxP site-specific recombination synapse. Proc Natl Acad Sci U S A 96:7143–7148
38. Doudna JA, Charpentier E (2014) Genome editing. The new frontier of genome engineering with CRISPR-Cas9. Science 346:1258096
39. Sternberg SH, LaFrance B, Kaplan M, Doudna JA (2015) Conformational control of DNA target cleavage by CRISPR-Cas9. Nature 527:110–113
40. Schwartz T, Rould MA, Lowenhaupt K, Herbert A, Rich A (1999) Crystal structure of the Zalpha domain of the human editing enzyme ADAR1 bound to left-handed Z-DNA. Science 284:1841–1845
41. Boyer LA, Lee TI, Cole MF, Johnstone SE, Levine SS, Zucker JP, Guenther MG, Kumar RM, Murray HL, Jenner RG et al (2005) Core transcriptional regulatory circuitry in human embryonic stem cells. Cell 122:947–956
42. Loh YH, Wu Q, Chew JL, Vega VB, Zhang W, Chen X, Bourque G, George J, Leong B, Liu J et al (2006) The Oct4 and Nanog transcription network regulates pluripotency in mouse embryonic stem cells. Nat Genet 38:431–440

43. Yamamoto S, De D, Hidaka K, Kim KK, Endo M, Sugiyama H (2014) Single molecule visualization and characterization of Sox2-Pax6 complex formation on a regulatory DNA element using a DNA origami frame. Nano Lett 14:2286–2292
44. Remenyi A, Lins K, Nissen LJ, Reinbold R, Scholer HR, Wilmanns M (2003) Crystal structure of a POU/HMG/DNA ternary complex suggests differential assembly of Oct4 and Sox2 on two enhancers. Genes Dev 17:2048–2059
45. Esadze A, Stivers JT (2018) Facilitated diffusion mechanisms in DNA Base excision repair and transcriptional activation. Chem Rev 118:11298–11323
46. Halford SE, Marko JF (2004) How do site-specific DNA-binding proteins find their targets? Nucleic Acids Res 32:3040–3052
47. Kamagata K, Murata A, Itoh Y, Takahashi S (2017) Characterization of facilitated diffusion of tumor suppressor p53 along DNA using single-molecule fluorescence imaging. J Photochem Photobiol C 30:36–50
48. Wang X, Chen X, Yang Y (2012) Spatiotemporal control of gene expression by a light-switchable transgene system. Nat Methods 9:266–269
49. Raghavan G, Hidaka K, Sugiyama H, Endo M (2019) Direct observation and analysis of the dynamics of the photoresponsive transcription factor GAL4. Angew Chem 58:7626–7630
50. Endo M, Tatsumi K, Terushima K, Katsuda Y, Hidaka K, Harada Y, Sugiyama H (2012b) Direct visualization of the movement of a single T7 RNA polymerase and transcription on a DNA nanostructure. Angew Chem Int Ed 51:8778–8782
51. Bacolla A, Wells RD (2004) Non-B DNA conformations, genomic rearrangements, and human disease. J Biol Chem 279:47411–47414
52. Bacolla A, Wells RD (2009) Non-B DNA conformations as determinants of mutagenesis and human disease. Mol Carcinog 48:273–285
53. Shirude PS, Okumus B, Ying L, Ha T, Balasubramanian S (2007) Single-molecule conformational analysis of G-quadruplex formation in the promoter DNA duplex of the proto-oncogene c-kit. J Am Chem Soc 129:7484–7485
54. Siddiqui-Jain A, Grand CL, Bearss DJ, Hurley LH (2002) Direct evidence for a G-quadruplex in a promoter region and its targeting with a small molecule to repress c-MYC transcription. Proc Natl Acad Sci U S A 99:11593–11598
55. Sannohe Y, Endo M, Katsuda Y, Hidaka K, Sugiyama H (2010) Visualization of dynamic conformational switching of the G-quadruplex in a DNA nanostructure. J Am Chem Soc 132:16311–16313
56. Xu Y, Sato H, Sannohe Y, Shinohara K, Sugiyama H (2008) Stable lariat formation based on a G-quadruplex scaffold. J Am Chem Soc 130:16470–16471
57. Rajendran A, Endo M, Hidaka K, Sugiyama H (2014a) Direct and single-molecule visualization of the solution-state structures of G-hairpin and G-triplex intermediates. Angew Chem Int Ed 53:4107–4112
58. Rajendran A, Endo M, Hidaka K, Tran PL, Mergny JL, Sugiyama H (2013c) Controlling the stoichiometry and strand polarity of a tetramolecular G-quadruplex structure by using a DNA origami frame. Nucleic Acids Res 41:8738–8747
59. Rajendran A, Endo M, Hidaka K, Tran PLT, Teulade-Fichou MP, Mergny JL, Sugiyama H (2014b) G-quadruplex-binding ligand-induced DNA synapsis inside a DNA origami frame. RSC Adv 4:6346–6355
60. Rajendran A, Endo M, Hidaka K, Tran PL, Mergny JL, Gorelick RJ, Sugiyama H (2013b) HIV-1 nucleocapsid proteins as molecular chaperones for tetramolecular antiparallel G-quadruplex formation. J Am Chem Soc 135:18575–18585
61. Huppert JL, Balasubramanian S (2007) G-quadruplexes in promoters throughout the human genome. Nucleic Acids Res 35:406–413
62. Gehring K, Leroy JL, Gueron M (1993) A tetrameric DNA structure with protonated cytosine. Cytosine base pairs. Nature 363:561–565

63. Dhakal S, Schonhoft JD, Koirala D, Yu Z, Basu S, Mao H (2010) Coexistence of an ILPR i-motif and a partially folded structure with comparable mechanical stability revealed at the single-molecule level. J Am Chem Soc 132:8991–8997
64. Dhakal S, Yu Z, Konik R, Cui Y, Koirala D, Mao H (2012) G-quadruplex and i-motif are mutually exclusive in ILPR double-stranded DNA. Biophys J 102:2575–2584
65. Sun D, Hurley LH (2009) The importance of negative superhelicity in inducing the formation of G-quadruplex and i-motif structures in the c-Myc promoter: implications for drug targeting and control of gene expression. J Med Chem 52:2863–2874
66. Xu Y, Sugiyama H (2006) Formation of the G-quadruplex and i-motif structures in retinoblastoma susceptibility genes (Rb). Nucleic Acids Res 34:949–954
67. Miyoshi D, Matsumura S, Nakano S, Sugimoto N (2004) Duplex dissociation of telomere DNAs induced by molecular crowding. J Am Chem Soc 126:165–169
68. Endo M, Xing X, Zhou X, Emura T, Hidaka K, Tuesuwan B, Sugiyama H (2015b) Single-molecule manipulation of the duplex formation and dissociation at the G-Quadruplex/i-motif site in the DNA nanostructure. ACS Nano 9:9922–9929
69. Moser HE, Dervan PB (1987) Sequence-specific cleavage of double helical DNA by triple helix formation. Science 238:645–650
70. Yamagata Y, Emura T, Hidaka K, Sugiyama H, Endo M (2016) Triple helix formation in a topologically controlled DNA Nanosystem. Chemistry 22:5494–5498
71. Beal PA, Dervan PB (1992) Recognition of double helical DNA by alternate Strand triple Helix formation. J Am Chem Soc 114:4976–4982
72. Jovin TM, Soumpasis DM, Mcintosh LP (1987) The transition between B-DNA and Z-DNA. Annu Rev Phys Chem 38:521–560
73. Mao CD, Sun WQ, Shen ZY, Seeman NC (1999) A nanomechanical device based on the B-Z transition of DNA. Nature 397:144–146
74. Rajendran A, Endo M, Hidaka K, Sugiyama H (2013a) Direct and real-time observation of rotary movement of a DNA nanomechanical device. J Am Chem Soc 135:1117–1123
75. Behe M, Felsenfeld G (1981) Effects of methylation on a synthetic polynucleotide: the B–Z transition in poly(dG-m5dC).Poly(dG-m5dC). Proc Natl Acad Sci U S A 78:1619–1623
76. Asanuma H, Liang X, Nishioka H, Matsunaga D, Liu M, Komiyama M (2007) Synthesis of azobenzene-tethered DNA for reversible photo-regulation of DNA functions: hybridization and transcription. Nat Protoc 2:203–212
77. Liang X, Mochizuki T, Asanuma H (2009) A supra-photoswitch involving sandwiched DNA base pairs and azobenzenes for light-driven nanostructures and nanodevices. Small 5:1761–1768
78. Endo M, Yang Y, Suzuki Y, Hidaka K, Sugiyama H (2012c) Single-molecule visualization of the hybridization and dissociation of photoresponsive oligonucleotides and their reversible switching behavior in a DNA nanostructure. Angew Chem Int Ed 51:10518–10522
79. Yang Y, Endo M, Suzuki Y, Hidaka K, Sugiyama H (2014) Direct observation of the dual-switching behaviors corresponding to the state transition in a DNA nanoframe. Chem Commun 50:4211–4213
80. Jash B, Muller J (2017) Metal-mediated base pairs: from characterization to application. Chemistry 23:17166–17178
81. Takezawa Y, Shionoya M (2012) Metal-mediated DNA base pairing: alternatives to hydrogen-bonded Watson-crick base pairs. Acc Chem Res 45:2066–2076
82. Ono A, Torigoe H, Tanaka Y, Okamoto I (2011) Binding of metal ions by pyrimidine base pairs in DNA duplexes. Chem Soc Rev 40:5855–5866
83. Tanaka Y, Kondo J, Sychrovsky V, Sebera J, Dairaku T, Saneyoshi H, Urata H, Torigoe H, Ono A (2015) Structures, physicochemical properties, and applications of T-hg(II)-T, C-Ag(I)-C, and other metallo-base-pairs. Chem Commun 51:17343–17360
84. Xing X, Feng Y, Yu Z, Hidaka K, Liu F, Ono A, Sugiyama H, Endo M (2018) Direct observation of the double-stranded DNA formation through metal ion-Mediated Base pairing in the nanoscale structure. Chemistry 25(6):1446–1450

85. Liu X, Lu CH, Willner I (2014) Switchable reconfiguration of nucleic acid nanostructures by stimuli-responsive DNA machines. Acc Chem Res 47:1673–1680
86. Wang F, Liu X, Willner I (2015) DNA switches: from principles to applications. Angew Chem Int Ed 54:1098–1129
87. Gu H, Furukawa K, Weinberg Z, Berenson DF, Breaker RR (2013) Small, highly active DNAs that hydrolyze DNA. J Am Chem Soc 135:9121–9129
88. Endo M, Takeuchi Y, Suzuki Y, Emura T, Hidaka K, Wang F, Willner I, Sugiyama H (2015a) Single-molecule visualization of the activity of a Zn(2+)-dependent DNAzyme. Angew Chem Int Ed 54:10550–10554
89. Van Melckebeke H, Devany M, Di Primo C, Beaurain F, Toulme JJ, Bryce DL, Boisbouvier J (2008) Liquid-crystal NMR structure of HIV TAR RNA bound to its SELEX RNA aptamer reveals the origins of the high stability of the complex. Proc Natl Acad Sci U S A 105:9210–9215
90. Grabow WW, Jaeger L (2014) RNA self-assembly and RNA nanotechnology. Acc Chem Res 47:1871–1880
91. Shukla GC, Haque F, Tor Y, Wilhelmsson LM, Toulme JJ, Isambert H, Guo P, Rossi JJ, Tenenbaum SA, Shapiro BA (2011) A boost for the emerging field of RNA nanotechnology. ACS Nano 5:3405–3418
92. Durand G, Lisi S, Ravelet C, Dausse E, Peyrin E, Toulme JJ (2014) Riboswitches based on kissing complexes for the detection of small ligands. Angew Chem Int Ed 53:6942–6945
93. Takeuchi Y, Endo M, Suzuki Y, Hidaka K, Durand G, Dausse E, Toulme JJ, Sugiyama H (2016) Single-molecule observations of RNA-RNA kissing interactions in a DNA nanostructure. Biomater Sci 4:130–135
94. Bath J, Turberfield AJ (2007) DNA nanomachines. Nat Nanotechnol 2:275–284
95. Wickham SFJ, Endo M, Katsuda Y, Hidaka K, Bath J, Sugiyama H, Turberfield AJ (2011) Direct observation of stepwise movement of a synthetic molecular transporter. Nat Nanotechnol 6:166–169
96. Bath J, Green SJ, Turberfield AJ (2005) A free-running DNA motor powered by a nicking enzyme. Angew Chem Int Ed 44:4358–4361
97. Wickham SFJ, Bath J, Katsuda Y, Endo M, Hidaka K, Sugiyama H, Turberfield AJ (2012) A DNA-based molecular motor that can navigate a network of tracks. Nat Nanotechnol 7:169–173
98. Lund K, Manzo AJ, Dabby N, Michelotti N, Johnson-Buck A, Nangreave J, Taylor S, Pei R, Stojanovic MN, Walter NG et al (2010) Molecular robots guided by prescriptive landscapes. Nature 465:206–210
99. Gu HZ, Chao J, Xiao SJ, Seeman NC (2010) A proximity-based programmable DNA nanoscale assembly line. Nature 465:202–205
100. Yang Y, Goetzfried MA, Hidaka K, You M, Tan W, Sugiyama H, Endo M (2015) Direct visualization of walking motions of photocontrolled nanomachine on the DNA nanostructure. Nano Lett 15:6672–6676
101. You M, Chen Y, Zhang X, Liu H, Wang R, Wang K, Williams KR, Tan W (2012) An autonomous and controllable light-driven DNA walking device. Angew Chem Int Ed 51:2457–2460
102. Yang Y, Tashiro R, Suzuki Y, Emura T, Hidaka K, Sugiyama H, Endo M (2017) A photoregulated DNA-based rotary system and direct observation of its rotational movement. Chem Eur J 23:3979–3985
103. Tashiro R, Iwamoto M, Morinaga H, Emura T, Hidaka K, Endo M, Sugiyama H (2015) Linking two DNA duplexes with a rigid linker for DNA nanotechnology. Nucleic Acids Res 43:6692–6700
104. Gardner TS, Cantor CR, Collins JJ (2000) Construction of a genetic toggle switch in *Escherichia coli*. Nature 403:339–342
105. Nielsen AA, Der BS, Shin J, Vaidyanathan P, Paralanov V, Strychalski EA, Ross D, Densmore D, Voigt CA (2016) Genetic circuit design automation. Science 352:aac7341

106. Masubuchi T, Endo M, Iizuka R, Iguchi A, Yoon DH, Sekiguchi T, Qi H, Iinuma R, Miyazono Y, Shoji S et al (2018) Construction of integrated gene logic-chip. Nat Nanotechnol 13:933–940
107. Endo M, Sugita T, Katsuda Y, Hidaka K, Sugiyama H (2010c) Programmed-assembly system using DNA jigsaw pieces. Chemistry 16:5362–5368
108. Rajendran A, Endo M, Katsuda Y, Hidaka K, Sugiyama H (2011a) Photo-cross-linking-assisted thermal stability of DNA origami structures and its application for higher-temperature self-assembly. J Am Chem Soc 133:14488–14491
109. Rajendran A, Endo M, Katsuda Y, Hidaka K, Sugiyama H (2011b) Programmed two-dimensional self-assembly of multiple DNA origami jigsaw pieces. ACS Nano 5:665–671
110. Endo M, Sugita T, Rajendran A, Katsuda Y, Emura T, Hidaka K, Sugiyama H (2011b) Two-dimensional DNA origami assemblies using a four-way connector. Chem Commun 47:3213–3215
111. Yoshidome T, Endo M, Kashiwazaki G, Hidaka K, Bando T, Sugiyama H (2012) Sequence-selective single-molecule alkylation with a pyrrole-imidazole polyamide visualized in a DNA nanoscaffold. J Am Chem Soc 134:4654–4660
112. Bando T, Sugiyama H (2006) Synthesis and biological properties of sequence-specific DNA-alkylating pyrrole-imidazole polyamides. Acc Chem Res 39:935–944
113. Nakata E, Fong L, Uwatoko C, Kiyonaka S, Mori Y, Katsuda Y, Endo M, Sugiyama H, Morii T (2012) Zinc-finger proteins for site-specific protein positioning on DNA-origami structures. Angew Chem Int Ed 51:2421–2424
114. Kuzyk A, Schreiber R, Fan Z, Pardatscher G, Roller EM, Hogele A, Simmel FC, Govorov AO, Liedl T (2012) DNA-based self-assembly of chiral plasmonic nanostructures with tailored optical response. Nature 483:311–314
115. Endo M, Yang Y, Emura T, Hidaka K, Sugiyama H (2011c) Programmed placement of gold nanoparticles onto a slit-type DNA origami scaffold. Chem Commun 47:10743–10745
116. Yang Y, Endo M, Hidaka K, Sugiyama H (2012) Photo-controllable DNA origami nanostructures assembling into predesigned multiorientational patterns. J Am Chem Soc 134:20645–20653
117. Suzuki Y, Endo M, Yang Y, Sugiyama H (2014c) Dynamic assembly/disassembly processes of photoresponsive DNA origami nanostructures directly visualized on a lipid membrane surface. J Am Chem Soc 136:1714–1717
118. Suzuki Y, Endo M, Sugiyama H (2015b) Mimicking membrane-related biological events by DNA origami nanotechnology. ACS Nano 9:3418–3420
119. Suzuki Y, Endo M, Sugiyama H (2015a) Lipid-bilayer-assisted two-dimensional self-assembly of DNA origami nanostructures. Nat Commun 6:8052
120. Liu W, Zhong H, Wang R, Seeman NC (2011) Crystalline two-dimensional DNA-origami arrays. Angew Chem Int Ed 50:264–267
121. Suzuki Y, Sugiyama H, Endo M (2018) Complexing DNA origami frameworks through sequential self-assembly based on directed docking. Angew Chem 57:7061–7065
122. Ohno H, Kobayashi T, Kabata R, Endo K, Iwasa T, Yoshimura SH, Takeyasu K, Inoue T, Saito H (2011) Synthetic RNA-protein complex shaped like an equilateral triangle. Nat Nanotechnol 6:116–120
123. Osada E, Suzuki Y, Hidaka K, Ohno H, Sugiyama H, Endo M, Saito H (2014) Engineering RNA-protein complexes with different shapes for imaging and therapeutic applications. ACS Nano 8:8130–8140
124. Jasinski D, Haque F, Binzel DW, Guo P (2017) Advancement of the emerging field of RNA nanotechnology. ACS Nano 11:1142–1164
125. Endo M, Yamamoto S, Tatsumi K, Emura T, Hidaka K, Sugiyama H (2013a) RNA-templated DNA origami structures. Chem Commun 49:2879–2881
126. Endo M, Takeuchi Y, Emura T, Hidaka K, Sugiyama H (2014a) Preparation of chemically modified RNA origami nanostructures. Chem Eur J 20:15330–15333

127. Endo M, Hidaka K, Kato T, Namba K, Sugiyama H (2009) DNA prism structures constructed by folding of multiple rectangular arms. J Am Chem Soc 131:15570–15571
128. Endo M, Hidaka K, Sugiyama H (2011a) Direct AFM observation of an opening event of a DNA cuboid constructed via a prism structure. Org Biomol Chem 9:2075–2077
129. Endo M, Yamamoto S, Emura T, Hidaka K, Morone N, Heuser JE, Sugiyama H (2014b) Helical DNA origami tubular structures with various sizes and arrangements. Angew Chem Int Ed 53:7484–7490
130. Endo M, Miyazaki R, Emura T, Hidaka K, Sugiyama H (2012a) Transcription regulation system mediated by mechanical operation of a DNA nanostructure. J Am Chem Soc 134:2852–2855
131. Douglas SM, Bachelet I, Church GM (2012) A logic-gated nanorobot for targeted transport of molecular payloads. Science 335:831–834
132. Takenaka T, Endo M, Suzuki Y, Yang Y, Emura T, Hidaka K, Kato T, Miyata T, Namba K, Sugiyama H (2014) Photoresponsive DNA nanocapsule having an open/close system for capture and release of nanomaterials. Chem Eur J 20:14951–14954
133. Tohgasaki T, Shitomi Y, Feng Y, Honna S, Emura T, Hidaka K, Sugiyama H, Endo M (2019) A photocaged DNA nanocapsule for controlled unlocking and opening inside the cell. Bioconjug Chem 30:1860–1863
134. Willner EM, Kamada Y, Suzuki Y, Emura T, Hidaka K, Dietz H, Sugiyama H, Endo M (2017) Single-molecule observation of the Photoregulated conformational dynamics of DNA origami Nanoscissors. Angew Chem 56:15324–15328
135. Gerling T, Wagenbauer KF, Neuner AM, Dietz H (2015) Dynamic DNA devices and assemblies formed by shape-complementary, non-base pairing 3D components. Science 347:1446–1452
136. Gopinath A, Miyazono E, Faraon A, Rothemund PW (2016) Engineering and mapping nanocavity emission via precision placement of DNA origami. Nature 535:401–405
137. Kuzyk A, Yang Y, Duan X, Stoll S, Govorov AO, Sugiyama H, Endo M, Liu N (2016) A light-driven three-dimensional plasmonic nanosystem that translates molecular motion into reversible chiroptical function. Nat Commun 7:10591
138. Koirala D, Shrestha P, Emura T, Hidaka K, Mandal S, Endo M, Sugiyama H, Mao H (2014) Single-molecule mechanochemical sensing using DNA origami nanostructures. Angew Chem Int Ed 53:8137–8141
139. Shrestha P, Emura T, Koirala D, Cui Y, Hidaka K, Maximuck WJ, Endo M, Sugiyama H, Mao H (2016) Mechanical properties of DNA origami nanoassemblies are determined by Holliday junction mechanophores. Nucleic Acids Res 44:6574–6582
140. Brinker A, Pfeifer G, Kerner MJ, Naylor DJ, Hartl FU, Hayer-Hartl M (2001) Dual function of protein confinement in chaperonin-assisted protein folding. Cell 107:223–233
141. Tang YC, Chang HC, Roeben A, Wischnewski D, Wischnewski N, Kerner MJ, Hartl FU, Hayer-Hartl M (2006) Structural features of the GroEL-GroES nano-cage required for rapid folding of encapsulated protein. Cell 125:903–914
142. Baumketner A, Jewett A, Shea JE (2003) Eff
ects of confinement in chaperonin assisted protein folding: rate enhancement by decreasing the roughness of the folding energy landscape. J Mol Biol 332:701–713
143. Jewett AI, Baumketner A, Shea JE (2004) Accelerated folding in the weak hydrophobic environment of a chaperonin cavity: creation of an alternate fast folding pathway. Proc Natl Acad Sci U S A 101:13192–13197
144. Takagi F, Koga N, Takada S (2003) How protein thermodynamics and folding mechanisms are altered by the chaperonin cage: molecular simulations. Proc Natl Acad Sci U S A 100:11367–11372
145. Shrestha P, Jonchhe S, Emura T, Hidaka K, Endo M, Sugiyama H, Mao H (2017) Confined space facilitates G-quadruplex formation. Nat Nanotechnol 12:582–588

146. Dhakal S, Cui Y, Koirala D, Ghimire C, Kushwaha S, Yu Z, Yangyuoru PM, Mao H (2013) Structural and mechanical properties of individual human telomeric G-quadruplexes in molecularly crowded solutions. Nucleic Acids Res 41:3915–3923
147. Koirala D, Dhakal S, Ashbridge B, Sannohe Y, Rodriguez R, Sugiyama H, Balasubramanian S, Mao H (2011) A single-molecule platform for investigation of interactions between G-quadruplexes and small-molecule ligands. Nat Chem 3:782–787
148. Jonchhe S, Pandey S, Emura T, Hidaka K, Hossain MA, Shrestha P, Sugiyama H, Endo M, Mao H (2018) Decreased water activity in nanoconfinement contributes to the folding of G-quadruplex and i-motif structures. Proc Natl Acad Sci U S A 115:9539–9544
149. Miyoshi D, Karimata H, Sugimoto N (2006) Hydration regulates thermodynamics of G-quadruplex formation under molecular crowding conditions. J Am Chem Soc 128:7957–7963
150. Kurokawa T, Kiyonaka S, Nakata E, Endo M, Koyama S, Mori E, Tran NH, Dinh H, Suzuki Y, Hidaka K et al (2018) DNA origami scaffolds as templates for functional tetrameric Kir3 K(+) channels. Angew Chem 57:2586–2591

Part III
Materials and Chemicals for Cell Control

Materials Designed for Biological Nitric Oxide Delivery

Shuhei Furukawa

1 Introduction

Living cells can be considered the most advanced and complex assembly of molecules. Cellular function is regulated by spatiotemporal molecular traffic known as signal transduction. Understanding these signaling pathways at the molecular level is essential for tackling significant biological issues such as treating intractable diseases, or establishing regenerative medicine. In contrast to traditional approaches of molecular biology, recent cross-disciplinary research works have drastically narrowed the boundary between cell biology and materials science, which allows new approaches for materials scientists to tackle biological issues by applying the state-of-the-art materials to living cells.

Nitric oxide (NO) and carbon monoxide (CO), considered the smallest among various signaling molecules, have attracted attention in recent years [1, 2]. These diatomic molecules are gaseous at ambient condition (room temperature and atmospheric pressure), and therefore, transduce signals much faster compared to other organic messengers. On the other hand, they are highly reactive, resulting in shorter half-life times (NO in a few seconds and CO in a few minutes). Highly reactive NO and CO gas molecules are toxic at high concentrations. In biological system, these gaseous signaling molecules, NO or CO, are produced by the enzymatic degradation of L-arginine or heme, respectively, at any time when necessary. In particular, the biological role of NO has been intensively studied and the enzyme essential to produce NO, nitric oxide synthase (NOS), has been identified in a variety of cells. For instance, neuronal NOS (nNOS) are expressed in neurons, and endogenously produced NO works as an intercellular signaling molecule. NOS is also expressed

S. Furukawa (✉)
Institute for Integrated Cell-Material Sciences, Kyoto University, Kyoto, Japan
e-mail: shuhei.furukawa@icems.kyoto-u.ac.jp

D. O. Wang, D. Packwood (eds.), *Cell-Inspired Materials and Engineering*, Fundamental Biomedical Technologies, https://doi.org/10.1007/978-3-030-55924-3_5

in endothelial cells (known to be eNOS) and induces vasodilation. There is an NOS induced by endotoxin or cytokines (inducible NOS; iNOS) in macrophages, which is related to the control of immune system. While a biological approach by activating or inhibiting NOS activity directly controls endogenous NO production, the on-demand generation of exogenous NO using chemicals or materials allows us to easily and precisely elucidate a physiological and pathological role of NO. Indeed, NO-releasing molecules (NO-donors) have been synthesized and applied to a variety of biological researches related to NO. In this chapter, I summarize the development of NO-donor molecules, assemblies of NO donor molecules as solid-state materials, and more functional NO donor materials in response to light stimulation and their usage in vitro and in vivo.

2 NO Donors with Spontaneous Release

Most of NO donor compounds developed so far spontaneously release NO by decomposing themselves under physiological condition. Among those, the most known compound is nitroglycerin that is commonly known as gunpowder. Since 1870s, the compound has been used as a medicine for angina in the UK; however, the physiological mechanism was not completely elucidated. In 1977, Murad et al. proposed that NO is involved in the process of vasodilation and works as endothelium-derived relaxation factor (EDRF) [3]. Thereafter, the mechanism of NO release from nitroglycerin in vivo was continuously studied and in 2002 Stamler et al. unveiled that mitochondrial aldehyde dehydrogenase (mtALDH) directly decomposes nitroglycerin to NO [4, 5].

Various organic nitrate ester compounds that spontaneously release NO have been developed; however, it is also known that the medical efficacy gradually decreases due to the resistance against nitric acid. In addition to the organic nitrate ester compounds, *N*-diazeniumdilates (NONOates) or S-nitrosothiols have been synthesized as organic NO donors [6, 7]. NONOates decompose under physiological condition (37 °C, pH = 7.4) and produce two NO molecules per functional group. Its releasing rate can be regulated in the range from a few seconds to a few days by modifying molecules by chemical functionality. S-nitrosothiols are known to be biological molecules and the derivatives are endogenously present in life. The release of NO is triggered by heat, light, and reaction with copper(I) ions. It is also known that the transnitrosation reaction occurs with a free thiol functionality and exchange NO between them. These two compound series do not require enzymes for the release of NO so that these compounds have been widely used for biological experiments to produce NO in target environments. In particular, (Z)-1-[*N*-(3-aminopropyl)-*N*-(n-propyl)amino]diazen-1-ium-1,2-diolate (PAPA/NO), 1-[N-(aminoethyl)-N-(2-aminoethyl)amino]diazen-1-ium-1,2-diolate (DETA/NO), S-nitrosoglutathione (GSNO), and S-nitroso-N-acetyl-penicillamine (SNAP) are well-known NO donors.

3 Assembling NO Donor Molecules into NO Donor Solid-State Materials

NO donor compounds described above are small molecules and should be dissolved in water or physiological buffer when being used, which means that the NO concentration depends on its solubility and homogeneous dispersion in solution. It is therefore difficult to localize NO molecules in desired tissues, cells, or organelles and to control the NO concentration. To overcome these shortcomings, a new approach of assembling these NO donors into solid-state materials has been recently introduced (Fig. 1) [8, 9]. This approach allows for greatly improving the NO payloads per volume. Furthermore, it is rather easy to handle solid-state materials and to shape them by microfabrication techniques, which leads to transplantation to animals and enables localized release of NO in target tissues.

For instance, there is an example to assemble NO donors into the surface of silica nanoparticles, which have been used as drug carriers due to their simple synthetic procedures, their stability under physiological condition, their high biocompatibility, and their size controllability. The immobilization of NO donors on the surface of silica nanoparticles restricts the dispersion of NO donors in solution. The release rate can be regulated by altering chemical functionality on the silica surfaces. However, the NO payloads and the actual release amount were not so high because only the surface of particles are used for assembling NO donors and its inner part were not accessible. Schoenfish et al. developed a procedure to introduce NONOate derivatives into the core of particles and demonstrated the high NO payloads [10]. The authors demonstrated an antibacterial effect that requires high concentration of

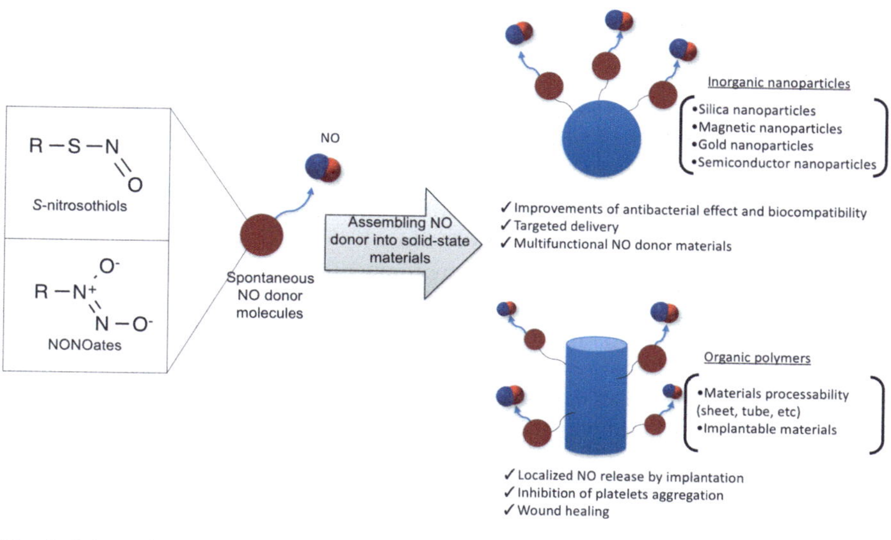

Fig. 1 Schematic illustration of spontaneous NO donor molecules and their assembly as solid-state materials such as inorganic nanoparticles and organic polymers

NO and confirmed that the NO donors decorated silica nanoparticles showed the better performance compared to NO donor itself. Interestingly, the cytotoxicity of nanoparticles was reduced compared to the NO donor solution. Besides silica nanoparticles, other inorganic nanoparticles (gold nanoparticles, magnetic nanoparticles, semiconductor nanoparticles) have been used to assemble NO donors on their surfaces and their hybrid materials have been studied as new NO carriers. At the current status, these materials were fabricated as proof-of-concept of new carriers and further synthetic studies are necessary to improve the performance. In the future, these materials are expected to show a new NO releasing system in response to magnetic field that has a high penetration depth in a body and a synergetic performance of NO releasing with bioimaging.

NO donor compounds can be integrated into a variety of solid-state materials by modifying chemical functionality. In particular, their integration into organic polymers has been intensively studied from the viewpoint of medical applications [11, 12]. For instance, NONOate derivatives were mixed into various type of organic polymers and the long term NO release in buffer solution over several weeks were achieved. Organic polymers here works to protect NO donors from solution and the release rate was drastically decreased. This hybrid polymers were processed into artificial blood vessel [13]. NO plays a significant role for preventing platelet thrombus formation in artificial blood vessel. Indeed, this synthetic system with NO donors were implanted into sheep and prevented the platelet aggregation even after 21 days. Organic polymers are also easily processed at macroscopic scale. A stent, a medical device to widen a blood vessel from inside the lumen, was coated with hybrid NO donors/organic polymers and it was confirmed that this stent suppressed the platelet aggregation [14]. As described in this section, assembling NO donor molecules into solid-state materials expands its usability in many biological and medical applications.

4 Photoactive NO Donors (Caged NO Donors)

NO donor compounds and materials described above spontaneously release NO in water or under physiological conditions due to self-decomposition. To elucidate the more precise role of NO in biological environments, it is essential to control the generation timing of NO, the localization and the concentration. Therefore, we need to develop a new releasing system that would be regulated by external stimuli. For instance, regarding glutamic acid known to be a neurotransmitter, an approach using caged compounds has been developed; using organic chemistry a photoactive protecting group is introduced into glutamic acid and this eliminates bioreactivity. Only under the light irradiation the photoactive group is deprotected and the bioactive glutamate would be generated on the position of light irradiation. This method allows for the spatiotemporal controlled production of glutamate. In the similar fashion, some of the caged NO compounds have been so far developed (Fig. 2) [15]. More specifically, aromatic nitro compounds (nitrobenzene) or *N*-nitrosamine are

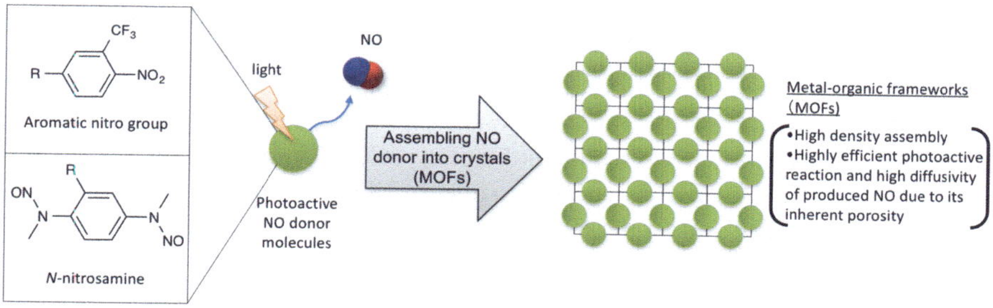

Fig. 2 Schematic illustration of photoactive NO donor molecules and their assembly as metal–organic frameworks

known to be caged NO compounds, which are stable under physiological conditions and do not spontaneously release NO. Although *S*-nitrosothiol or a metal nitrosyl complexes has been reported as photoactive NO donors but spontaneously and gradually release NO in the buffer solution. The performance of these photoactive NO donor compounds can be improved by organic chemistry approach and more sophisticated system have been developed, for example, mitochondria-specific NO donor compounds [16] and two-photon absorption NO donor compound [17]. More recently, the photoactive NO donor compounds were used to ex vivo control vasodilation by light in [18] and localized NO release in vivo [19].

5 Porous Materials with Photoactive NO Donors

Similar to the approach of assembling NO donors into solid-state materials discussed in the Sect. 3, researchers integrated photoactive NO donor molecules with other solid-state materials [15]. Whereas several materials science approaches have been so far reported such as a monolayer film formed on a substrate, organic polymer hybrid materials or the immobilization on nanoparticle surfaces, the applications of those materials in cell biology or in vivo medicine are yet to be achieved. This is because the photoactive materials are rather new compared to spontaneous NO donors and there is still a lack of the methods to precisely introduce a light source in cells or in vivo and to quantitatively correlate the light intensity with the estimation of locally NO produced.

Furukawa et al. recently developed new solid-state materials, into which photoactive NO donors are assembled, and demonstrated its application in cell biology [20]. Here porous materials, the so-called metal–organic frameworks (MOFs), assembled from metal ions/clusters and organic linkers, are used as a material platform [21]. MOFs are crystalline solids and have three-dimensional molecular scaffolds with permanent voids therein. Thanks to their nanosized porosity, into which a variety of gases can be accommodated, this type of materials have been developed for energy applications such as the storage of methane or hydrogen and for environmental applications such as separation of carbon dioxide from air. In order to use MOFs for gas delivery applications, it is important to design materials with no leakage at the ambient condition and the on-demand release of NO by external stimuli. The authors took the approach to use photoactive NO donors as a part of scaffold; aromatic nitro group with coordination ability, 2-nitroimidazole, was chosen as photoactive NO donors expected to release NO by the atomic rearrangement of nitro group to nitrite under the light irradiation. The reaction of 2-nitroimidazole with zinc ions led to the formation of MOF. X-ray crystallography revealed the imidazole coordinated to zinc ions and formed a three-dimensional framework structure with sodalite topology (Fig. 3a) and the material was named as NOF-1 (NOF = nitric oxide framework: Fig. 3a). When the ultraviolet light at 370 nm was irradiated, the compound immediately released NO molecules, and the release was stopped just right after the irradiation was terminated (Fig. 3b). Furthermore, the release amount can be regulated by the light intensity. The total release amount from NOF-1 was estimated to be 3.4 mol mg^{-1}, which was the highest value among the conventional photoactive NO donor materials. This is because of its crystallinity that allows for densely assembling photoactive NO donors and of its internal porosity that enhance the efficiency of the photochemical reaction. Note that the the solid powder of 2-nitroimidazole does not release NO even under the light irradiation, which suggests the importance of internal porosity to inhibit the self-quenching process at the excited state.

To apply NOF-1 to cell biology applications, a novel functional cell culture substrate was fabricated. First, microcrystals of NOF-1 was spin-coated onto a glass substrate, and then, polydimethylsiloxane (PDMS), a polymer material with high

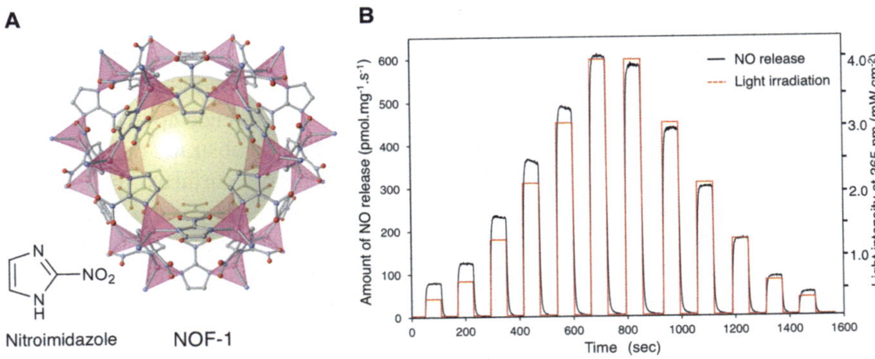

Fig. 3 Chemical structure and NO release property of NOF-1 in response to light irradiation. (**a**) Crystal structure and nitroimidazole chemical structure of a representative pore of NOF-1. Pink indicates a zinc ion. Yellow spheres shows an internal space of NOF-1. (**b**) ON / OFF release of NO from NOF-1 in response to light irradiation. NO molecules were released only under ultraviolet light irradiation. The amount of released NO depends on the light intensity (Adapted with permission from Ref. [20]. Copyright Nature Publishing Group)

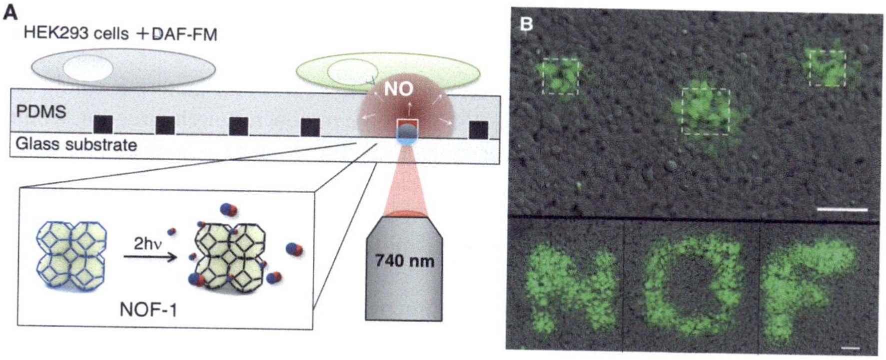

Fig. 4 Cell biology application of NOF-1 as functional cellular substrate and the localized NO release using two-photon excitation microscopy. (**a**) Schematic illustration of a cross-sectional view of prepared cell culture substrate. NOF-1 crystals were completely embedded into PDMS. A laser beam of 740 nm was used as light irradiation using two-photon excitation microscope, to reduce cell damage as much as possible. (**b**) Localized cellular stimulation by NO. White square area surrounded by dotted line indicates the selected area of light irradiation. Green fluorescence indicates intracellular NO uptake due to DAF-FM, an intracellular NO probe. By controlling the selected area of light irradiation using a laser beam, writing letters of "N," "O," "F" using HEK293 cells (Adapted with permission from Ref. [20]. Copyright Nature Publishing Group)

biocompatibility and gas permeability, was further spin-coated on the top of NOF-1 microcrystals (Fig. 4a). In this configuration, NOF-1 crystals were totally embedded into PDMS and have no chance to be exposed into cell culture media, but NO produced by NOF-1 would penetrate through PDMS and chemically stimulate living cells. By selecting one of the microcrystals using laser irradiation, the NO release is localized to target cells. When NOF-1 crystals in the selected area were irradiated using two-photon excitation microscopy with the 740 nm laser, only HEK293 cells on the top of selected area showed a green emission, which originated from DAF-FM, a green NO fluorescent indicator. This experiments confirmed the localized NO stimulation using NOF-1 materials (Fig. 4b). Furthermore, to confirm that these exogenously produced NO from NOF-1 would influence on the physiological effects, the authors used another HEK293 cell with genetically overexpressed TRPC5 (transient receptor potential channel) [22], which is a transmembrane channel protein that introduce Ca2+ ions into intracellular domain in response to NO. Instead of DAM-FM, a Ca2+ chemical probe, Fluo-4 AM was used to monitor the change of Ca2+ concentration. In the similar way, two-photon excitation microscopy was used to irradiate selected area. Only HEK cells on the top of the selected area transiently increased intracellular Ca2+ concentration. As described here, the photoactive NO donor materials have been applied to a real cell biology experiment and using such functional substrates further study to unveil a secret role of NO in various types of cells would be expected.

6 Conclusion

Recent developments of photoactive NO donor molecules and materials allow for spatiotemporal stimulation of living cells. This type of new research direction would lead to more sophisticated NO delivery materials, for instance, localized release of NO toward intracellular targets and the elucidation of an intracellular role of NO using superresolution microscopy. Moreover, in the similar manner to spontaneous NO donors, it is highly expected to use solid-state NO donor materials toward in vivo applications.

From the viewpoint of materials development, it would be expected to synthesize a novel molecule or material that would release another gaseous signaling molecule, for example, CO; H_2S (that is recently intensively studied as a redox active transmitter); and the most known gaseous molecule, O_2. Most of the compounds have been targeted to release gaseous signaling molecules; however, toward the perfect control of gases in living system the synthesis of compounds or materials that reduce or trap gases in cells or physiological environment would become an important target in the next decade. In particular, decreasing O_2 concentration would be essential, and porous materials like MOFs would contribute to this research direction.

References

1. de Mel A, Murad F, Seifalian AM (2011) Nitric Oxide: A Guardian for Vascular Grafts?. Chem Rev 111:5742–5767
2. Motterlini R, Otterbein LE (2010) The therapeutic potential of carbon monoxide. Nat Rev Drug Disc 9:728–743
3. Murad F (1999) Discovery of Some of the Biological Effects of Nitric Oxide and Its Role in Cell Signaling (Nobel Lecture). Angew Chem Int Ed 38:1856–1868
4. Chen Z, Zhan J, Stamler JS (2002) Identification of the enzymatic mechanism of nitroglycerin bioactivation. Proc Natl Am Soc 99:8306–8311
5. Chen Z, Foster MW, Zhan J et al (2005) An essential role for mitochondrial aldehyde dehydrogenase in nitroglycerin bioactivation. Proc Natl Am Soc 102:12159–12164
6. Wang PG, Xian M, Tang X et al (2002) Nitric Oxide Donors: Chemical Activities and Biological Applications. Chem Rev 102:1091–1134
7. Williams DLH (2003) A chemist's view of the nitric oxide story. Org Biomol Chem 1:441–449
8. Nichols SP, Storm WL, Koh A, Schoenfisch MH (2012) Local delivery of nitric oxide: targeted delivery of therapeutics to bone and connective tissues. Adv Drug Deliv Rev 64:1177–1188
9. Kim J, Saravanakumar G, Choi HW et al (2014) A platform for nitric oxide delivery. J Mater Chem B 2:341–356
10. Hetrick EM, Shin JH, Stasko NA et al (2008) Bactericidal Efficacy of Nitric Oxide-Releasing Silica Nanoparticles. ACS Nano 2:235–246
11. Jen MC, Serrano MC, van Lith R, Ameer GA (2012) Polymer-Based Nitric Oxide Therapies: Recent Insights for Biomedical Applications. Adv Funct Mater 22:239–260
12. Naghavi N, de Mel A, Alavijeh OS et al (2013) Nitric oxide donors for cardiovascular implant applications. Small 9:22–35
13. Batchelor M, Reoma SL, Fleser PS et al (2003) More Lipophilic Dialkyldiamine-Based Diazeniumdiolates: Synthesis, Characterization, and Application in Preparing Thromboresistant Nitric Oxide Release Polymeric Coatings. J Med Chem 46:5153–5161

14. Weng Y, Song Q, Zhou Y et al (2011) Immobilization of selenocystamine on TiO2 surfaces for in situ catalytic generation of nitric oxide and potential application in intravascular stents. Biomaterials 32:1253–1263
15. Sortino S (2010) Light-controlled nitric oxide delivering molecular assemblies. Chem Soc Rev 39:2903–2913
16. Horinouchi T, Nakagawa H, Suzuki T et al (2011) Photoinduced Nitric Oxide Release from a Nitrobenzene Derivative in Mitochondria. Chem Eur J 17:4809–4813
17. Hishikawa K, Nakagawa H, Furuta T (2009) Photoinduced Nitric Oxide Release from a Hindered Nitrobenzene Derivative by Two-Photon Excitation. J Am Chem Soc 131:7488–7489
18. Ieda N, Hotta Y, Miyata N (2014) Photomanipulation of Vasodilation with a Blue-Light-Controllable Nitric Oxide Releaser. J Am Chem Soc 136:7085–7091
19. Nakagawa H, Hishikawa K, Eto K et al (2013) Fine Spatiotemporal Control of Nitric Oxide Release by Infrared Pulse-Laser Irradiation of a Photolabile Donor. ACS Chem Biol 8:2493–2500
20. Diring S, Wang DO, Kim C et al (2013) Localized cell stimulation by nitric oxide using a photoactive porous coordination polymer platform. Nat Commun 4:2684
21. Kitagawa S, Kitaura R, Noro S (2004) Functional Porous Coordination Polymers. Angew Chem Int Ed 43:2334–2375
22. Yoshida T, Inoue R, Morii T et al (2006) Nitric oxide activates TRP channels by cysteine S-nitrosylation. Nat Chem Biol 2:596–607

Designing Biomimicking Synthetic Transcription Factors for Therapeutic Gene Modulation

Ganesh N. Pandian and Hiroshi Sugiyama

1 Introduction

Living organisms regulate their biological information through a set of genetic and epigenetic codes [1]. Deoxyribonucleic acid (DNA) is the informational molecule that stores genetic information inside a living cell. Although DNA was identified as nuclein in 1869, it gained recognition in 1953 after Watson and Crick characterized it as a double-helical molecule [2]. Subsequently, there was an exponential increase in the biological knowledge database, especially with the rise of genomics, proteomics, metabolomics, and other modern technologies. Several therapeutically important genes associated with specific diseases can now be understood at the level of DNA sequences [3–5]. Despite this technical progress and accumulated biological information, there is a vast gap in the number of artificial tools that could be developed by harnessing the power of the data repositories available currently. In particular, there is a massive void in the available synthetic tools that mimic the structure and function of the natural biological factors that serve as ON and OFF marks [6–8].

Transcription factors (TFs) are sequence-specific DNA-binding elements that govern the rate of transcription of the biological information from DNA to messenger RNA. TFs operate by switching ON and OFF the endogenous expression of genes at a specific location, time, and amount inside living cells and organisms. About 2600 factors have been characterized as TFs in the human genome. In the

G. N. Pandian (✉)
Institute for Integrated Cell-Material Sciences (WPI-iCeMS), Kyoto University, Kyoto, Japan
e-mail: ganesh@kuchem.kyoto-u.ac.jp

H. Sugiyama (✉)
Department of Chemistry, Graduate School of Science, Kyoto University, Kyoto, Japan
e-mail: hs@kuchem.kyoto-u.ac.jp

D. O. Wang, D. Packwood (eds.), *Cell-Inspired Materials and Engineering*, Fundamental Biomedical Technologies, https://doi.org/10.1007/978-3-030-55924-3_6

natural cellular environment, TFs coordinate biological processes such as the growth, division, and death of cells throughout life. Moreover, TFs play an important role in embryonic development, cellular differentiation, and the modulation of the key signaling pathways associated with cell-fate specification. Recently, clinicians have been shifting their interest toward TFs because of the recent surge in evidence linking TF mutations with specific rare diseases and yet-to-be-cured disorders [9]. Furthermore, uncontrolled function, such as the overexpression and repression exerted by TFs, results in a faulty transcription machinery, which in turn switches cells from a normal to a diseased state. A closer look at the coordinated control exerted by TFs may help develop synthetic TF mimics that partake in the precise programmable functionality of natural TFs.

The structures of natural TFs vary and are adaptable. In general, TFs should necessarily contain at least one component, the DNA-binding domain (DBD), to precisely read the specific key sequence of DNA that regulates gene expression. TFs operate alone via the DBD or by assembling in complexes with other functional components through a linker domain (LD), to either promote (activation domain, AD) or block (repressor domain, RD) the recruitment of RNA polymerase, which transcribes specific DNA sequences to RNA [10]. However, the presence of the DBD alone is sometimes insufficient to regulate the expression of certain genes. The role of epigenetics is evidenced by the distinctive features of humans compared with the fruit fly, *Drosophila melanogaster*, the genome of which contains half the DNA sequences that are present in the human genome. Therefore, mimicking the function of natural TFs artificially is not straightforward because of the existence of several layers of control [11, 12]. It is known that, through an unknown epigenetic code, DNA strands wrap around proteins called histones and are packaged into extremely condensed structures called nucleosomes. Two significant phenomena control the ON and/or OFF status of the expression of specific genes at a particular location and at a particular time. The first one encompasses the structural changes and conformational flexibility of the base sequence of DNA, such as the A-form, Z-form, and G-quadruplex form, which mediate. The other aspect is related to (1) proteins associated with signaling pathways, such as kinases, and (2) epigenetic enzymes, such as histone acetyltransferases and histone deacetylases, which catalyze acetylation, methylation, and other histone modifications in the nucleosome structure [13].

Accordingly, several synthetic small molecules that are capable of restoring the expression of disease-associated genes have been developed to mimic the function of TFs partially. However, small molecules targeting signaling proteins and epigenetic enzymes cannot be classified as synthetic TF mimics because they lack the defining feature of a TF, that is, the DBD. Consequently, there is an increasing demand to develop DNA-based small molecules that could gain chemical control over the intricate transcription machinery. DNA-based small molecules and/or those with a functional moiety, such as epigenetic activity, are claimed to have the advantage of inducing specific transcriptional activation. We have successfully developed hairpin pyrrole–imidazole polyamides (PIPS) as synthetic transcription factor mimics that are capable of switching ON and OFF therapeutically important genes in a preprogrammed manner. PIPs could also be harnessed as fluorescent probes to clarify the function of key TFs and other factors. Here, we detail the ongoing research on how PIPs and their conjugates cause the targeted expression/suppression of a specific gene of interest.

2 Natural Transcription Factors: A Brief Introduction

Brent and Ptashne first revealed the modular organization of transcription factors (TFs) in 1985 by swapping the DNA-binding domain (DBD) of LexA with the activation domain (AD) of GAL4 [14]. Subsequently, Hollenberg and Evans interchanged the DBD, AD, and ligand-binding (LD) functions in estrogen and glucocorticoid nuclear receptors to validate the remarkable modular nature of TFs [15]. Moreover, the flexible nature of TFs was further verified by showing that the intermediary proteins could either interact with the modules or covalently attach to them [16]. Response elements are DNA sequences to which TFs bind explicitly using combined interactions involving electrostatic and van der Waals forces. Interestingly, although TFs operate by binding to specific DNA sequences, this binding is not limited to one particular DNA sequence and could be expanded to several other related DNA sequences with varied strengths of interaction [17]. One example is the binding of the TATA-binding protein to either TATAAAA or TATATAT or TATATAA elements. It is important to note here that TFs do not bind to all compatible sequences, as nucleosome assembly and the accessibility of cofactors affect the recognition properties of TFs. In addition, tandem DBDs bestow different specificities to TFs. The major proteins involved in the coregulation of transcription in mammals include p53, NFAT, NF-κB, and VP16 [18]. TFs are also classified based on their regulatory function as being either constitutively active in cells at all times or conditionally active at sites where they are activated depending upon developmental, signal-dependent, and cell-membrane receptor factors [19]. While genomic DNA serves as the blueprint of life, as it carries a vast amount of biological information using simply four nucleobases (bases; A, T, G, and C), TFs function as an operating system by reading and interpreting this information in either an increased or decreased manner. Eukaryotic transcription requires a set of general transcription factors (i.e., TFIIA, B, D, E, F, and H) that do not bind to DNA but are the essential components of the transcription preinitiation complex, which interacts directly with RNA polymerase to regulate basal gene expression (Fig. 1) [20]. Other TFs differentially control the enhancement of gene transcription by binding to enhancer regions in a particular gene, thus inducing or suppressing them at the right level, depending on the requirement [21]. The major families of DBD-containing proteins include basic helix-loop-helix, helix-turn-helix, winged helix, zinc finger, basic-leucine zipper, C-terminal effector domain, lambda repressors, and homeodomain proteins. In the past decade, synthetic transcription factors developed from naturally occurring DNA-binding proteins were shown to modulate important genes therapeutically [22].

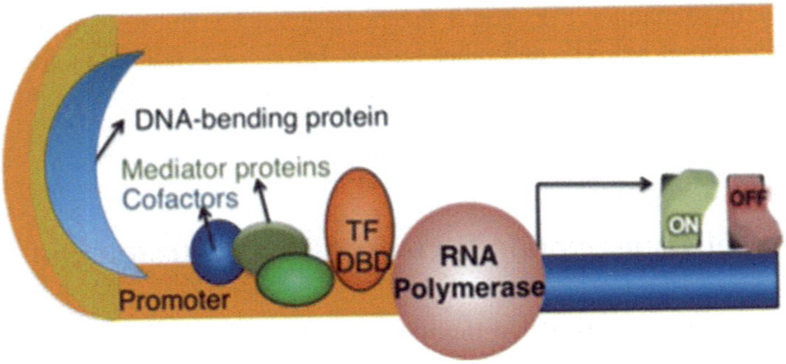

Fig. 1 Illustration of the eukaryotic transcription machinery

3 Natural DNA-Binding Proteins for Therapeutic Gene Modulation

Zinc fingers (ZFs) are natural DNA-binding proteins that are encoded by about 3% of the genes known in the human genome. Synthetic polydactyl ZFs supplemented with nucleases, transcription effectors, or epigenetic enzymes alter gene transcription by modifying the local DNA structure in a site-specific manner [23].

Similarly, the homing endonuclease family of proteins has been harnessed to recognize and cleave particular DNA sequences with enhanced specificity [24]. A programmed endonuclease with increased recognizing capacity was used to repair deleterious mutations in the human *RAG1* gene [25]. The plant pathogen *Xanthomonas*-derived transcription activator-like effectors (TALEs) contain highly conserved repeat variable diresidues (RVDs) and can be tailored to recognize specific DNA sequences [26, 27]. Designer TALEs initiated the cellular reprogramming of human cells by selectively modulating the endogenous expression of the cell-fate-regulating genes *SOX2* and *KLF4* [28]. TALEs have been preprogrammed to target diverse DNA sequences; however, their lax specificity and their efficacy toward methylated DNA sequences are significant concerns [29, 30]. Furthermore, TALEs supplemented with epigenetic enzyme modulators exhibited a lower efficiency compared with that of the small-molecule modulators [31]. Recently, progress has been made regarding the improvement of the specificity of TALE nucleases using nonconventional RVDs [32], which have also been used in T-cell immunotherapies [33, 34].

Clustered regularly interspaced short palindromic repeats (CRISPR) represent a family of DNA sequences that contain snippets of viral DNA, leading to the acquisition of resistance toward foreign genetic elements. In this artificial gene-regulation system adapted from the natural bacterial defense mechanism, CRISPR act as the DNA-recognition module, while the Cas9 protein acts as the functional module. The CRISPR-Cas9 system has been revolutionizing the genome-editing field in recent years and is claimed to cure the incurable disorders. The CRISPR-Cas9 system has been successfully used to modulate therapeutically important genes associated with several diseases, including neurological disorders, cancer, polycystic kidney disease, and focal segmental glomerulosclerosis [35, 36]. Recently, a CRISPR system encompassing the Cpf1 nuclease, which causes staggered cutting, was developed and its mechanistic variations were shown [37–39]. Despite the remarkable success of this natural toolbox in the genome engineering of therapeutically significant genes, this technique still requires a plasmid or vector transfection system. Furthermore, the effect of the CRISPR-Cas9 system is permanent and the process remains too expensive for translation to the clinic. Therefore, there is a need for a simple, transgene-free, and cost-effective strategy that could partake in the structural and functional capacity of its natural counterparts.

4 Development of DNA-Based Synthetic Ligands as Synthetic Transcription Factor Mimics

DNA-binding natural products that specifically interact with the four nucleobases could artificially alter gene transcription inside living cells. Correspondingly, the oligopeptide antibiotics netropsin and distamycin A, with different base compositions, were shown to hinder nucleic acid synthesis [40]. In 1985, the crystal structure of netropsin was solved and revealed that a new type of small molecule capable of reading G/C-containing base sequences could be generated by substituting imidazole (I) with pyrrole (P) in netropsin [41]. The use of a DNA minor-groove-binding peptide that recognizes the 5′-(A,T)GCGC(A,T)-3′ sequence led to the formation of a 2:1 positively cooperative complex with a four-ligand (IPPI). The pairing of aliphatic and aromatic amino acids gave rise to hairpin polyamides, which could be designed to have versatile sequence-recognition capacity together with the incorporation of γ-aminobutyric acid [42]. Consequently, a binding rule asserting that the GC base pair (bp) and the AT or TA bp could be recognized by an antiparallel pairing of I/P and P/P pair, respectively, was generated (Fig. 2a) [43]. Hairpin pyrrole–imidazole polyamides (PIPs) exhibited the potential to be developed as synthetic transcription factor mimics because of their remarkable binding property, which is equivalent to that observed in natural transcription factors. Moreover, PIPs can be harnessed as sequence-specific probes because they can permeate and localize inside the nucleus of living cells to recognize any desired DNA sequence without requiring transfection agents (Fig. 2b). PIPs supplemented with fluorescent properties have been harnessed as intrinsic probes because of their remarkable maintenance of specificity and affinity, even after binding to the target DNA sequences [44]. However, various obstacles, such as their molecular size, P/I content, and the choice of dye, can hamper the effective nuclear localization of PIPs [45]. Consequently, several fluorophore conjugates with different photochemical properties, uptake properties, and components (such as heterocycles) were developed to have bioactivity as anticancer compounds [46]. Intriguingly, the PIP-fluorophore conjugate excimers exhibited an increase in emission with the increasing number of

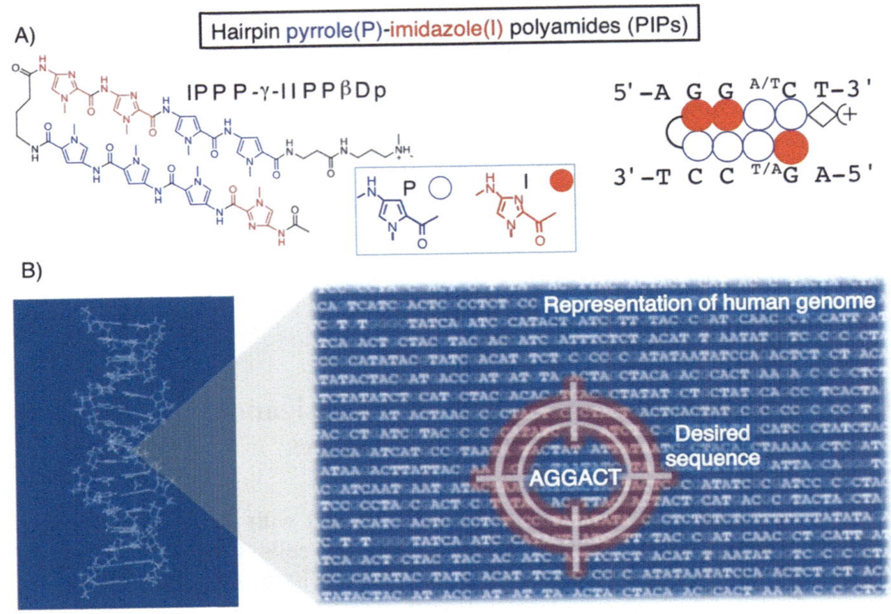

Fig. 2 (**a**) Chemical architecture of hairpin pyrrole (blue circle)–imidazole (red circle) polyamides, β-alanine, γ-aminobutyric acid, and the proposed binding rule. (**b**) PIPs could be predesigned to target a specific DNA sequence in the human genome

CAG repeats, which suggested their scope to quantify triplet nucleotide sequences, such as the CAG repeats associated with trinucleotide expansion disorders [47, 48]. PIP-fluorophore conjugates were further developed by altering their chemical architecture, namely the γ-turn position and the β-alanine linker [49]. A PIP-fluorescein isothiocyanate conjugate successfully targeted the promoter region of the gene encoding the human transforming growth factor-β1, which is a key transcription factor associated with cell-fate regulation in human vascular smooth muscle cells [50].

Transcriptional regulation of telomerase activity plays a fundamental role in several physiological processes and is associated with diseases such as cancer. PIP-fluorophore conjugates have been developed to detect the human telomere oligonucleotide sequence (5′-(GGGTTA)4GGG-3′/3′-(CCCAAT)4CCC-5′) [51]. Recently, a tandem hairpin PIP mimicking the tandem DBD observed in natural TFs was developed to successfully target and easily label both human and mouse telomeres under mild conditions, while keeping their structures intact. Characterization and microheterogeneity studies using tandem PIP-fluorophore conjugates revealed that the level of the TRF1 protein in telomeres is associated with telomere length at the single-telomere level [52]. Subsequently, we developed fluorescent tandem trimer PIP probes comprising three hairpins, termed TT59, which were capable of discriminating mismatches and of recognizing 18 bp dsDNA in human telomeric repeats [53]. Recently, we successfully developed a tandem tetramer PIP that targets 24 bp within telomere repeat sequences, and successfully verified its sequence specificity using large-scale sequencing studies [54].

4.1 Gene Regulation Using Designer PIPs Mimicking TF DBDs

Designer PIPs targeting a cognate DNA sequence in a gene-coding region display bioactivity only for the unique target DNA sequences. Moreover, the off-target recognition of flanking sequences and nonspecific TFs is a significant concern. PIPs designed to target key sequences of therapeutically important genes, including those encoding aurora kinases, TGF-β1, human lectin-like oxidized low-density lipoprotein, HER2, and EBNA1, successfully altered their endogenous expression in different cell types (Fig. 3a) [55–59]. A designer PIP was successfully introduced into human breast adenocarcinoma (MDA-MB-231) cells and recognized the specific bp of the REL/ELK1-binding site in the promoter region of the human ectopic viral integration site 1 (EVI1), which is a well-known oncogenic TF associated with various types of cancer [60]. Microarray studies revealed that the designer PIP for EVI1 successfully inhibited the genes associated with breast cancer cell migration. Similarly, a PIP targeting the activator protein-1-binding site in the promoter region of the gene encoding the matrix metalloprotein 9 (MMP9), which is a tumor-invasion- and metastasis-associated factor, successfully inhibited the expression of the MMP9 gene and protein and altered the migration and invasion of MDA-MB-231 cells [61]. Pharmacokinetic studies performed after their intravenous administration in rats suggested that designer PIPs have notable absorption, distribution, metabolism, and excretion (ADME) properties [62, 63]. A designer PIP also successfully recognized a cognate dsRNA sequence of TGF-β1 and the influenza A virus, and not the mismatched sequences. However, it had a lower binding affinity than that observed for PIP–DNA binding [64]. Designer PIPs were also shown to bind to methylated 5′-CpG-3′ sequences and their bioefficacy over the natural DNA-binding proteins was demonstrated [65].

Recently, the epoch-making application of a designer PIP in human induced pluripotent stem cells (hiPSCs) was reported. Targeted differentiation of hiPSCs using chemicals alone is claimed to have the potential to generate clinically useful terminal cell types, such as cardiomyocytes, which are the muscle cells of the heart. Several small-molecule inhibitors targeting the key proteins associated with signaling pathways were shown to be useful in differentiating hiPSCs into cardiomyocytes. However, the off-target effect of this system was a major concern. Considering the potential of DNA-binding inhibitors in obviating the existing difficulties associated with signaling-protein inhibitors, we designed PIPs to target key cell-fate-regulating genes, such as *SOX2*. By exploiting the available knowledge about the intertwined transcription machinery associated with cardiomyocyte differentiation,

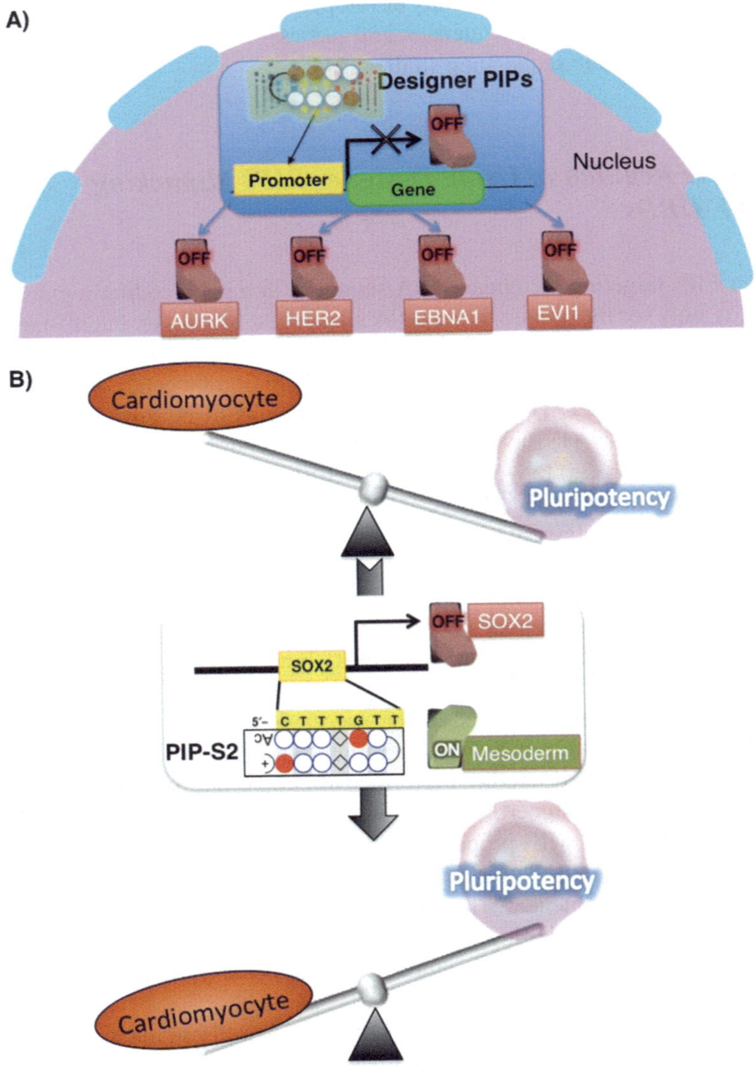

Fig. 3 (**a**) Illustration of designer PIPs causing promoter-specific transcription suppression of therapeutically important genes. (**b**) A designer PIP termed PIP-S2 targeting the consensus motif of the SOX2-binding region switched OFF the expression of SOX2, which in turn activated the mesodermal genes and shifted 201B7-iPS cells from the pluripotent state to the cardiomyocyte state

a designer PIP (termed **PIP-S2**) was constructed as a first-ever synthetic DNA-binding inhibitor. **PIP-S2** recognized the 5′-CTTTGTT-3′ sequence and inhibited the SOX2–DNA interaction. **PIP-S2** caused promoter-specific transcription suppression of the *SOX2* gene and inhibited the expression of the encoded protein (Fig. 3b). Whole-transcriptome analyses suggested that **PIP-S2** specifically targeted and altered *SOX2*-associated gene regulatory networks that are involved in mesoderm induction. Furthermore, the combination of **PIP-S2** with a Wnt/β-catenin inhibitor effectively generated spontaneously contracting cardiomyocytes, which provided new opportunities for differentiating stem cells into a desired cell type [66]. Recently, we showed that **PIP-S2** yielded targeted suppression in various cancer cell lines and inhibited the progression of metastasis [67].

4.2 Designer PIPs with DNA Alkylating Agents and Their Bioactivity

PIPs supplemented with a duocarmycin A (Du) or distamycin A moiety exhibit improved alkylation efficiency, particularly at the G residues of GC-rich sequences [68]. This unique property aids the design of PIP conjugates aimed at recognizing the coding region of the target gene. Alterations in the chemical architecture of PIPs via the introduction of vinyl linkers, L-moieties, β-alanine, and indole linkers improved their sequence-specific recognition capacity [69]. To extend the length of the target DNA sequence of the hairpin PIP, a Y-shaped designer PIP retaining 1-(chloromethyl)-5-hydroxy-1,2-dihydro-3H-benz[e]indole (Seco-CBI) DNA-alkylating moieties is used, thus extending the recognition of the target DNA sequence [70]. PIP–Seco-CBI conjugates showed superior bioefficacy compared with the PIP–chlorambucil conjugates [71]. Designer PIP-chlorambucil conjugates capable of 11 bp recognition that were constructed by introducing an amino group into GABA (γ-turn) inhibited the progression of RNA polymerase by causing alkylation at the cognate site [72]. The biological evaluation of PIPs conjugated with alkylating agents revealed their anticancer activity [73]. Heterotrimeric PIP–Seco-CBI and PIP–chlorambucil conjugates recognized the 11 bp sequence in human telomere repeats and triggered cooperative alkylation [74]. Designer PIP–indole-Seco-CBI conjugates encompassing a tandem hairpin motif exhibited enhanced alkylating efficacy and better affinity toward the TTAGG sequence in the telomere repeat regions [75]. Similarly, a PIP–Seco-CBI conjugate showed potent alkylating activity and improved sequence specificity in the histone *H4* gene fragment and caused apoptosis in K562 cells [76]. Furthermore, biological evaluation of 7-bp-recognizing PIP conjugates with alkylating agents not only exhibited potent cytotoxicity against A549 cells, but also suppressed tumors in nude mice transplanted with DU145 cells [77].

Because PIP-alkylating conjugates recognize specific DNA sequences in the coding region, they could be harnessed to recognize point mutations in disease-associated genes in a sequence-specific manner. Recently, PIP–Seco-CBI conjugates were synthesized and developed to recognize mutated sequences at codons 12 and 13 of the *KRAS* gene, which occur frequently in different types of cancer. Screening studies identified a PIP–Seco-CBI conjugate with high reactivity toward the *KRAS* codon 13 mutation site that could alkylate the A base in the 5′-ACGTCACCA-3′ sequence [78]. Subsequently, an advanced PIP–Seco-CBI conjugate called KR12 (Fig. 4a) was developed to recognize selectively oncogenic *KRAS* codon 12 mutations and alkylate adenine N3 at the target sequence (Fig. 4b). KR12 induced strand cleavage and apoptosis by suppressing the growth of human colon cancer cells carrying G12D or G12V mutations. KR12 significantly suppressed tumor growth with low host toxicity in *KRAS*-mutated, but not wild-type, xenograft models (Fig. 4c) [79]. The remarkable targeting efficacy of KR12 toward the mutated and not the wild-type sequence was demonstrated using a computer-minimized model and Bind-n-Seq [80]. Furthermore, KR12-binding targets were validated in the LS180 colorectal cancer genome via a comparative analysis using next-generation sequencing and gene expression data [81]. The successful demonstration of the designer PIP KR12 represents a new opportunity to target point mutations of any oncogene. Similarly, a designer alkylating PIP successfully repressed the runt-related transcription factor (RUNX), a notable TF that is associated with a tumor suppressor in the p53-mediated cell death pathway. Biological evaluation studies showed that the alkylating PIP conjugates modulated RUNX-cluster genes in acute myeloid leukemia cells and were effective even against poor-prognosis solid tumors in a xenograft mouse model of AML [82]. This study not only demonstrated the bioefficacy of PIP alkylating agents but also clarified the role of the RUNX family of proteins in the maintenance and progression of cancer cells.

4.3 Alteration of the Chemical Architecture of PIPs for Enhanced Bioefficacy

In vitro binding assays using distinct PIPs designed to contain a varying number of imidazole rings revealed that the planarity of PIPs caused by imidazole decreased the binding association rate of PIPs to their target DNA sequences [83]. Similarly, the incorporation of a chiral β-hydroxyl-γ-aminobutyric acid–β-alanine pair aided in discriminating T/A from A/T base pairs [84]. Moreover, the replacement of a single pyrrole in a PIP with β-alanine changed its DNA-binding affinity to acquire improved selectivity toward the 10 bp target DNA sequence [85]. Similarly, PIP conjugates that were altered to carry methoxy polyethylene glycol 750 (PEG-750) exhibited a notable improvement in their aqueous solubility [86]. A new synthetic route for the introduction of a vinylpyrrole unit into the C terminus via (fluorenyl-methoxy)carbonyl solid-phase peptide synthesis and subsequent liquid-phase coupling with Seco-CBI generated a vinyl-linker-encompassing PIP–Seco-CBI conjugate that was capable of recognizing a 7 bp DNA sequence [87]. PIPs with a

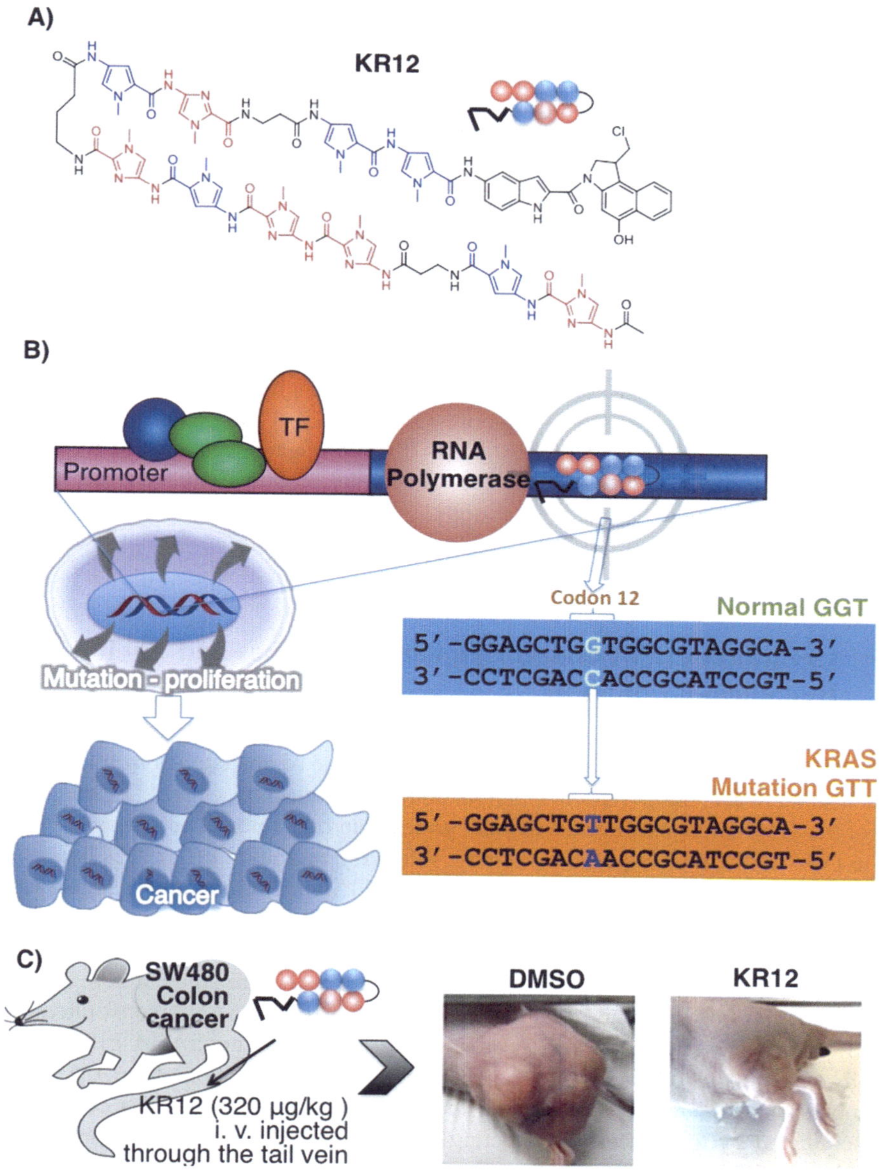

Fig. 4 (**a**) Chemical structure and design of KRAS point mutations targeting the alkylating PIP conjugate KR12. (**b**) Illustration of KR12 binding to the codon 12 region of the KRAS gene carrying a GTT mutation, leading to the proliferation of normal cells into cancer cells. (**c**) Injection of KR12, but not DMSO, reversed the cancer progression in a mouse model carrying SW480 colon cancer cells (Modified and reproduced from Ref. [79])

new design (called cyclic PIPs) were synthesized using a novel method involving a cysteine and chloroacetyl residue. In vitro binding assay studies suggested that the cyclic PIPs with two β-alanine molecules had higher affinity and specificity than those observed for their hairpin PIP counterparts, which indicated their potential as next-generation DNA-binding agents [88]. Subsequently, this molecular design was extended to attain a large dimeric PIP that was capable of recognizing a 14 bp sequence [89]. Together, these studies demonstrated that even a small alteration in the chemical architecture of PIPs modifies their binding efficacy and, hence, their bioactivity. Although a pseudocomplementary peptide nucleic acid (pcPNA) has successfully recognized physiologically important DNA structures, such as G-quadruplex and telomeric DNA, their poor invasion efficacy under physiological ionic-strength conditions is a primary concern in their development as a synthetic transcription factor mimic. The conjugation of pcPNA with PIP led to the selective recognition of the target DNA sequence (5′-AGTCCT-3′) and caused sequence-specific scission [90]. Interestingly, the PIP–PNA conjugate overcame the technical difficulties associated with pcPNA, as the conjugate bound to its target DNA sequence even under high salt conditions, which is relevant to the physiological conditions.

5 Creation of DNA-Based Epigenetic Switches and Their Biological Evaluation

Synthetic TF mimics often overlook the essential role of epigenetic enzymes, such as histone deacetylases (HDACs) and histone acetyltransferases (HATs), that control the acetylation status of the histone proteins and, therefore, the accessibility of the DNA to the transcription machinery. While the inhibition of HDACs by small molecules like suberoylanilide hydroxamic acid (SAHA) modulates gene expression inside living cells, the lack of a DBD (the essential component of TFs) in the mimics hampers their use in gene regulation [11, 12, 52]. Because PIPs can be tuned to have variable activity, we conjugated them with SAHA to create a new type of synthetic TF mimic termed SAHA-PIP, in which PIP and SAHA constituted the DBD and functional modules, respectively (Fig. 5a). A designer SAHA-PIP specifically triggered histone H3 Lys9 acetylation in the promoter region of the *p16* tumor suppressor gene and transformed the morphological features of HeLa cells [91]. Encouraged by the success of targeting SAHA-PIP, we designed a library of 16 distinct SAHA-PIPs (A to P) that were capable of recognizing a 6 bp sequence abiding the general binding rule of PIPs. Because targeted histone acetylation can reprogram somatic cells into pluripotent stem cells, we screened this library of 16 SAHA-PIPs to evaluate their effect on the endogenous expression of the pluripotency-associated TFs Oct-4, Sox2, Klf4, and C-Myc in mouse embryonic fibroblasts (MEFs). qRT-PCR studies showed that SAHA-PIPs D, E, J, and O distinctively induced the *c-Myc*, *Nanog*, *Sox2*, and *Klf4* genes, respectively. A chroma-

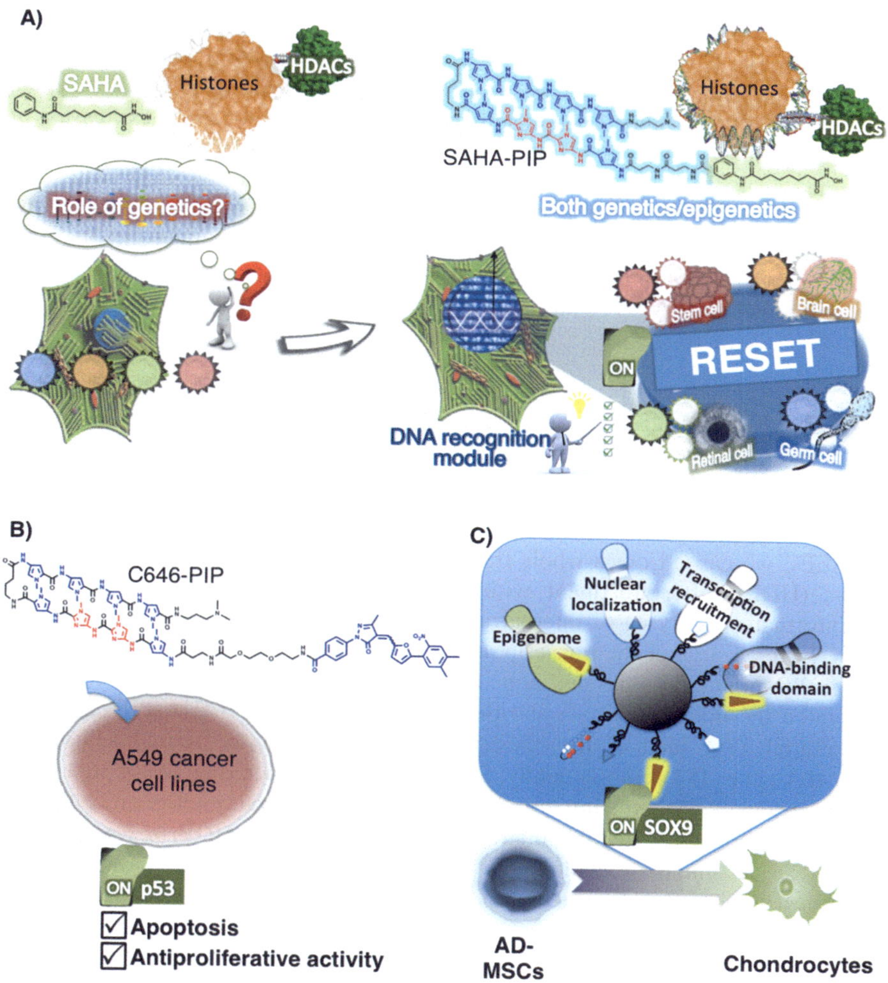

Fig. 5 (**a**) Graphical illustration depicting that the target of a histone deacetylase (HDAC) inhibitor is unclear, as it overlooks the role of genetics (left), whereas complementing SAHA with a DNA-recognition module, to create SAHA-PIP, bestows them with genetic and epigenetic access, which in turn aids in switching ON and resetting therapeutically important genes. (**b**) Chemical structure of the sequence-specific histone acetyltransferase (HAT) C646-PIP and its effect on the A549 cancer cell line, to induce p53-dependent apoptosis and antiproliferative activity. (**c**) NanoScript encompassing different functions and a DNA-recognition module for SOX9-mediated differentiation of adipose-derived mesenchymal stem cells (AD-MSCs) into chondrocytes with enhanced bioefficacy

tin immune precipitation analysis revealed that SAHA-PIP induced gene expression by establishing transcriptionally permissive chromatin modifications in histone H3, including the acetylation of Lys9 and Lys14 and the trimethylation of Lys4 [92]. However, the level of endogenous gene expression was low (fivefold) and the combination of SAHA-PIPs did not result in cellular reprogramming. Therefore, a second library of 16 SAHA-PIPs (termed Q to Φ) with improved capacity to recognize GC-rich sequences was synthesized. Screening studies using qRT-PCR revealed that SAHA-PIP δ exhibited potent activity to induce the endogenous expression of *OCT-3/4* and *NANOG*, by about 30-fold, even at nanomolar concentrations. A genome-wide gene analysis verified the remarkable capacity of SAHA-PIP δ to switch ON multiple pluripotency TFs inside fibroblasts [93]. In just 24 h, SAHA-PIP δ rapidly overcame the rate-limiting step of the dedifferentiation process by shifting the transcriptional network from the fibroblast state to the mesenchymal–epithelial transition (MET) state.

5.1 *Distinct DNA-Based Epigenetic Switches for Therapeutically Important TFs*

Considering the notable effect of SAHA-PIP δ on the genome-wide gene expression in MEFs, we carried out a whole-transcriptome analysis of human dermal fibroblasts (HDFs) after treatment with both libraries of SAHA-PIPs (A to Φ). Bioinformatics analyses of microarray data validated the remarkable capacity of SAHA-PIPs to trigger the transcriptional activation of particular clusters of therapeutically important genes and noncoding RNAs (Fig. 5a). qRT-PCR studies corroborated the microarray data and the finding that distinct SAHA-PIPs activated differential TFs, such as the obesity-related *KSR2* and the retinal cell-related *SEMA6A* genes [94]. Subsequently, SAHA-PIP K was identified as the first small molecule capable of inducing unusual transcriptional activation of germ cell genes in a human somatic cell [95]. Interestingly, the K-induced *MOV10L1* and *piRNA* factors are essential in maintaining postmeiotic genome integrity and are not expressed in somatic cells. It is important to note here that SAHA-PIP I, which contains a similar number of P and I components as does K, displayed distinctive bioactivity and switched ON core pluripotency TFs, including the miR-302 family [96]. A next-generation sequencing analysis confirmed that SAHA-PIP I, but not K, induced hyperacetylation in the *OCT-3/4* promoter region. Prolonged incubation of I (for 21 days) generated partially reprogrammed alkaline phosphatase-positive cells with an induction efficiency of 0.06%. Similarly, SAHA-PIP X triggered transcriptional activation of retinal cell-specific therapeutic genes associated with ocular disorders, including *CERKL*, *PAX6*, *RS1*, *USH2A*, *CRYBB3*, and *STRA6* [97]. A ChiP-Seq analysis verified that the significantly enriched motif matched the cognate binding site (5′-WCGGWW-3′). Moreover, SAHA-PIP L was characterized as the first-ever multitarget small molecule that could epigenetically induce neural-system and brain-synapse TFs in BJ human foreskin fibroblasts and 201B7-iPS cells. An ingenuity pathway analysis showed that the bioactivity of SAHA-PIP L was correlated with the signaling of synaptic receptors, such as glutamate and γ-aminobutyric acid, which are linked with autism spectrum disorders. We also successfully differentiated 201B7-iPS cells into a neural progenitor state, which suggested their potential for acquiring desired neural cell types by targeting the corresponding TFs.66e.

5.2 Next-Generation Synthetic-TF Mimics for Gene Regulation

To identify the optimal chemical architecture for enhanced bioefficacy, the number of β-alanine linkers in the PIP architecture of SAHA-PIP E was changed. Biological evaluation studies revealed that the presence of three β-alanine linkers in SAHA-PIP moderately increased the *OCT-4* expression levels [98]. Alteration of the structure of SAHA in SAHA-PIP δ generated a new type of synthetic-TF mimic termed JAHA-PIP δ (or Jδ, a JAHA-a derivative of SAHA lacking its surface-recognition domain). Bioassays demonstrated that Jδ had higher HDAC8 activity than did δ and activated the skull-morphogenesis-governing *Otx2* and *Lhx1* genes [99]. Similarly, the incorporation of isophthalic acid (IPA) at the C terminus of δ improved its aqueous solubility and overcame aggregation issues. Consequently, the chemically modified δ boosted bioefficacy significantly ($P < 0.05$) compared with the unmodified δ, as the chemically modified δ yielded an elevated expression of *Oct-3/4* [100]. A Bind-n-Seq analysis provided vital information about the sequence bias observed for the β-β-PIPs and aided in the construction of next-generation SAHA-PIPs with improved binding specificity and bioefficacy [101, 102]. A chromatin immunoprecipitation analysis verified this result by extrapolating the model oligonucleotide duplex to the corresponding sequence within chromatin. Together, these results proved that SAHA-PIPs can be tailored for the improvement of their bioefficacy.

A transcriptionally permissive chromatin architecture, which can be achieved by activating epigenetic activators such as HATS, rather than by blocking the HDAC epigenetic suppressor, provides a more direct way to activate gene(s) of interest. However, unlike the widely studied HDAC inhibitors, there is a considerable gap in the study of HAT modulators. Because HAT activation provides a more direct way to induce transcriptionally permissive marks, we substituted the SAHA in SAHA-PIP I (SAHA-I) with a known HAT activator [*N*-(4-chloro-3-trifluoromethyl-phenyl)-2-ethoxy-benzamide (CTB)] to generate CTB-PIP I (CTB-I). A transcriptome analysis of SAHA-I- and CTB-I-induced genes indicated that CTB-I remarkably boosted the bioefficacy of CTB to activate a similar set of genes, similar to SAHA-I, in HDFs. A chromatin immunoprecipitation-sequencing analysis revealed the importance of specific sequences in the putative promoter region, to which the DBD in both CTB-I and SAHA-I binds to trigger transcriptionally permissive chromatin marks. This proof-of-concept study clearly indicated that distinct epigenetic modulators can be transformed upon conjugation with a PIP DBD to yield identical bioactivity [103]. Moreover, we conjugated PIPs with a HAT inhibitor termed C646, to construct sequence-specific HAT inhibitors that could induce antiproliferative and apoptotic activity in cancer cell lines (A549 and MV4–11) (Fig. 5b). A microarray analysis and studies using mutant cell lines revealed that this sequence-specific HAT inhibitor operated by triggering the tumor suppressor p53-dependent apoptosis pathway [104]. The integration of CTB, a designer PIP, and other functional modules onto a nanoparticle resulted in a new type of gene-regulation platform termed NanoScript. NanoScript targeting SOX9, a master TF of chondrogenic differentiation, successfully activated the endogenous expression of *Sox9* in adipose-derived mesenchymal stem cells and differentiated them into chondrocytes with enhanced bioefficacy (Fig. 5c) [105].

6 Creation of a Designer PIP for Mitochondrial Gene Modulation

Considering the recent surge in evidence linking the mitochondrial genome and human diseases, there is an increasing demand to devise new strategies to modulate mitochondrial gene transcription. The mitochondrial genome has its own ~16.6 kbp double-stranded, circular DNA (mtDNA), in which each DNA strand is categorized based on its guanine content as the heavy (H) strand and the light (L) strand. Mitochondrial transcription occurs polycistronically from three transcription initiation sites in mtDNA: the L-strand promoter (LSP), the H-strand promoter 1 (HSP1), and HSP2. Despite the accumulation of information regarding mitochondrial transcription factors and the resolved fundamental mechanisms, chemical control of the mitochondrial transcription machinery has not been achieved, because no synthetic ligands have been developed that are adept at (1) mitochondrial localization, (2) DNA recognition, and (3) targeted suppression of endogenous gene expression.

Therefore, we transformed the nuclear-localizing property of PIPs by conjugating them with a mitochondria-penetrating peptide composed of cyclohexylalanine (Cha) and D-arginine (Arg), to create a new type of mitochondria-specific synthetic ligands, termed MITO-PIPs. MITO-PIPs were designed to target the known binding site of a mitochondrial transcription factor, TFAM, in LSP and HSP1. Biological evaluation studies using HeLa cells showed that MITO-PIP-LSP triggered targeted suppression of the mitochondrially encoded NADH dehydrogenase 6 (*ND6*) gene, which is located downstream of LSP (Fig. 6a). Interestingly, MITO-PIP-HSP1 recognizing different DNA sequences did not yield notable transcriptional change, suggesting the sequence-specific bioactivity of MITO-PIP-LSP inside living cells. A T_m analysis indicated that the affinity of MITO-PIP-HSP1 is weaker than that of MITO-PIP-LSP (Fig. 6b). Live-cell imaging studies using MITO-PIP labeled with TAMRA showed its efficient localization only in the mitochondria, and not the nuclei (Fig. 6c). This result further validates the contention that the changes in gene expression caused by MITO-PIPs arise because of their localization in mitochondria and their targeted recognition of mitochondrial DNA. To our knowledge, MITO-PIPs are the first DNA-based programmable small molecules that are capable

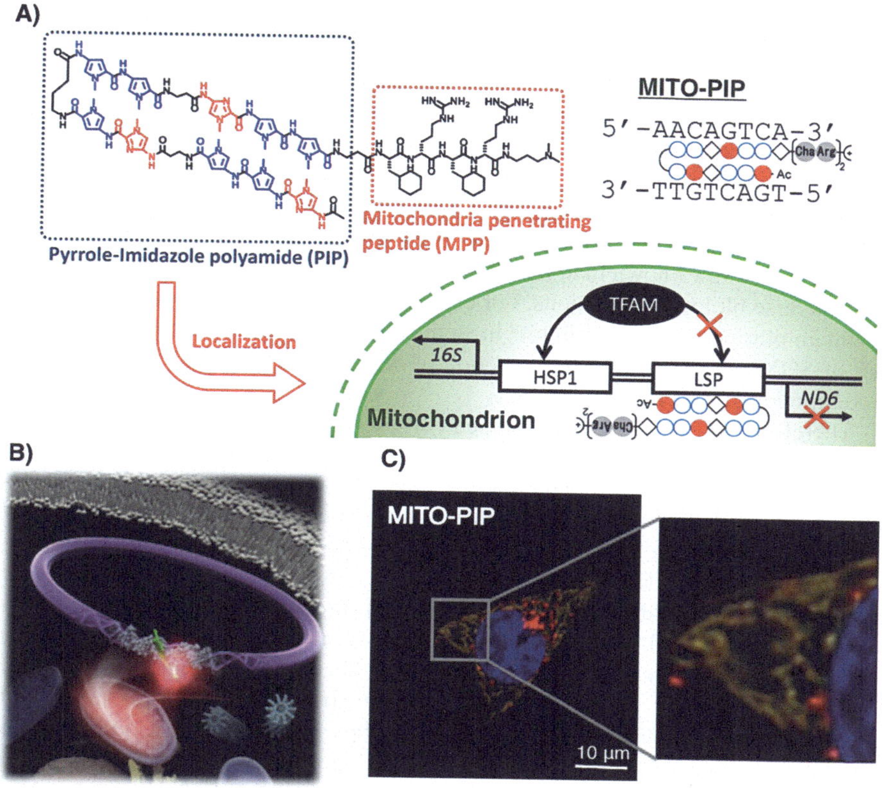

Fig. 6 (**a**) Chemical components of the mitochondrial gene switch termed MITO-PIP and their localization inside mitochondria to target the light strand promoter (LSP) region, containing the mitochondrial transcription factor A (TFAM), and not the heavy strand promoter 1 (HSP1) region. (**b**) Illustration of MITO-PIP selectively binding to the circular DNA of mitochondria inside the complex cellular environment. (**c**) Live-cell imaging of the fluorescent agent TAMRA-labeled MITO-PIP (Modified and reproduced from Ref. [106])

of inducing targeted suppression of *ND6*, an essential factor of the respiratory chain complex in mitochondria that is associated with complex genetic disorders.

7 Advanced System to Mimic the Synergistic TF Pair–DNA Interaction with Epigenetic Modulating Activity

In higher organisms, transcription factor (TF) binding occurs in dense clusters along DNA, suggesting that most TFs coordinate widespread cooperative interactions. A study showed that 50–70% of >1000 TFs expressed in mammalian cells function cooperatively, which allows the specific manipulation of the expression of downstream genes [107]. To mimic or inhibit the TF pair–DNA interaction synergistically using synthetic DNA binders, such as PIPs, three issues should be considered [108]. First, the TF pair–DNA interaction contains two TF-binding sites in which the binding sequences are predetermined. However, gap sequences are not conserved, making the conventional strategy of covalent conjugation of two individual PIPs ineffective for targeting a TF pair. Second, the gap distances are flexible and usually vary from −1 to 5 bp [107]. Third, the most important point to consider concerns the use of two separate short PIPs for targeting the two binding sites. However, it is now known that, if a TF pair partner is displaced, the TFs might generate diverse biological functions. For example, the SOX2–OCT4 pair functions to maintain pluripotency gene circuits in stem cells. However, the change of the partner of SOX2 to PAX6 shifts the function of the pair toward the regulation of lens development [109, 110].

Therefore, we sequentially developed two exemplary cooperative DNA-binding systems, that is, PIP-HoGu [111] and PIP-NaCo [112], to mimic TF pair binding (Fig. 7a, b). The cooperative interaction domains (CIDs), such as cyclodextrin–adamantane in PIP-HoGu, assure the synergistic binding of two PIPs at the nearby DNA sites and with a flexible gap distance between two PIPs. The PIP-NaCo system utilizes left-handed PNA as the CID and exhibits orthogonality to cellular nucleic acids and tunability of cooperation [113]. In an in vitro and cell study, both the PIP-HoGu and PIP-NaCo systems exhibited valuable improvement in sequence selectivity, binding flexibility, and cooperation. Specifically, PIP-HoGu showed a thermal stability of 7.2 °C and a minimum free energy of interaction of −2.32 kcal mol^{-1} with a targeting length of 14 bp, whereas PIP-NaCo exhibited a minimum energetics of cooperation of −3.26 kcal mol^{-1}. The orthogonal PIP–NaCo system has the potential to provide triple to multiple heterobinding systems to target further complicated natural TF cluster networks.

Recently, we moved one step forward to construct an advanced genetic switch by introducing epigenetic functionality into a cooperative DNA binding system, and demonstrated its effectiveness in an in vitro histone acetylation assay. We first upgraded the cooperation domain in the first-generation PIP–HoGu system by replacing cyclodextrin (Cyd) with CB7 [114, 115], because an advanced DNA-binding system requires a tighter CID interaction (Fig. 7c). We constructed a biomimetic epigenetic code with a synthetic compound called Bi-PIP by conjugating PIP with an epigenetic drug that was capable of inducing targeted acetylation at a particular locus of interest [116]. Subsequently, a small molecule epidrug was tethered to the CB7-assisted PIP–HoGu, to construct the advanced system termed ePIP–HoGu (Fig. 7d) [117]. In fact, the incorporation of a cooperative dimer system into PIP–epidrug conjugates increases the DNA recognition length, reinforces reasonable sequence selectivity, and allows versatile binding modes. As a proof-of-concept study, the ePIP–HoGu system was shown to be adept at synergistically augmenting proximate histone acetylation with good efficiency and selectivity [118].

These cooperative systems possess huge potential for in vitro and cell-based assays, and further optimization would render them feasible for versatile applications

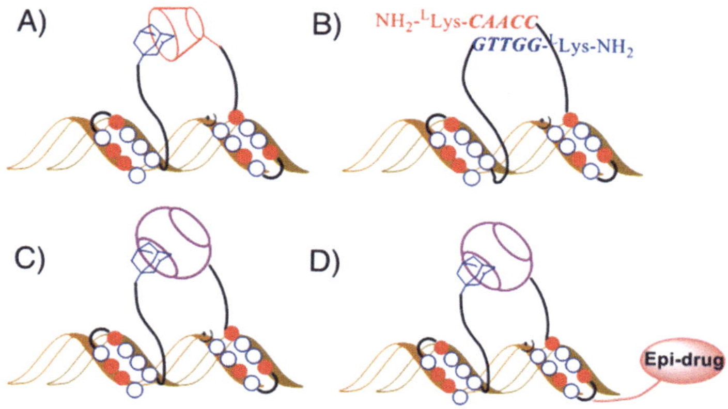

Fig. 7 Schematic view of the (**a**) PIP–HoGu, (**b**) PIP–NaCo, (**c**) CB7-assisted PIP–HoGu, and (**d**) ePIP–HoGu systems

[119]. Such concept of cooperative DNA can also be applied in other research studies, such as nucleic acid-based DNA binders, RNA binders, protein-based DNA binding systems, and even protein targeting.

8 Summary and Outlook

Transcription factors (TFs) play a fundamental role in maintaining cellular identity, homeostasis, and function by either operating alone or in coordination with other protein complexes. Contemporary analytical techniques have been revealing the cell-fate-regulating TFs and the association of defective TFs with several complex diseases. However, there remains a considerable gap between the accumulated knowledge of these molecules and the available tools to modulate them. Although small molecules partially function in resetting the gene transcription machinery inside living cells, they lack the essential component of TFs, that is, a DBD. One approach to gain artificial control over TFs is the artificial construction of TF mimics that maintain the structure and function of their natural counterparts. Natural DNA-binding proteins, such as ZFs and TALEs, were developed after gaining a closer look at the components of natural TFs and exhibited a notable success in modulating key TFs inside living cells. However, several obstacles, such as efficiency, lax specificity, and epigenome, hamper their widespread use. The CRISPR-Cas9 system was developed by mimicking the natural defense system of bacteria and is, to date, the most widely used genetic toolbox. This system has been successfully employed for editing several key TFs in both the prokaryotic and eukaryotic genomes.

Considering that both patients and clinicians favor the use of small molecules over biological drugs, there is a growing demand to create genetic-knowledge-based

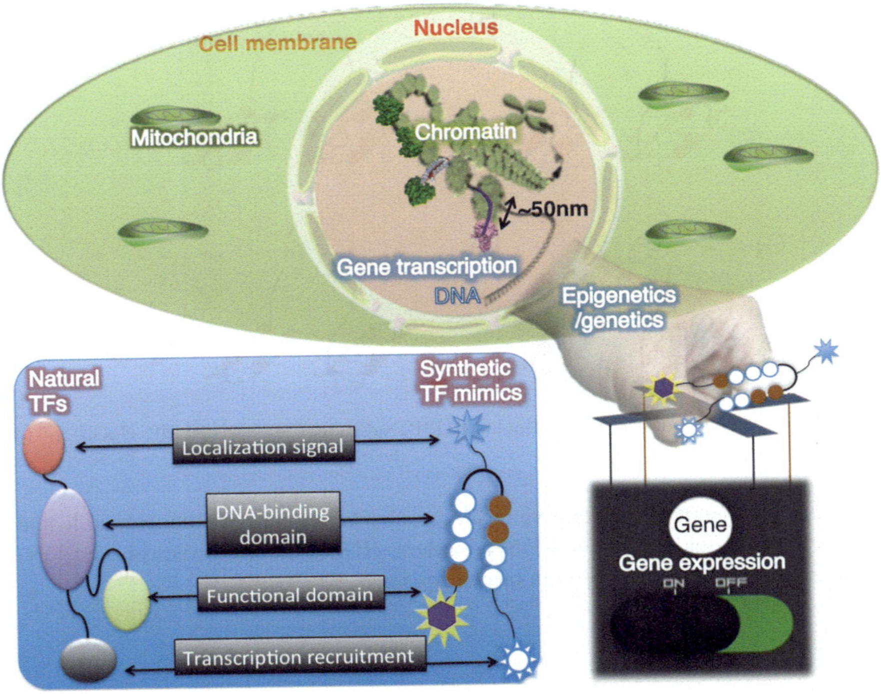

Fig. 8 Graphical illustration of synthetic transcription factor (TF) mimics retaining the structure and function of natural TFs (blue inset), which could aid in gaining artificial control over gene expression in living cells

small molecules, such as synthetic TF mimics. Hairpin pyrrole–imidazole polyamides (PIPs) remain at the forefront of the synthetic gene regulators. In our lab, we have been developing designer PIPs and their conjugates as synthetic TF mimics. PIP DBDs have been used to probe therapeutically important TFs associated with telomeres, cancer, and trinucleotide genetic disorders. Moreover, they have been used to cause the promoter-specific transcription suppression of therapeutically important genes, including oncogenes, such as *EVI1*, and cell-fate-regulating genes, such as *SOX2*. PIP and alkylating conjugates successfully targeted the point mutations in oncogenes such as *KRAS* and reversed tumor growth. By mining the critical role of the epigenome, we also constructed PIP-epigenetic modulator conjugates termed SAHA-PIPs that could trigger the transcriptional activation of therapeutically important genes and noncoding RNAs. As a paradigm-shifting study, recently, we delivered PIPs inside mitochondria and successfully demonstrated the promoter-specific transcription suppression of genes involved in the electron transport chain.

Advancing our synthetic TF mimics to construct knowledge-based and tailor-made gene-regulation tools could help overcome the current challenges in therapeutic gene modulation and cellular reprogramming (Fig. 8) [120]. To achieve this complex feat, several bottleneck issues, such as water solubility, limitation in the number of recognition sequences by DBDs, structural limitation to encompass multifunctional components, imidazole-content-dependent cell membrane permeability, and biodegradability, need to be addressed. Nevertheless, the tunable nature of PIPs and the recent success in incorporating them together with other functional nanoparticles suggest that these issues could be resolved in the near future.

References

1. Pandian GN, Sugiyama H (2013) Strategies to modulate heritable epigenetic defects in cellular machinery: lessons from nature. Pharmaceuticals 2013(6):1–24
2. Watson JD (1990) The human genome project: past, present, and future. Science 248:44
3. Chin L, Anderson JN, Futrea PA (2011) Cancer genomics: from discovery science to personalized medicine. Nat Med 17:297–303
4. Pandian GN, Taniguchi J, Sugiyama H (2014) Cellular reprogramming for pancreatic β-cell regeneration: clinical potential of small molecule control. Clin Transl Med 3:6
5. Pandian GN, Syed J, Sugiyama H (2015) Synthetic strategies to identify and regulate noncoding RNAs. In: Long noncoding RNAs. Springer, Tokyo, pp 23–43. https://doi.org/10.1007/978-4-431-55576-6_2
6. Pandian GN, Taylor RD, Junetha S, Saha A, Anandhakumar C, Vaijayanthi T, Sugiyama H (2014) Alteration of epigenetic program to recover memory and alleviate neurodegeneration: prospects of multi-target molecules. Biomater Sci 2:1043–1056
7. Masuda S, Wu J, Hishida T, Pandian GN, Sugiyama H, Belmonte JCI (2013) Chemically induced pluripotent stem cells (CiPSCs): a transgene-free approach. J Mol Cell Biol 5:354–355
8. Wu YL, Pandian GN, Ding YP, Zhang W, Tanaka Y, Sugiyama H (2013) Clinical grade iPS cells: need for versatile small molecules and optimal cell sources. Chem Biol 20:1311–1322
9. Pandian GN, Sugiyama H (2014) Targeted editing of therapeutic genes using DNA-based transcriptional activators: scope and challenges. In: Chemical biology of nucleic acids: fundamentals and clinical applications. Springer, pp 347–365. https://doi.org/10.1007/978-3-642-54452-1_19.
10. Patel S, Jung D, Yin PT, Carlton P, Yamamoto M, Bando T, Sugiyama H, Lee KB (2014) NanoScript: a nanoparticle-based artificial transcription factor for effective gene regulation. ACS Nano 8:8959–8967
11. Pandian GN, Sugiyama H (2012) Programmable genetic switches to control transcriptional machinery of pluripotency. Biotechnol J 7:798–809
12. Pandian GN, Sugiyama H (2016) Nature-inspired design of smart biomaterials using the chemical biology of nucleic acids. Bull Chem Soc Jp 89:843–868
13. Xu Y, Sugiyama H (2006) Photochemical approach to probing different DNA structures. Angew Chem Int Ed 45:1354–1362
14. Brent Ft, Ptashne M (1985) A eukaryotic transcriptional activator bearing the DNA specificity of a prokaryotic repressor. Cell 43:729–736
15. Hollenberg SM, Evans RM (1998) Multiple and cooperative trans-activation domains of the human glucocorticoid receptor. Cell 55:899–906
16. Liu F, Green MR (1990) A specific member of the ATF transcription factor family can mediate transcription activation by the adenovirus E1a protein. Cell 61:1217–1224
17. Wang JC (2015) Finding primary targets of transcriptional regulators. Cell Cycle 4:356–358
18. Piskacek S, Gregor M, Nemethova M, Grabner M, Kovarik P, Piskacek M (2007) Nineamino-acid transactivation domain: establishment and prediction utilities. Genomics 89:756–768
19. Brivanlou AH, Darnell JE (2002) Signal transduction and the control of gene expression. Science 295(5556):813–818

20. Thomas MC, Chiang CM (2006) The general transcription machinery and general cofactors. Crit Rev Biochem Mol Biol 41:105–178
21. Lemons D, McGinnis W (2006) Genomic evolution of hox gene clusters. Science 313(5795):1918–1922
22. Blancafort P, Magnenat L, Barbas CF (2003) Scanning the human genome with combinatorial transcription factor libraries. Nat Biotechnol 21:269–274
23. Gonzalez B, Schwimmer LJ, Fuller RP, Ye Y et al (2010) Modular system for the construction of zinc-finger libraries and proteins. Nat Protoc 5:791–810
24. Rosen LE, Morrison HA, Masri S, Brown MJ et al (2006) Homing endonuclease I-CreI derivatives with novel DNA target specificities. Nucleic Acids Res 34:4791–4800
25. Grizot S, Smith J, Daboussi F, Prieto J et al (2009) Efficient targeting of a SCID gene by an engineered single-chain homing endonuclease. Nucleic Acids Res 37:5405–5419
26. Bogdanove AJ, Schornack S, Lahaye T (2010) TAL effectors: finding plant genes for disease and defense. Curr Opin Plant Biol 13:394–401
27. Christian M, Cermak T, Doyle EL, Schmidt C et al (2010) Targeting DNA double-strand breaks with TAL effector nucleases. Genetics 186:757–761
28. Zhang F, Cong L, Lodato S, Kosuri S et al (2011) Efficient construction of sequence-specific TAL effectors for modulating mammalian transcription. Nat Biotechnol 292:149–153
29. Bogdanove AJ, Voytas DF (2011) TAL effectors: customizable proteins for DNA targeting. Science 333:1843–1846
30. Moscou MJ, Bogdanove AJ (2009) A simple cipher governs DNA recognition by TAL effectors. Science 326:1501
31. Kouzarides T (2007) Chromatin modifications and their function. Cell 128:693–705
32. Alexandre J, Coline P, Gwendoline D, Valérie G, Alan M, Julien V, Fayza D, Laurent P, Aymeric D (2015) Optimized tuning of TALEN specificity using non-conventional RVDs. Sci Rep 5:e8150
33. Valton J, Guyot V, Marechal A, Jean-Marie J, Alexandre DA, Philippe D, Poirot L (2015) Multiplex TALEN treatment of primary T cells enables efficient depletion of TCR membrane expression. Mol Ther 23:1507–1518
34. Poirot L, Philip B, Schiffer-Mannioui C, Clerre DL, Chion-Sotinel I, Derniame S, Bas C, Potrel P, Lemaire L (2015) Multiplex genome-edited T-cell manufacturing platform for "Off-the-Shelf" adoptive T-cell immunotherapies. Cancer Res 75:3853–3864
35. Cruz NM, Song X, Czerniecki SM, Gulieva RE, Churchill AJ, Kim YK, Winston K, Tran LM, Diaz MA, Fu H, Finn LS, Pei Y, Himmelfarb J, Freedman BS (2017) Organoid cystogenesis reveals a critical role of microenvironment in human polycystic kidney disease. Nat Materials 16:1112–1119
36. Sandweiss AJ, McIntosh MI, Moutal A, Davidson-Knapp R, Hu J, Giri AK, Yamamoto T, Hruby VJ, Khanna R (2018) Genetic and pharmacological antagonism of NK1 receptor prevents opiate abuse potential. Mol Psych 23:1745–1755
37. Liu Y, Zhan Y, Chen Z, He A, Li J, Wu H, Liu L, Zhuang C, Lin J, Guo X, Zhang Q, Huang W, Cai Z (2016) Directing cellular information flow via CRISPR signal conductors. Nat Methods 13:938–944
38. Fonfara I, Richter H, Bratovič M, Le Rhun A, Charpentier E (2016) The CRISPR-associated DNA-cleaving enzyme Cpf1 also processes precursor CRISPR RNA. Nature 532:517–521
39. Mohanraju P, Makarova KS, Zetsche B, Zhang F, Koonin EV, van der Oost J (2016) Diverse evolutionary roots and mechanistic variations of the CRISPR-Cas systems. Science 353:6299
40. Wähnert U, Zimmer C, Luck G, Pitra C (1975) (dA.dT)-dependent inactivation of the DNA template properties by interaction with netropsin and distamycin a. Nucleic Acids Res 2:391
41. Kopka ML, Yoon C, Goodsell D, Pjura P, Dickerson RE (1985) Binding of an antitumor drug to DNA: Netropsin and C-G-C-G-A-A-T-T-BrC-G-C-G. J Mol Biol 183:553–563
42. Trauger JW, Baird EE, Dervan PB (1996) Recognition of DNA by designed ligands at subnanomolar concentrations. Nature 382:559–561

43. White S, Szewczyk JW, Turner JM, Baird EE, Dervan PB (1998) Recognition of the four Watson–Crick base pairs in the DNA minor groove by synthetic ligands. Nature 391:468–471
44. Hsu CF, Dervan PB (2008) Quantitating the concentration of Py-Im polyamide-fluorescein conjugates in live cells. Bioorg Med Chem Lett 18:5851–5815
45. Nishijima S, Shinohara K, Bando T, Minoshima M, Kashiwazaki G, Sugiyama H (2010) Cell permeability of Py–Im-polyamide-fluorescein conjugates: influence of molecular size and Py/Im content. Bioorg Med Chem 18:978–983
46. Vaijayanthi T, Bando T, Pandian GN, Sugiyama H (2012) Progress and prospects of pyrroleimidazole polyamide–fluorophore conjugates as sequence- selective DNA probes. Chembiochem 13:2170
47. Bando T, Fujimoto J, Minoshima M, Shinohara K, Sasaki S, Kashiwazaki G, Mizumura M, Sugiyama H (2007) Detection of CAG repeat DNA sequences by pyrene-functionalized pyrroleimidazole polyamides. Bioorg Med Chem 15:6937–6942
48. Fujimoto J, Bando T, Minoshima M, Uchida S, Iwasaki M, Shinohara K, Sugiyama H (2008) Detection of triplet repeat sequences in the double-stranded DNA using pyrene-functionalized pyrrole-imidazole polyamides with rigid linkers. Bioorg Med Chem 16:5899–5907
49. Vaijayanthi T, Bando T, Hashiya K, Pandian GN, Sugiyama H (2013) Design of a new fluorescent probe: pyrrole/imidazole hairpin polyamides with pyrene conjugation at their γ-turn. Bioorg Med Chem 21:852–855
50. Lai Y-M, Fukuda N, Ueno T, Matsuda H, Saito S, Matsumoto K, Ayame H, Bando T, Sugiyama H, Mugishima H, Serie K (2005) Synthetic pyrrole-imidazole polyamide inhibits expression of the human transforming growth factor-beta1 gene. J Pharmacol Exp Ther 315:571–575
51. Minoshima M, Bando T, Sasaki S, Shinohara K, Shimizu T, Fujimoto J, Sugiyama H (2007) DNA alkylation by pyrrole–imidazole seco-CBI conjugates with an indole linker: sequence-specific DNA alkylation with 10-base-pair recognition through heterodimer formation. J Am Chem Soc 129:5384–5390
52. Kawamoto Y, Bando T, Kamada F, Li Y, Hashiya K, Maeshima K, Sugiyama H (2013) Development of a new method for synthesis of tandem hairpin pyrrole–imidazole polyamide probes targeting human telomeres. J Am Chem Soc 135:16468–16477
53. Kawamoto Y, Sasaki A, Hashiya K, Ide S, Bando T, Maeshima K, Sugiyama H (2015) Tandem trimer pyrrole–imidazole polyamide probes targeting 18 base pairs in human telomere sequences. Chem Sci 6:2307–2312
54. Kawamoto Y, Sasaki A, Chandran A, Hashiya K, Ide S, Bando T, Maeshima K, Sugiyama H (2016) Targeting 24 bp within telomere repeat sequences with tandem tetramer pyrrole–imidazole polyamide probes. J Am Chem Soc 138:14100–14107
55. Takahashi T, Asami Y, Kitamura E, Suzuki T, Wang X, Igarashi J, Morohashi A, Shinojima Y, Kanou H, Saito K, Takasu T, Nagase H, Harada Y, Kuroda K, Watanabe T, Kumamoto S, Aoyama T, Matsumoto Y, Bando T, Sugiyama H, Yoshida-Noro C, Fukuda N, Hayashi N (2008) Development of pyrrole-imidazole polyamide for specific regulation of human aurora kinase-A and -B gene expression. Chem Biol 15:829–841
56. Ueno T, Fukuda N, Tsunemi A, Yao EH, Matsuda H, Tahira K, Matsumoto T, Matsumoto K, Matsumoto Y, Nagase H, Sugiyama H, Sawamura T (2009) A novel gene silencer, pyrroleimidazole polyamide targeting human lectin-like oxidized low-density lipoprotein receptor-1 gene improves endothelial cell function. J Hypertens 2009(27):508–516
57. Suzuki T, Asami Y, Takahashi T, Wang X, Watanabe T, Bando T, Sugiyama H, Fukuda N, Nagase H (2009) Development of a molecule-recognized promoter DNA sequence for inhibition of HER2 expression. J Antibiot 62:339–341
58. Matsuda H, Fukuda N, Ueno T, Katakawa M, Wang X, Watanabe T, Matsui S, Aoyama T, Saito K, Bando T, Matsumoto Y, Nagase H, Matsumoto K, Sugiyama H (2011) Transcriptional inhibition of progressive renal disease by gene silencing pyrrole–imidazole polyamide targeting of the transforming growth factor-b1 promoter. Kidney Int 79:46–56

59. Yasuda A, Noguchi K, Yasuda A, Minoshima M, Kashiwazaki G, Kanda T, Katayama K, Mitsuhashi J, Bando T, Sugiyama H, Sugimoto Y (2011) DNA ligand designed to antagonize EBNA1 represses Epstein–Barr virus- induced immortalization. Cancer Sci 102:2221–2230
60. Syed J, Pandian GN, Sato S, Taniguchi J, Chandran A, Hashiya K, Bando T, Sugiyama H (2014) Targeted suppression of EVI1 oncogene expression by sequence-specific pyrroleimidazole polyamide. Chem Biol 21:1370–1380
61. Wang X, Nagase H, Watanabe T, Nobusue H, Suzuki T, Asami Y, Shinojima Y, Kawashima H, Takagi K, Mishra R, Igarashi J, Kimura M, Takayama T, Fukuda N, Sugiyama H (2010) Inhibition of MMP- 9 transcription and suppression of tumor metastasis by pyrrole- imidazole polyamide. Cancer Sci 101:759–766
62. Nagashima T, Aoyama T, Yokoe T, Fukasawa A, Fukuda N, Ueno T, Sugiyama H, Nagase H, Matsumoto Y (2009) Pharmacokinetic modeling and prediction of plasma pyrrole-imidazole polyamide concentration in rats using simultaneous urinary and biliary excretion data. Biol Pharm Bull 32:921–927
63. Fukasawa A, Aoyama T, Nagashima T, Fukuda N, Ueno T, Sugiyama H, Nagase H, Matsumoto Y (2009) Pharmacokinetics of pyrrole–imidazole polyamides after intravenous administration in rat. Biopharm Drug Dispos 30:81–89
64. Iguchi A, Fukuda N, Takahashi T, Watanabe T, Matsuda H, Nagase H, Bando T, Sugiyama H, Shimizu K (2013) RNA binding properties of novel gene silencing pyrrole-imidazole polyamides. Biol Pharm Bull 36:1152–1158
65. Minoshima M, Bando T, Sasaki S, Fujimoto J, Sugiyama H (2008) Pyrrole-imidazole hairpin polyamides with high affinity at 5′–CGCG–3′ DNA sequence; influence of cytosine methylation on binding. Nucleic Acids Res 36:2889–2894
66. Taniguchi J, Pandian GN, Hidaka T, Hashiya K, Bando T, Kim KK, Sugiyama H (2017) A synthetic DNA-binding inhibitor of SOX2 guides human induced pluripotent stem cells to differentiate into mesoderm. Nucleic Acids Res 45:9219–9228
67. Malinee M, Kumar A, Hidaka T, Horie M, Hasegawa K, Pandian GN, Sugiyama H (2020) Targeted suppression of metastasis regulatory transcription factor SOX2 in various cancer cell lines using a sequence-specific designer pyrrole–imidazole polyamide. Bioorg Med Chem 28:115248
68. Sugiyama H, Lian C, Isomura M, Saito I, Wang AHJ (1996) Distamycin a modulates the sequence specificity of DNA alkylation by duocarmycin A. Proc Natl Acad Sci U S A 93:14405–14410
69. Bando T, Sugiyama H (2006) Synthesis and biological properties of sequence-specific DNAalkylating pyrrole–imidazole polyamides. Acc Chem Res 39:935–944
70. Sasaki S, Bando T, Minoshima M, Shinohara K, Sugiyama H (2008) Sequence-specific alkylation by Y-shaped and tandem hairpin pyrrole-imidazole polyamides. Chem Eur J 14:864–870
71. Minoshima M, Bando T, Shinohara K, Kashiwazaki G, Nishijima S, Sugiyama H (2010) Comparative analysis of DNA alkylation by conjugates between pyrrole–imidazole hairpin polyamides and chlorambucil or seco-CBI. Bioorg Med Chem 18:1236–1243
72. Asamitsu S, Kawamoto Y, Hashiya F, Hashiya K, Yamamoto M, Kizaki S, Bando T, Sugiyama H (2014) Sequence-specific DNA alkylation and transcriptional inhibition by longchain hairpin pyrrole–imidazole polyamide-chlorambucil conjugates targeting CAG/CTG trinucleotide repeats. Bioorg Med Chem 22:4646–4657
73. Shinohara K, Bando T, Sugiyama H (2010) Anticancer activities of alkylating pyrroleimidazole polyamides with specific sequence recognition. Anti-Cancer Drugs 21:228–242
74. Kashiwazaki G, Bando T, Shinohara K, Minoshima M, Kumamoto H, Nishijima S, Sugiyama H (2010) Alkylation of a human telomere sequence by heterotrimeric chlorambucil PI polyamide conjugates. Bioorg Med Chem 18:2887–2893
75. Guo C, Kawamoto Y, Asamitsu S, Sawatani Y, Hashiya K, Bando T, Sugiyama H (2015) Rational design of specific binding hairpin py-im polyamides targeting human telomere sequences. Bioorg Med Chem 23:855–860

76. Minoshima M, Chou JC, Lefebvre S, Bando T, Shinohara K, Gottesfeld JM, Sugiyama H (2010) Potent activity against K562 cells by polyamide-seco-CBI conjugates targeting histone H4 genes. Bioorg Med Chem 18:168–174
77. Kashiwazaki G, Bando T, Yoshidome T, Masui S, Takagaki T, Hashiya K, Pandian GN, Yasuoka J, Akiyoshi K, Sugiyama H (2012) Synthesis and biological properties of highly sequence-specific-alkylating N- Methylpyrrole- N- Methylimidazole polyamide conjugates. J Med Chem 55:2057–2066
78. Taylor RD, Asamitsu S, Takenaka T, Yamamoto M, Hashiya K, Kawamoto Y, Bando T, Nagase H, Sugiyama H (2014) Sequence-specific DNA alkylation targeting for Kras codon 13 mutation by pyrrole-imidazole polyamide seco-CBI conjugates. Chem Eur J 20:1310–1317
79. Hiraoka K, Inoue T, Taylor RD, Watanabe T, Koshikawa N, Yoda H, Shinohara K, Takatori A, Sugimoto K, Maru Y, Denda T, Fujiwara K, Balmain A, Ozaki T, Bando T, Sugiyama H, Nagase H (2015) Inhibition of KRAS codon 12 mutants using a novel DNA-alkylating pyrroleimidazole polyamide conjugate. Nat Commun 6:6706
80. Taylor RD, Chandran A, Kashiwazaki G, Hashiya K, Bando T, Nagase H, Sugiyama H (2015) Selective targeting of the KRAS codon 12 mutation sequence by pyrrole-imidazole polyamide seco-CBI conjugates. Chem Eur J 21:14996–15003
81. Lin J, Hiraoka K, Watanabe T, Kuo T, Shinozaki Y, Takatori A, Koshikawa N, Chandran A, Otsuki J, Sugiyama H, Horton P, Nagase H (2016) Identification of binding targets of a pyrroleimidazole polyamide KR12 in the LS180 colorectal genome. PLoS One 11:e0165581
82. Morita K, Suzuki K, Maeda S et al (2017) Genetic regulation of the RUNX transcription factor family has antitumor effects. J Clin Invest 127(7):2815–2828
83. Han YW, Matsumoto T, Yokota H, Kashiwazaki G, Morinaga H, Hashiya K, Bando T, Harada Y, Sugiyama H (2012) Binding of hairpin pyrrole and imidazole polyamides to DNA: relationship between torsion angle and association rate constants. Nucleic Acids Res 40:11510–11517
84. Taylor RD, Kawamoto Y, Hashiya K, Bando T, Sugiyama H (2014) Sequence-specific DNA alkylation by tandem py-im polyamide conjugates. Chem Asian J 9:2527–2533
85. Han YW, Kashiwazaki G, Morinaga H, Matsumoto T, Hashiya K, Bando T, Sugiyama H (2013) Effect of single pyrrole replacement with β-Alanine on DNA binding affinity and sequence specificity of hairpin pyrrole/imidazole polyamides targeting 5′-GCGC-3′. Bioorg Med Chem 21:5436–5441
86. Takagaki T, Bando T, Kitano M, Hashiya K, Kashiwazaki G, Sugiyama H (2011) Evaluation of PI polyamide conjugates with eight-base pair recognition and improvement of the aqueous solubility by PEGylation. Bioorg Med Chem 19:5896–5902
87. Takagaki T, Bando T, Sugiyama H (2012) Synthesis of pyrrole-imidazole polyamide seco-1-Chloromethyl-5-hydroxy-1,2-dihydro-3H-benz[e]indole conjugates with a vinyl linker recognizing a 7 bp DNA sequence. J Am Chem Soc 134:13074–13081
88. Morinaga H, Bando T, Takagaki T, Yamamoto M, Hashiya K, Sugiyama H (2011) Cysteine cyclic pyrrole-imidazole polyamide for sequence-specific recognition in the DNA minor groove. J Am Chem Soc 133:18924–18930
89. Yamamoto M, Bando T, Morinaga N, Kawamoto Y, Hashiya K, Sugiyama H (2014) Sequence-specific DNA recognition by cyclic pyrrole-imidazole cysteine-derived polyamide dimers. Chem Eur J 20:752–759
90. Kameshima W, Ishizuka T, Minoshima M, Yamamoto M, Sugiyama H, Xu Y, Komiyama M (2013) Conjugation of peptide nucleic acid with a pyrrole/imidazole polyamide to specifically recognize and cleave DNA. Angew Chem Int Ed 52:13681–13684
91. Ohtsuki A, Kimura MT, Minoshima M, Suzuki T, Ikeda M, Bando T, Nagase H, Shinohara K, Sugiyama H (2009) Synthesis and properties of PI polyamide-SAHA conjugate. Tetrahedron Lett 50:7288–7292
92. Pandian GN, Shinohara K, Ohtsuki A, Nakano Y, Minoshima M, Bando T, Nagase H, Yamada Y, Watanabe A, Terada N, Sato S, Morinaga H, Sugiyama H (2011) Synthetic small molecules

for epigenetic activation of pluripotent genes in mouse embryonic fibroblasts. Chembiochem 12:2822–2828
93. Pandian GN, Nakano Y, Sato S, Morinaga H, Bando T, Nagase H, Sugiyama H (2012) A synthetic small molecule for rapid induction of multiple pluripotency genes in mouse embryonic fibroblast. Sci Rep 2:e544
94. Pandian GN, Taniguchi J, Junetha S, Sato S, Han L, Saha A, AnandhaKumar C, Bando T, Nagase H, Vaijayanthi T, Taylor RD, Sugiyama H (2014) Distinct DNA-based epigenetic switches trigger transcriptional activation of silent genes in human dermal fibroblasts. Sci Rep 4:e3843
95. Han L, Pandian GN, Junetha S, Sato S, AnandhaKumar C, Taniguchi J, Saha A, Bando T, Nagase H, Sugiyama H (2013) A synthetic small molecule for targeted transcriptional activation of germ cell genes in a human somatic cell. Angew Chem Int Ed 52:13410–13413
96. Pandian GN, Sato S, AnandhaKumar C, Taniguchi J, Takashima K, Syed J, Han L, Saha A, Bando T, Nagase J, Sugiyama H (2014) Identification of a small molecule that turns `ON` the pluripotency gene circuitry in human fibroblasts. ACS Chem Biol 9:2729–2736
97. Syed J, Chandran A, Pandian GN, Taniguchi J, Sato S, Hashiya K, Kashiwazaki G, Bando T, Sugiyama H (2015) A synthetic transcriptional activator of genes associated with the retina in human dermal fibroblasts. Chembiochem 16:1497–1501
98. Pandian GN, Ohtsuki A, Bando T, Sato S, Hashiya K, Sugiyama H (2012) Development of programmable small DNA-binding molecules with epigenetic activity for induction of core pluripotency genes. Bioorg Med Chem 20:2656–2660
99. Saha A, Pandian GN, Sato S, Taniguchi J, Hashiya K, Bando T, Sugiyama H (2013) Synthesis and biological evaluation of a targeted DNA-binding transcriptional activator with HDAC8 inhibitory activity. Bioorg Med Chem 21:4201–4209
100. Saha A, Pandian GN, Sato S, Taniguchi J, Kawamoto Y, Hashiya K, Bando T, Sugiyama H (2014) Chemically modified synthetic small molecule boosts its biological efficacy against pluripotency genes in mouse fibroblast. ChemMedChem 9:2374–2380
101. AnandhaKumar C, Li Y, Kizaki S, Pandian GN, Hashiya K, Bando T, Sugiyama H (2014) Next-generation sequencing studies guide the design of pyrrole-imidazole polyamides with improved binding specificity by the addition of β-alanine. Chembiochem 15:2647–2651
102. AnandhaKumar C, Kizaki S, Pandian GN, Bando T, Sugiyama H (2015) Advancing smallmolecule-based chemical biology with next-generation sequencing technologies. Chembiochem 16:20–38
103. Han L, Pandian GN, Chandran A, Sato S, Taniguchi J, Kashiwazaki G, Sawatani Y, Hashiya K, Bando T, Xu Y, Qian X, Sugiyama H (2015) A synthetic DNA-binding domain guides distinct chromatin-modifying small molecules to activate an identical gene network. Angew Chem Int Ed 54:8700–8703
104. Yu Z, Taniguchi J, Wei Y, Pandian GN, Hashiya K, Bando T, Sugiyama H Antiproliferative and apoptotic activities of sequence-specific histone acetyltransferase inhibitor. Eur J Med Chem 138:320–327
105. Patel S, Pongkulapa T, Yin PT, Pandian GN, Rathnam C, Bando T, Vaijayanthi T, Sugiyama H, Lee KB (2015) Integrating epigenetic modulators into NanoScript for enhanced chondrogenesis of stem cells. J Am Chem Soc 137:4598–4601
106. Hidaka T, Pandian GN, Taniguchi J, Nobeyama T, Hashiya K, Bando T, Sugiyama H (2017) Creation of a synthetic ligand for mitochondrial DNA sequence recognition and promoter-specific transcription suppression. J Am Chem Soc 139:8444–8447
107. Jolma A, Yin Y, Nitta KR, Dave K, Popov A, Taipale M, Enge M, Kivioja T, Morgunova E, Taipale J (2015) DNA-dependent formation of transcription factor pairs alters their binding specificity. Nature 527(7578):384–388
108. Yu Z, Pandian GN, Hidaka T, Sugiyama H (2019) Therapeutic gene regulation using pyrrole-imidazole polyamides. Adv Drug Deliv Rev 147:66–85

109. Tapia N, MacCarthy C, Esch D, Marthaler AG, Tiemann U, Araúzo-Bravo MJ, Jauch R, Cojocaru V, Schöler HR (2015) Dissecting the role of distinct OCT4-SOX2 heterodimer configurations in pluripotency. Sci Rep 5:e13533
110. Kamachi Y, Uchikawa M, Tanouchi A, Sekido R, Kondoh H (2001) Pax6 and SOX2 form a co-DNA-binding partner complex that regulates initiation of lens development. Genes Dev 15(10):1272–1286
111. Yu Z, Guo C, Wei Y, Hashiya K, Bando T, Sugiyama H (2018) Pip-HoGu: an artificial assembly with cooperative DNA recognition capable of mimicking transcription factor pairs. J Am Chem Soc 140(7):2426–2429
112. Yu Z, Hsieh WC, Asamitsu S, Hashiya K, Bando T, Ly DH, Sugiyama H (2018) Orthogonal γPNA dimerization domains empower DNA binders with cooperativity and versatility mimicking that of transcription factor pairs. Chem Eur J 24(53):14183–14188
113. Sacui I, Hsieh WC, Manna A, Sahu B, Ly DH (2015) Gamma peptide nucleic acids: as orthogonal nucleic acid recognition codes for organizing molecular self-assembly. J Am Chem Soc 137(26):8603–8610
114. Liu S, Ruspic C, Mukhopadhyay P, Chakrabarti S, Zavalij PY, Isaacs L (2005) The cucurbit[n]uril family: prime components for self-sorting systems. J Am Chem Soc 127(45):15959–15967
115. Jeon WS, Moon K, Park SH, Chun H, Ko YH, Lee JY, Lee ES, Samal S, Selvapalam N, Rekharsky MV, Sindelar V, Sobransingh D, Inoue Y, Kaifer AE, Kim K (2005) Complexation of ferrocene derivatives by the cucurbit[7]uril host: a comparative study of the cucurbituril and cyclodextrin host families. J Am Chem Soc 127:12984–12989
116. Taniguchi J, Feng Y, Pandian GN, Hashiya F, Hidaka T, Hashiya K, Park S, Bando T, Ito S, Sugiyama H (2018) Biomimetic artificial epigenetic code for targeted acetylation of histones. J Am Chem Soc 140:7108–7115
117. Yu Z, Ai M, Samanta SK, Hashiya F, Taniguchi J, Asamitsu S, Ikeda S, Hashiya K, Bando T, Pandian GN, Isaacs L, Sugiyama H (2020) A synthetic transcription factor pair mimic for precise recruitment of an epigenetic modifier to the targeted DNA locus. Chem Comm 56:2296–2299
118. Zou T, Hashiya F, Wei Y, Yu Z, Pandian GN, Sugiyama H (2018) Direct observation of H3–H4 octasome by high- speed AFM. Chem Eur J 24(60):15998–16002
119. Yu Z (2020) Synthetic DNA binding assembly: architecture, application and perspectives. In: Artificial assemblies with cooperative DNA recognition. Springer, Singapore, pp 1–39
120. Vaijayanthi T, Pandian GN, Sugiyama H (2018) Chemical control system of epigenetics. Chem Rec 18:1833–1853

Part IV
Physical Methods for Cell Control

Magnetic Nanoparticles and Alternating Magnetic Field for Cancer Therapy

Harutaka Mekaru, Yuko Ichiyanagi, and Fuyuhiko Tamanoi

1 Introduction

The era of nanotechnology was opened in 1960, and since then a dramatic progress has been made. This technology uses engineered materials and devices that have a nanometer scale in at least one dimension. A variety of nanomaterials have been developed over the years that have found application in various fields such as biomedicine, energy, and environmental preservation [1]. Nanotechnology is having a major impact on cancer therapy, as nanoparticles, nanosize particles that range from 40 to 400 nm diameter [2], can accumulate in tumors and alter biological processes. One of the mechanisms for this tumor accumulation is the so-called EPR (enhanced permeability and retention effect); due to breakage in the tumor vessel walls, nanosize materials will leak out from the vasculature and accumulate in tumor [3]. In

The original version of this chapter was revised. The correction to this chapter is available at https://doi.org/10.1007/978-3-030-55924-3_12

H. Mekaru
National Institute of Advanced Industrial Science and Technology, Tsukuba, Japan

Y. Ichiyanagi
Department of Physics, Graduate School of Engineering, Yokohama National University, Yokohama, Japan

F. Tamanoi (✉)
Department of Microbiology, Immunology and Molecular Genetics, University of California, Los Angeles, CA, USA

Institute for Integrated Cell-Material Sciences, Institute for Advanced Study, Kyoto University, Kyoto, Japan
e-mail: Tamanoi.Fuyuhiko.2c@kyoto-u.ac.jp

D. O. Wang, D. Packwood (eds.), *Cell-Inspired Materials and Engineering*, Fundamental Biomedical Technologies, https://doi.org/10.1007/978-3-030-55924-3_7

addition, ligands for cancer-cell specific receptors overexpressed on cell surface can be used to preferentially take up nanosize materials. This tumor targeting capability of nanoparticles has been used to selectively deliver anticancer drugs into tumor thus reducing side effects associated with current chemotherapy.

One of the exciting developments concerning nanomaterials is that they are opening ways to treat cancer by using external control such as light, magnetic field, and ultrasound [4]. This potential therapeutic strategy became possible as a variety of nanomaterials that respond to external stimuli such as light and magnetic field have been developed. Once they accumulate in the tumor, external stimuli can be applied to tumor that contains the nanomaterials. For example, magnetic nanomaterials such as iron oxide nanoparticles can be exposed to alternating magnetic field (AMF), leading to heat generation. This is because the materials yield thermal energy upon the application of the external AMF. We have proposed "magnetic hyperthermia" and developed the nanomaterials for biomedical applications over recent years [5–8].

In this review, we first discuss magnetic nanomaterials such as iron oxide nanoparticles. We will then describe in vitro and in vivo experiments carried out using these materials on biological systems. Finally, we will discuss progress made concerning devices used to carry out hyperthermia treatments. The potential use of magnetic field manipulation in clinical settings will be discussed.

2 Magnetic Nanoparticles (MNPs)

Magnetic nanoparticles (MNPs) consisting of iron oxide have been studied for magnetic hyperthermia treatment [9–13]. Iron oxide nanoparticles are desirable because of their biocompatibility and chemical stability. However, there are issues with the nanoscale magnetic materials. First, bulk ferromagnetic materials form multiple domains in order to keep energy low. However, small particles typically form only single domains. Consequently, magnetic spins in the particle fluctuate due to the thermal energy, causing the ferromagnetism to disappear. Accordingly, it is difficult to maintain ferromagnetism on the nanometer scale. It is therefore necessary to optimize magnetic parameters in order to make meaningful applications in biomedical fields.

One of the approaches to improve the magnetic properties of iron oxide is to dope them with other transition metal ions to obtain ferrites. Ferrite essentially consists of iron oxide including ferric ions. It is a curious and interesting phenomenon that when zinc ions are added to spinel ferrites, the magnetization increases even though zinc ion is nonmagnetic [14]. This is because the spinel structure (Fig. 1a) has two sublattices which are antiferromagnetically coupled each other. On doping, Zn^{2+} ions prefer to distribute at the A-sites, causing an increase in magnetic ions at the B-sites. The difference in magnetic moments of the A and B-sites causes a net increase in magnetization of the material.

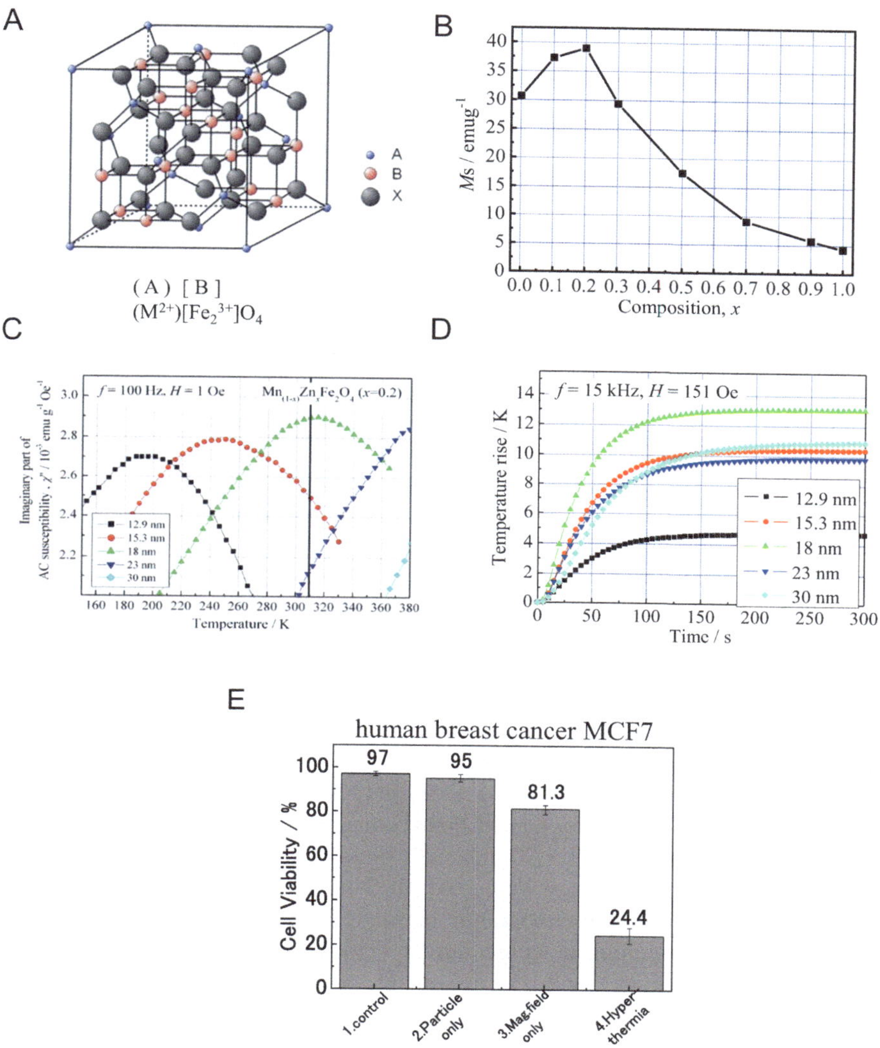

Fig. 1 Magnetic nanoparticles and their physical/chemical property. (**a**) Spinel structure of iron oxide doped with a metal. (**b**) Maximum magnetization of direct current of Mn-Zn ferrite for each composition. (**c**) Heating mechanism by magnetic relaxation loss. (**d**) Increase in temperature after AMF exposure. (**e**) Effect of magnetic hyperthermia on human breast cancer cells. Human breast cancer cells MCF7 were cultured and incubated with magnetic nanoparticles. They are then exposed to oscillating magnetic field for hyperthermia experiments. Cell viability is shown

As an example, we have prepared Mn-Zn ferrite nanoparticles and estimated Zn doping effect. Various Zn content of $Mn_{1-x}Zn_xFe_2O_4$ (x = 0, 0.1, 0.2, 0.3, 0.4, 0.5, 0.6, 0.7, 0.8) nanoparticles were prepared by our original wet chemical method [5–7]. Aqueous solutions of $MnCl_2 \cdot 4H_2O$, $ZnCl_2$, and $FeCl_2 \cdot 4H_2O$ were mixed with a solution of $Na_2SiO_3 \cdot 9H_2O$. The molar ratio of the prepared reagent was Mn:Zn:Fe:Si = $1-x$:x:2:3. The precipitates were washed several times with distilled water and dried at about 350 K in a thermostat. The as-prepared samples were then subjected to heat treatment in a furnace in an argon environment. Particle sizes were controlled by adjusting temperature. All samples were examined by CuKa X-ray powder diffraction $\lambda = 0.154$ nm and X-ray fluorescence analyses.

Figure 1b shows the maximum magnetization at direct current (DC) of Mn-Zn ferrite ($Mn_{1-x}Zn_xFe_2O_4$) for each composition. The magnetization (*M*s) value was largest at $x = 0.2$ of Zn content in this particle system. As an agent of hyperthermia treatment, large magnetization value *M*s and without hysteresis as well as coercive force *H*c (coercive force is a measure of the ability of a ferromagnetic material to withstand an external magnetic field without becoming demagnetized) in magnetization curves would be required. We obtained ideal superparamagnetic nanoparticles with large magnetization and no hysteresis in Mn-Zn ferrite systems.

Heating mechanism of magnetic nanoparticles in this system would be due to the relaxation loss. Heat dissipation in alternating current (AC) can be expressed by the following equation [15].

$$P = \mu^0 \pi \chi'' f h^2, \tag{1}$$

where μ_0 is the permeability in vacuum, χ'' is the imaginary part of the magnetic susceptibility, f is the frequency of the AMF, and h is the strength of the applied AC magnetic field. The imaginary part of the AC magnetic susceptibility χ'' was measured under a 1 Oe, 100 Hz AC magnetic field, for different particle sizes of $Mn_{0.8}Zn_{0.2}Fe_2O_4$. Particle sizes were controlled to be between 12.9 and 30 nm, depending on the annealing temperature between 1123 and 1223 K. It is assumed that an acceptable biological temperature is 310 K (37 °C, the average normal human body temperature), and heat dissipation can be predicted for this temperature. Hence, the heating mechanism by magnetic relaxation loss is most efficient for the sample with 18 nm particles as estimated in Fig. 1c.

In order to confirm the temperature increase using the samples in AMF, a measurement system involving the heating of particles by AMF was constructed and the temperature increase was examined. Figure 1d shows the increase in temperature of the samples as a function of time. The temperature was measured after placing the samples in a 151 Oe, 15 kHz AMF at 310 K. As expected from the AC magnetic susceptibility measurements, particle size of 18 nm had the highest heating effect.

3 Magnetic Hyperthermia

Magnetic nanoparticles (MNPs) were used to increase cell temperature and cause cell death. Using the above nanoparticles to induce hyperthermia, the following in vitro experiments were carried out [7]. Human breast cancer cells (MCF7) were cultured in a 35 mm dish for 24 h and incubated with magnetic particles at a concentration of 1 mg/ml. After application of AMF of 31 kHz and 90 Oe for 30 min, the cells were stained with trypan blue, and the live cells were counted. Figure 1e shows the result of the experiment for breast cancer cells. Conditions were 1. Control, 2. Particle only, 3. Magnetic field only, and 4. Magnetic Hyperthermia. Results of the "4. Magnetic Hyperthermia" condition confirmed that the viability of cancer cells was significantly decreased. Results of the "2. Particle only" condition confirmed the low toxicity of the nanomaterials. Ferrites are generally not considered to be biocompatible; however, our particles showed superior biocompatibility. This fact could be due to the characteristic configuration of the nanoparticles encapsulated with amorphous SiO_2. The results for different cancer cells such as prostate cancer (DU145) and breast cancer cells (KPL4, MDA-MB231), under this condition were similar [7]. A decrease in the cell viability of about 80% was observed for all four types of cancer cells, thus confirming that this sample is an effective agent for hyperthermia treatment.

4 Nanovalves and Controlled Release of Anticancer Drugs

Temperature increase due to an AMF can be used to release anticancer drugs from magnetic nanoparticles (MNPs). This approach uses mechanized nanoparticles based on mesoporous silica [4]. These nanoparticles have an average diameter of 100 nm and contain thousands of pores where anticancer drugs can be stored [16] in addition to the iron oxide core (Fig. 2a). Nanovalves are placed at the openings of the pores, thus providing on/off function for drug release. Pseudorotaxanes with cucurbituril are used as a material for making the nanovalves (Fig. 2b). When this nanoparticle is placed in an AMF, the nanoparticle is heated due to superparamagnetic property of iron oxide. This temperature increase causes the nanovalve to open as cucurbituril comes off the stalk.

Experimental setup for the operation of the nanovalve by AMF is shown in Fig. 2c. In this experiment, doxorubicin was loaded onto the nanoparticle that has nanovalves attached to pore openings. They were taken up into breast cancer MDA-MB231 cells and then the cells were exposed to AMF at the frequency of 500 kHz and Amplitude of 37.4 K/Am. The release of doxorubicin was detected by red fluorescence inside the cell after exposure (Fig. 2d). This was associated with increased cell killing after exposure (Fig. 2e). The advantage of this method is that the nanovalve opening and the drug release does not require high temperature; thus, adverse side effects on surrounding cells can be minimized.

Dong and Zink developed nanothermometer to detect the temperature of the interior of magnetically heated nanoparticles [17]. This involved the use of the upconversion emission spectrum of $NaYF_4$:Yb^{3+}, Er^{3+}. The temperature-dependent intensity ratio of two luminescence bands can be detected. This technology should contribute to finding optimum exposure conditions of magnetic nanoparticles to AMF.

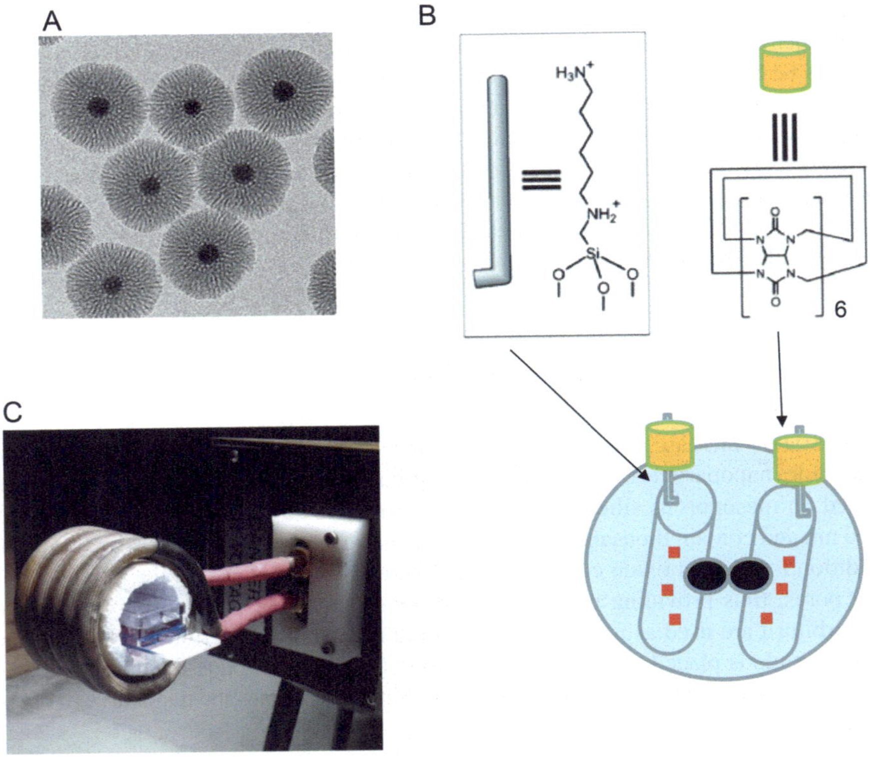

Fig. 2 Nanovalves that respond to AMF. (**a**) Mesoporous silica nanoparticles with iron oxide core. (**b**) Nanovalves attached to the opening of the pore provides open and close function for drug release. (**c**) Coils used for oscillating magnetic field exposure to cancer cells that have taken up magnetic nanoparticles. (**d**) Magnetic field dependent release of anticancer drugs and cell killing. Breast cancer cells MDA-MB231 were cultured and incubated with magnetic core silica nanoparticles (MCSN) equipped with magnetic field sensitive nanovalves (Fig. 2b). MCSNs were loaded with doxorubicin. Exposure to AMF with a frequency of 500 kHz and a current amplitude of 37.4 kAm^{-1} resulted in the release of doxorubicin as detected by red fluorescence. E: Cell killing was observed by the combined treatment of magnetic nanoparticles loaded with doxorubicin and AMF exposure

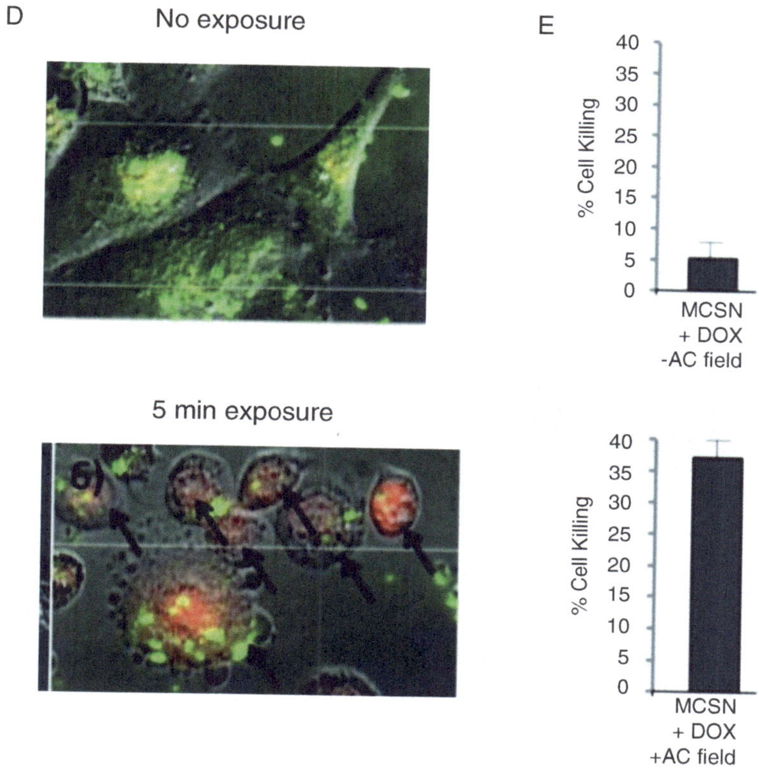

Fig. 2 (continued)

5 Localizing Magnetic Nanoparticles in Living Tissues Using External Magnetic Field

Magnetic field can also be used to move MNPs inside of a tissue. Figure 3 describes an experiment to examine the movement of MNPs through tissue [18]. In this experiment (Fig. 3), rhodamine (red fluorescent material) labeled MNPs were applied on the surface of a mouse's ear. A magnet was attached as an external field. For control nonmagnetic material, aluminum was also applied. After 24 h, the magnet was removed and the cross section of the sample was observed by confocal microscopy. Figure 3 shows magnetic area, the MNPs passed through the densely packed cells such as the epidermis and tissues, but were incapable of passing through areas of higher cell density, such as cartilage. Thus, MNPs could be concentrated in a localized area. Figure 3 shows outside of magnetic area and III is nonmagnetic area. The MNPs could not pass through the layers of epidermis, dermis, and the subcutaneous tissue in areas outside the influence of the magnetic field or when no magnetic field was present. Thus, we can localize NMPs to a targeted region.

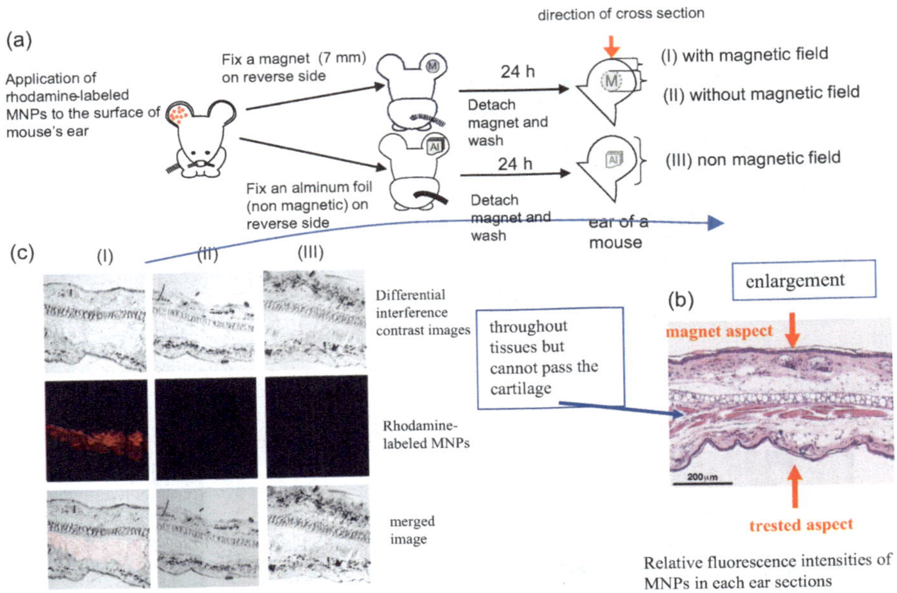

Fig. 3 Preparation of mouse external ear section for external magnetic field specific internalization of NMPs (a). Hematoxylin–eosin stained external ear section (b). Confocal laser scanning microscopy images of mouse external ear sections (c): differential interference contrast images (upper panel), rhodamine fluorescence images (middle panel), and merged images (lower panel). Localization of the NMPs ($d = 3$ nm) was investigated in the presence (c-I) and absence (c-II) of an external magnetic field and without the permanent magnet (aluminum foil control) (c-III).The relative fluorescence intensity was determined by dividing the sum of all pixel intensities

6 Apparatus and Instruments That Generate Alternating Magnetic Fields (AMF)

For research and clinical applications of MNPs, it is essential to use instruments that generate AMF. Here, we would like to describe advance in designing these instruments. Currently, various labs use magnetic coils to generate AMF. An AMF-generator is a simple system comprising of a high-frequency power supply and a solenoid coil. Commercial apparatus that generate AMFs are available from a number of sources. Therefore, in laboratory experiments, many research groups use self-assembled AMF systems. A transistor inverter LTG-100-05 (Dai-ichi High Frequency Co., Japan) incorporates an inverter circuit for inverting high and low levels of a digital signal by the switching operation of a transistor. This device was used as a high-frequency power source, and an AMF frequency of 118 kHz was obtained [19, 20]. Another device uses an AC power supplied to a solenoid coil by combining a programmable power supply EC1000S (NF Co., Japan) and a digital function generator DF1906 (NF Co., Japan) to generate an AMF of 20 mT [21]. In another case, an AMF generator with a frequency of 500 kHz and a magnetic field strength of 46.9 mT was developed by Taeyang Instrument Company [16].

Several AMF apparatuses are commercially available for studies of hyperthermia. One type of hyperthermic apparatus is developed and sold by nanoScale Biomagnetics (nB). The company nB was established in 2008 as a spin-off from the University of Zaragoza, based on technologies for magnetic heating of nanomaterials developed at the Magnetic Hyperthermia Group of the Institute of Nanoscience of Aragón (Spain). In 2010, the DM100 series with a frequency range of 50–1000 kHz and a maximum magnetic field strength of 50 mT was marketed as a nanoparticle induction heating analyzer, and was used to develop magnetic nanoparticles for hyperthermia treatment. It has been reported that an AMF with an operating frequency of 260 kHz and magnetic field intensity of 16 mT was applied to iron oxide magnetic nanoparticles incorporated in primary, monocyte-derived dendritic cells, and selectively induced cell death [22]. The influence of AMF at a frequency of 260 kHz and a field strength of 15.9 mT has been investigated on dendritic cells loaded with a magnetite core (Fe_3O_4) with a carboxyl-functionalized dextran shell at the surface [23]. Iron oxide nanocube-modified microcapsules were heated by the AMF at a frequency of 300 kHz and a magnetic field strength of 30.1 mT, eventually destroying the microcapsule walls to release the inclusions [24]. An engineered thermosensitive random copolymer, which exhibits a linear (soluble) to globular (insoluble) transition at 40–43 °C, was coated on the surface of mesoporous silica nanoparticles with superparamagnetic iron oxide cores. The subsequent application of an AMF at frequencies in the range of 424–838 kHz and a magnetic field strength of 25.1–29.9 mT caused the cargo to be released by contraction of the polymer network [25]. FePt core–shell nanoparticles, which were neutron activated and coated with an amorphous tungsten oxide were heated by an AMF at 831 kHz with a 25 mT field strength [26]. Magnetic hyperthermia at a frequency of 409 kHz and a magnetic field strength of 18 mT through the conjugation of carboxyl-modified DNA20 onto the aminated magnetic mesoporous silica nanoparticles has been reported [27]. The hyperthermic effect of magnetic nanocapsules prepared by sequential layer-by-layer adsorption of polyelectrolytes and magnetite nanoparticles has been demonstrated by applying an AMF at frequencies up to 430 kHz and maximum fields strengths up to 25 mT [28]. Submicrometer and magnetic polymer nanocomposite capsules, combined with a biodegradable polymer, poly-ε-caprolactone, and oleic acid-functionalized magnetite nanoparticles, were heated by AMF at frequencies in the range of 337–869 kHz and a magnetic field strength of 19.9 mT [29]. As a successor to the DM100 series, the company nB released the D5 series in 2017 (Fig. 4a), which improved the control technology offering an unlimited number of frequency modes and power levels, thus, enabling greater flexibility and scalability. The maximum frequency can now be selected from 450 to 900 kHz depending on the choice of CoilSets, and the maximum field amplitude can be output from 60 to 120 mT by selection of drivers, which exceeds the range of the DM100 series [30]. In the near future, it is expected that a considerable amount of research on magnetic nanoparticles based on the D5 series will be reported.

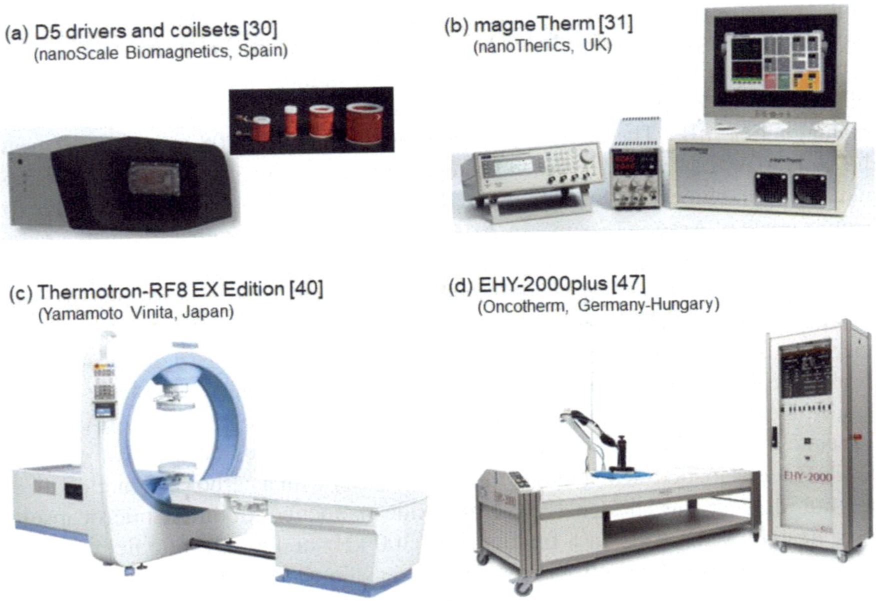

Fig. 4 Commercial hyperthermic apparatus for laboratory and clinical use with nanoparticles (Reprinted with permission from [30, 31, 40, 47])

Another hyperthermia apparatus used in laboratories, is the magneTherm (Fig. 4b), manufactured and sold by nanoTherics [31]. The company nanoTherics was established in the UK in 2007 based on technologies resulting from over 30 years of research into the use of magnetic nanoparticles for biomedical applications, at Keele University and the University of Florida. The magneTherm series specialize in research on hyperthermic therapy using magnetic nanoparticles and research on novel magnetic nanoparticles. Because of its compact size, it is possible to change the direction of the solenoid coil connected to the main unit from vertical to horizontal. The frequency of the AMF that can be applied ranges from 100 to 1000 kHz, and the maximum magnetic field strength reaches 25 mT, although this varies depending on the frequency. It has been demonstrated that internalization of both CMDx-coated and EGF-conjugated nanoparticle heaters of 47.0 mT at 233 kHz conjugated to epidermal growth factor (EGF) on MCF-7 and MDA-MB-468 cells. This system was used to target the epidermal growth factor receptor (EGFR) and resulted in a significant (up to 99.9%) reduction in cell viability and clonogenic survival of tumor cells depending on the heat dose [32]. The therapeutic effects of magnetic fluid hyperthermia in immortalized T lymphocyte (Jurkat) cells were investigated with monodisperse 16-nm-diameter magnetite nanoparticles at a frequency of 373 kHz and a magnetic field strength of 17.6 mT [33]. The AMF (334.5 kHz, 9–17 mT) triggered a distance-dependent release of the drug as demonstrated in a cytotoxicity assay of KB cancer cells with iron oxide nanoparticles [34].

Successful conjugation of a latent TGF-β (prodrug) to the surface of poly(ethylene glycol)-coated iron oxide nanoparticles and subsequent triggered release of an active TGF-β upon application of an external radio frequency magnetic field (350 kHz, 22.6 mT) have been demonstrated [35]. Micro/nanoparticles containing ciprofloxacin (CIP) and magnetic nanoparticles encapsulated in poly (lactic-co-glycolic acid) PLGA were prepared. Significantly, magnetic nanoparticles containing systems all showed the ability to trigger drug release when exposed to an external AMF (355 kHz, 0.7 mT) [36]. An iron oxide nanoparticle/wax-coated composite capsule protected the capsule contents from the highly variable chemical conditions of the GI tract. The coating completely melted within 2 min of hyperthermia treatment (521.3 kHz, 24 mT) under biologically relevant conditions of temperature, pH, buffer and external field strength, allowing the delivery and dispersal of the capsule contents [37]. A magnetic field (110.6 kHz, 20.8 mT) triggered the drug release, and different poly(methacrylic acid)-g-poly(ethyleneglycol methacrylate) polymers were investigated as in situ coating agents for magnetite nanocrystallites [38]. Chitosan microbeads embedded with magnetic nanoparticles were loaded with the antibiotic vancomycin and stimulated by a 109.9-kHz frequency AMF of 25 mT [39].

7 Clinical Potential

Soft heating by magnetic nanoparticles still remains at the stage of basic research and clinical studies and practice are yet to be realized. Hyperthermia apparatuses are RF dielectric heating devices, which are currently most widely used as heating therapy for cancer patients. These devices were developed with an aim of selectively killing (inducing necrosis) cancer cells with a heat of 42–43 °C, because human cells rapidly die at temperatures above 42.5 °C. However, it is technically very difficult to selectively heat only cancer cells located deep inside the human body, such as internal organs, by sending heat energy from outside the human body. Therefore, the accumulation of magnetic nanoparticles in a diseased tissue through nanotargeting technologies are considered to be an effective way of achieving local heating [40].

Yamamoto Vinita Co., Ltd. (Japan) was founded in 1953, and currently manufactures and sells high-frequency dielectric heating applied equipment. On the basis of collaborative research on hyperthermia equipment for cancer hyperthermia with Kyoto University led to the development of the Thermotron-RF 8. The Thermotron-RF 8 EX Edition (Fig. 4c) features a capacitive method that sets a pair of electrodes on the front and back surface of the human body capable of applying an 8-MHz frequency electric field with a maximum output power of 1500 W. As a result, a high-frequency current flows inside the human body, and the temperature of the affected part rises to 42 °C or higher owing to Joule heating [41]. Generally, the pH of cancerous tissue is lower than that of normal tissues, and cells at low pH are more sensitive to heat. Cancer tissue cannot increase blood flow in response to heating and blood flow may actually decrease. Owing to the lack of a cooling effect from blood flow, cancer cells are heated to temperatures of 42 °C or higher in a relatively short time. Several cases have been reported, verifying the hyperthermic effect of magnetic nanoparticles using the Thermotron-RF8. Blood vessels were placed on magnetite nanoparticles-containing agar piece and treated with a Thermotron-RF for 30 min. As a result, no serious changes were noted in the vascular endothelial cells or the vascular wall elastic fibers [42]. $CoFe_2O_4$ 10-nm nanoparticles were incorporated into human breast cancer MCF-7 cells. When the cells containing nanoparticles were subjected to an AMF, MCF-7 cell death was induced by heat generated from the $CoFe_2O_4$ nanoparticles [43]. Safety of regional 8-MHz RF capacitive hyperthermia combined with magnetite nanoparticles coated with positively charged liposomes in patients with castration-resistant prostate cancer has been demonstrated [44].

Local heating by the oscillating magnetic field is important for enhancing the positional accuracy for cancer therapy, and minimizing the region over which AMF is applied is important. Deep hyperthermia induced by fractal-modulated 10-MHz RF waves in cancer cells is called oncothermia [45]. Unlike normal cells, cancer cells form disordered and independent tissues, so an oncothermia apparatus is based on one of the characteristics of cancer cells, that is, absorbing fractal-modulated radio waves at a high rate. This phenomenon allows for selection of cellular tissues on a micro scale. Furthermore, as cancer cells ingest adenosine triphosphate (ATP) at a low rate, they try to incorporate ATP by activating metabolism. Hence, the conductivity around cancer cells is typically higher and cells creating a region of low potential around cancer cells. Owing to the increase in dielectric caused by changes of the outer matrix structures, the heat resistance of cancer cells is lowered. Thus, an environment where the heat generated by high-frequency waves is intensively transferred to the tumor tissue is achieved. This technique is patented by Oncotherm and is sold as Oncotherm EHY-2000 (Fig. 4d). Oncotherm (Hungary) was established in 1988, and received investment from a German company in 2002 and was then reorganized as a German-Hungarian company. Approximately 400 Oncotherm EHY-2000 units are currently in operation, mainly in Hungary and Germany, which have contributed to more than 200,000 treatments [46]. The output carrier frequency of Oncotherm EHY-2000 plus is 13.56 MHz and the maximum output power is 150 W [47]. An advantage is that the output power is 1/8–1/10 lower than that of conventional hyperthermic apparatus, which lowers the risk of burns. However, attempts to combine this device and magnetic nanoparticles have not yet been reported..

8 Challenges Related to MNPs and AMF

To confirm whether a sufficient amount of nanomaterial for treatment is accumulated in cancer cells, imaging techniques based on electromagnetic waves are necessary. Ideally, it is desirable to control the behavior of nanomaterial distribution while monitoring the effects of drug administration, and to perform diagnosis and therapy together. However, clinical inspection equipment is very large and there is usually insufficient space to add new functionalities. To protect tissues other than the target site from unnecessary heating, the application area of AMF should be as small as possible. Therefore, miniaturization or portable AMF generating devices are required that do not physically and electromagnetically interfere with existing inspection equipment, to enable local heating of the tumor site. This will likely become a major area for development in future hyperthermic devices. Furthermore, considering the influence of magnetic fields on the human body, stronger magnetic fields are difficult to use over large areas owing to potential side effects, despite the highly beneficial therapeutic effects. Therefore, the development of magnetic nanoparticles, that can efficiently generate heat, even under a relatively weak magnetic field, is required, together with new device developments.

Focusing AMF field by combining with another coil has been investigated. Murase et al. [48] used an external static magnetic field (SMF) with a field-free point generated by two solenoid coils on top of the AMF. This approach is similar to the one proposed by Tsai et al. [49]. This will decrease the risk of damaging surrounding tissues and could make the MNP/AMF approach more desirable for cancer therapy.

Acknowledgments This work was supported by a grant from JSPS KAKENHI, grant number JP15K21764. This work was partially supported by the Precursory Research for Embryonic Science and Technology of the Japan Science and Technology Agency, JST-Mirai Program No. JPMJMI17D7, and Grand-in Aid for Science Research (No. 25286041, 23656013) from the Japan Society for the Promotion of Science (JSPS) by Y.I.

References

1. Maruya A, Singh AK, Mishra G, Kumari K, Rai A, Sharma B, Kulkarni GT, Awasthi R (2019) Strategic use of nanotechnology in drug targeting and its consequences on human health: a focused review. Interv Med Appl Sci 11:38–54
2. Perez-Herrero E, Fernandez-Medarde A (2015) Advanced targeted therapies in cancer: drug nanocarriers, the future of chemotherapy. Eur J Pharm Biopharm 93:52–790
3. Maeda H, Wu J, Sawa T, Matsumura Y, Hori K (2000) Tumor vascular permeability and the EPR effect in macromolecular therapeutics: a review. J Control Release 65:271–284
4. Mekaru H, Lu J, Tamanoi F (2015) Development of mesoporous silica-based nanoparticles with controlled release capability for cancer therapy. Adv Drug Deliv Rev 95:40–49
5. Ichiyanagi Y, Moritake S, Taira S, Setou M (2007) Functional magnetic nanoparticles for medical application. J Magn Magn Mater 310:2877–2879
6. Ichiyanagi Y, Shigeoka D, Hiroki T, Mashino T, Kimura S, Tomitaka A, Ueda K, Takemura Y (2012) Study on increase in temperature of Co-Ti ferrite nanoparticles for magnetic hyperthermia treatment. Thermochim Acta 532:123–126
7. Kondo T, Mori K, Hachisu M, Yamazaki T, Okamoto D, Watanabe M, Gonda K, Tada H, Hamada Y, Takano M, Ohuchi N, Ichiyanagi Y (2015) Alternating current magnetic susceptibility and heat dissipation by $Mn_{1-x}Zn_xFe_2O_4$ nanoparticles for hyperthermia treatment. J Appl Phys 117:17D157

8. Kamzin AS, Ichiyanagi Y (2016) Study of ferrite $Co_{1+x} Ti_x Fe_{2-2x} O_4$ ($0.2 < x < 0.5$) nanoparticles for magnetic hyperthermia. Phys Solid State 58:2101–2108
9. Jordan A, Scholz R, Wust P, Fähling H, Felix R (1999) Magnetic field hyperthermia: cancer treatment with AC magnetic field induced excitation of biocompatible superparamagnetic nanoparticles. J Magn Magn Mater 201:413–419
10. Suto M, Hirota Y, Mamiya H, Fujita A, Kasuya R, Tohji K, Jeyadevan B (2009) Hear dissipation mechanism of magnetic nanoparticles in magnetic field hyperthermia. J Magn Magn Mater 321:1493–1496
11. Li Z, Kawashita M, Araki N, Mitsumori M, Hiraoka M, Doi M (2010) Magnetic SiO_2 gel microspheres for arterial embolization hyperthermia. Biomed Mater 5:065010
12. Laurent S, Dutz S, H€afeli, U.O. and Mahmoudi, M. (2011) Magnetic fluid hyperthermia: focus on superparamagnetic ion oxide nanoparticles. Adv Colloid Interface Sci 166:8–23
13. Lin TC, Lin FH, Lin JC (2012) In vitro feasibility study of the use of a magnetic electrospun chitosan nanofiber composite for hyperthermia treatment of tumor cells. Acta Biomater 8:2704–2711
14. Ichiyanagi Y, Uehashi T, Yamada S (2004) Magnetic properties of Ni-Zn ferrite nanoparticles. Phys Status Solidi C 1:3485
15. Rosensweig RE (2002) Heating magnetic fluid with alternating magnetic field. J Magn Magn Mater 252:370–374
16. Thomas CR, Ferris DP, Lee J-H, Choi E, Cho MH, Kim ES, Stoddart JF, Shin J-S, Cheon J, Zink JI (2010) Noninvasive remote-controlled release of drug molecules in vitro using magnetic actuation of mechanized nanoparticles. J Am Chem Soc 132:10623–10625
17. Dong J, Zink JI (2014) Taking the temperature of the interiors of magnetically heated nanoparticles. ACS Nano 8:5199–5207
18. Moritake S, Taira S, Ichiyanagi Y, Morone N, Song S-Y, Hatanaka T, Yuasa S, Setou M (2007) Functionalized nano-magnetic particles for an in vivo delivery system. J Nanosci Nanotechnol 7:937–944
19. Ito A, Tanaka K, Kondo K et al (2003) Tumor regression by combined immunotherapy and hyperthermia using magnetic nanoparticles in an experimental subcutaneous murine melanoma. Cancer Sci 94:308–313
20. Kikumori T, Kobayashi T, Sawaki M et al (2009) Anti-cancer effect of hyperthermia on breast cancer by magnetite nanoparticle-loaded anti-HER2 immunoliposomes. Breast Cancer Res Treat 113:435–441
21. Ohki A, Tanoue M, Kobayashi S et al (2017) Magnetic particle imaging for quantitative evaluation of tumor response to magnetic hyperthermia treatment combined with chemotherapy using cisplatin. Thermal Med 33:39–51
22. Marcos-Campos I, Asín L, Torres TE et al (2011) Cell death induced by the application of alternating magnetic fields to nanoparticle-loaded dendritic cells. Nanotechnology 22:205101
23. Asín L, Ibarra MR, Tres A et al (2012) Controlled cell death by magnetic hyperthermia: effects of exposure time, field amplitude, and nanoparticle concentration. Pharm Res 29:1319–1327
24. Carregal-Romero S, Guardia P, Yu X et al (2015) Magnetically triggered release of molecular cargo from iron oxide nanoparticle loaded microcapsules. Nanoscale 7:570–576
25. Guisasola E, Baeza A, Talelli M et al (2015) Magnetic-responsive release controlled by hot spot effect. Langmuir 31:12777–12782
26. Seemann KM, Luysberg M, Révay Z et al (2015) Magnetic heating properties and neutron activation of tungsten-oxide coated biocompatible FePt core-shell nanoparticles. J Control Release 197:131–137
27. Xu Y, Zhu Y, Kaskel S (2015) A smart magnetic nanosystem with controllable drug release and hyperthermia for potential cancer therapy. RSC Adv 5:99875–99883
28. Cristofolini L, Szczepanowicz K, Orsi D et al (2016) Hybrid polyelectrolyte/Fe3O4 nanocapsules for hyperthermia applications. ACS Appl Mater Interfaces 8:25043–25050

29. Grillo R, Gallo J, Stroppa DG et al (2016) Sub-micrometer magnetic nanocomposites: insights into the effect of magnetic nanoparticles interactions on the optimization of SAR and MRI performance. ACS Appl Mater Interfaces 8:25777–25787
30. Nanoscale Biomagnetic (2020) Products D5 series. https://www.nbnanoscale.com/d5-series-equipment-magnetic-hyperthermia/. Accessed 20 Nov 2020
31. nanoTherics (2020) magneTherm Range. http://www.nanotherics.com/magnetherm-products/. Accessed 20 Nov 2020
32. Creixell M, Bohórquez AC, Torres-Lugo M et al (2011) EGFR-targetedmagneticnanoparticle heaters kill cancer cells without a perceptible temperature rise. ACS Nano 5:7124–7129
33. Khandhar AP, Ferguson RM, Simon JA (2012) Enhancing cancer therapeutics using size-optimized magnetic fluid hyperthermia. J Appl Phys 111:07B306
34. Riedinger A, Guardia P, Curcio A et al (2013) Subnanometer local temperature probing and remotely controlled drug release based on Azo-functionalized Iron oxide nanoparticles. Nano Lett 13:2399–2406
35. Monsalve A, Bohórquez C, Rinaldi C et al (2015) Remotely triggered activation of TGF-β with magnetic nanoparticles. IEEE Magn Lett 6:1700204
36. Hua X, Tan S, Bandara HMHN et al (2014) Externally controlled triggered-release of drug from PLGA micro and nanoparticles. PLoS One 9:e114271
37. Rose LC, Bear JC, Southern P et al (2016) On-demand, magnetic hyperthermia-triggered drug delivery: optimisation for the GI tract. J Mater Chem B 4:1704–1711
38. Voulgari E, Bakandritsos A, Galtsidis S et al (2016) Synthesis, characterization and *in vivo* evaluation of a magnetic cisplatin delivery nanosystem based on PMMA-graft-PEG copolymers. J Control Release 243:342–356
39. Mohapatra A, Harris MA, LeVine D et al (2017) Magnetic stimulus responsive vancomycin drug delivery system based on chitosan microbeads embedded with magnetic nanoparticles. J Biomed Mater Res B Appl Biomater 106:2169–2176
40. Vinita Y (2020) Medical system, to health professionals. http://www.vinita.co.jp/en/medical_div/medical/product/. Accessed 20 Nov 2020
41. Szigeti G, Lee DY, Hegyi G (2017) What is on the horizon for hyperthermic cancer therapy? J Tradit Med Clin Nat 6:10000217
42. Kawai N, Kobayashi D, Yasui T et al (2014) Evaluation of side effects of radiofrequency capacitive hyperthermia with magnetite on the blood vessel walls of tumor metastatic lesion surrounding the abdominal large vessels: an agar phantom study. Vasc Cell 6:15
43. Matuda S, Nakanishi T, Kaneko K et al (2015) Synthesis of cobalt ferrite nanoparticles using spermine and their effect on death in human breast cancer cells under an alternating magnetic field. Electrochem Acta 183:153–159
44. Kawai N, Kobayashi T, Kobayashi D et al (2016) Safety of regional 8-MHz radiofrequency capacitive hyperthermia combined with magnetic cationic liposomes in patients with castration-resistant prostate cancer: a phase I clinical study. J Int Transl Med 4:258–267
45. Szasz P, Szasz MA, Minnaar C et al (2017) Heating preciosity—trends in modern oncological hyperthermia. Open J Biophys 7:116–144
46. Baronzio G, Parmar G, Ballerini M et al (2014) A brief overview of hyperthermia in cancer treatment. J Integr Oncol 3:115
47. Oncothrm (2020) EHY-2000plus, The Oncothermia method & devices. https://www.medica.de/vis-content/event-medcom2018.MEDICA/exh-medcom2018.2586286/MEDICA-2018-OncoTherm-GmbH-Paper-medcom2018.2586286-MS8nDjPjTw2lq8lJVeZ6fQ.pdf. Accessed 20 Nov 2020
48. Murase K, Takata H, Takeuchi Y, Saito S (2013) Control of the temperature rise in magnetic hyperthermia with use of an external static magnetic field. Phys Med 29:624–630
49. Tasci, TO, Vargel I, Arat A, Guzel E, Korkusuz P, Atalar E (2009) Focused RF hyperthermia using magnetic fluids. Med Phys 36:1906–1912

Light-Control of Cell Membrane Potential and Its Environment

Yuta Takano

1 Targets for Light Control of Cell Membrane

Cell membrane is a compartment that separates two different environments. Main components of the cell membrane are lipid molecules and proteins. There is an ion gradient between the inside and outside of cell membranes in living cells. This gradient is led by the impermeable nature of ions to the lipid bilayer of the cell membrane, and active or passive ion transportation by membrane proteins, which are called ion channels [1]. Ion channels are proteins which locate in the cell membrane, and possess a pore allowing for selective flux of ions, such as K^+, Ca^{2+}, and Na^+. Because ions are charged, uneven distribution of the ions between the inside and outside of the membrane induces local electric field. This is called cell membrane potential.

To control cell membrane potential, straightforward approaches are changing the nature of lipid molecules consisting of the lipid bilayer, and modulation of the ion transportation governed by ion channels. The present molecules (i.e., lipids and proteins) vary from the location of cell membranes [2]. Therefore, different approaches may be taken for controlling cell membrane by light in a different place in the cell.

Y. Takano (✉)
Laboratory of Molecular Photonics, Division of Material and Molecular Sciences, Research Institute for Electronic Science, Hokkaido University, Sapporo, Hokkaido, Japan
e-mail: tak@es.hokudai.ac.jp

D. O. Wang, D. Packwood (eds.), *Cell-Inspired Materials and Engineering*, Fundamental Biomedical Technologies, https://doi.org/10.1007/978-3-030-55924-3_8

1.1 Outer Cell Membrane

Outer cell membrane is a compartment separating the inside of the cell from the outside environment. It is the most important part which constructs the structure of cells. Proteins in or on the membrane are known to modulate various cellular activities such as ion transportation [3], and molecular transportation [4], resulting in cell signaling [5]. The membrane proteins, ion channels, which can control ion transportation play fundamental roles to maintain and change the cell membrane potentials and its surrounding environment [1] (Fig. 1). For instance, neural activities are precisely controlled by the transition between a static state and an excited state of cell membrane potentials. Generally, the static cell membrane potential in typical nerve cells are around −70 mV, which is a potential induced between the inside and outside of the membrane [6]. In an excited state of the cell, it may reach to +40 mV instantaneously. This event is finished within several hundred milliseconds and this pulsed-signal, which is called action potential or neuron firing, triggers various cellular activities [7].

The static membrane potential is maintained by the total balance of the ions surrounding the membrane. This balance is precisely controlled by the activities of the ion channels [1]. Change in the activities caused by a certain stimulus can lead to a change in membrane potentials. Representative cation channels, as a target for light control of cell membrane potentials, are listed below;

- Potassium (K^+) channels.
- Sodium (Na^+) channels.
- Calcium (Ca^{2+}) channels.

Not limited to these ion channels, ion channels modulate inward and outward flux of various ions, inducing cell membrane potentials.

To date, a vast number of families and subfamilies have been reported as these types of ion channels [8]. For simplicity, however, this chapter does not go further about its categorization. An important point here is such ion channels govern the ion transportations thorough the cell membrane, and thus permit fine control and tuning of cellular activities which relate to the cell membrane potentials [3–5]. Therefore, targeting and controlling the activities and functionalities of ion channels can be promising approaches to control cell membrane potential [9].

Lipid molecules and structure of the lipid bilayer can also be a target for controlling cell membrane potentials. Lipid bilayer is normally quite flexible at physiological

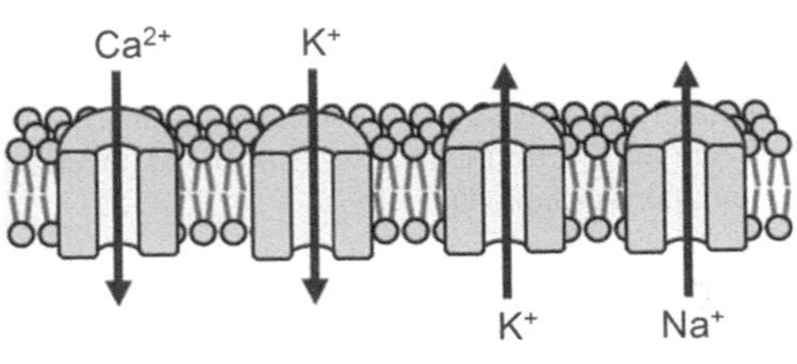

Fig. 1 Cell membrane consisting of lipid bilayers and representative ion channels

temperature and it is not permeable to charged hydrophilic molecules and ions [10]. However, when a lipid in the bilayer is degraded or altered, the permeability of ions would change, resulting in the change of membrane potentials [11]. It is caused by the change in polarity of lipid molecule and/or change in the intermolecular interaction of the lipid molecules constructing the lipid bilayer.

In summary, the following events on lipids can be an approach to control cell membrane potentials by light.

- Local alteration or degradation of lipid molecules.

Although we should not ignore the possible interplay between the proteins and the lipid molecules [12], so far, most reports are based on the consideration of their effects independently. The examples to be shown in this chapter are mostly such studies.

Mitochondrial Membrane

Organelles also possess cell membranes to separate their components from the outer environment, cytosol. Among the organelles, mitochondrion is an indispensable cellular organelle that is responsible for the production of the majority of the cell's energy supply and for the induction of cell death, such as apoptosis or necrosis [13]. A characteristic of mitochondrial membrane is the existence of two membranes, an outer membrane and an inner membrane. Particularly, the inner membrane is of importance for unique functionalities of mitochondria. A main feature of the inner membrane is construction and modulation of the important part of the respiratory chain to generate energy for the body [14]. A series of mitochondrial membrane proteins involving an adenosine triphosphate (ATP) synthase is called supercomplexes. In it, redox reactions promote the proton gradient between the inside and outside of the inner mitochondrial membrane. Consequently, inside of the inner membrane is much negatively charged and typically its membrane potential is about −180 mV relative to the outside of the outer cell membrane. This homogeneous distribution of protons drives an ATP synthase, which synthesizes ATP from ADP.

In this regard, mitochondrial membrane potential controls the respiratory chain [15], which is essential for living things. When it increases positively, or it is depolarized in other words, the activity is suppressed, and then the rate of ATP synthesis would descend. With the opposite change in the membrane potential, the ATP synthesis can be accelerated.

Large depolarization of the mitochondrial membrane leads to severe effects on mitochondria. It may cause fatal damage to cells because of the irregular generation of reactive oxygen species, which leaks from the respiratory chain, resulting in cell death. To remove such abnormal mitochondria, mitochondria are able to lead to apoptotic cell death [16]. When the mitochondrial membrane potential is largely depolarized, cytochrome *c*, which is membrane binding signaling protein, is released to the cytosol and trigger the cellular response to induce apoptosis [17].

Therefore, artificial control methods of mitochondrial membrane, particularly the inner membrane, potential potentially enable us to control life (i.e., ATP synthesis) and death (i.e., cell death) of cells. To control the membrane potential, the following approaches are taken on the basis of the unique structure of the membrane.

- Activation or deactivation of the supercomplex.
- Decrease or vanish the proton gradient around the membrane (called "uncoupling").

Several studies are known to achieve an artificial control of it by light, and some of them will be mentioned elsewhere in the next part of this chapter.

2 Light-Control Methods for Cell Membrane Potentials and Surrounding Environments

2.1 Photoexcited States of Molecules for Photocontrol

When a molecule absorbs light, it gets excited by the induced light energy. The excited molecule will take a transition path in excited states depending on the nature of the molecule. Fluorescence is one of the most known phenomena resulting in the photoexcitation (Fig. 2). A molecule which was excited by light first takes singlet excited state, and then fluorescence emits when it causes irradiative deactivation.

In certain molecules, which generally possess pi-conjugated system of electrons, such as porphyrins and fullerenes, intersystem crossing leads the singlet excited state to a triplet state [18, 19]. If the energy level of the triplet state is higher than that of molecular oxygen (O_2), intermolecular energy transfer from the molecule in the triplet state to O_2 is feasible in the presence of O_2 [20]. This generates singlet oxygen ($^1O_2$), which is a reactive oxygen species possessing large cytotoxicity [21].

If an appropriate combination of an electron donating molecule (D) and an electron accepting molecule (A) is used in the same system, photoexcitation remarkably promotes an electron transfer reaction from D to A, resulting in one-electron reduced

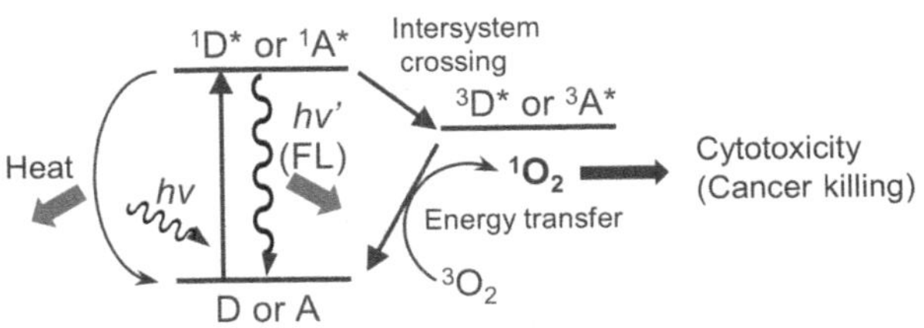

Fig. 2 Possible transient paths of a photoexcited molecule (D or A). The molecules first absorb light (*hν*). The excited molecule ($^1D^*$ or $^1A^*$) can demonstrate fluorescence (*hν′*, FL), can generate heat, or can occur intersystem crossing. In the presence of molecular oxygen ($^3O_2$), the excited molecule in the triplet excited state ($^3D^*$ or $^3A^*$) cause energy transfer to generate singlet oxygen ($^1O_2$). High cytotoxicity of $^1O_2$ is used for cancer therapy to kill cancer cells

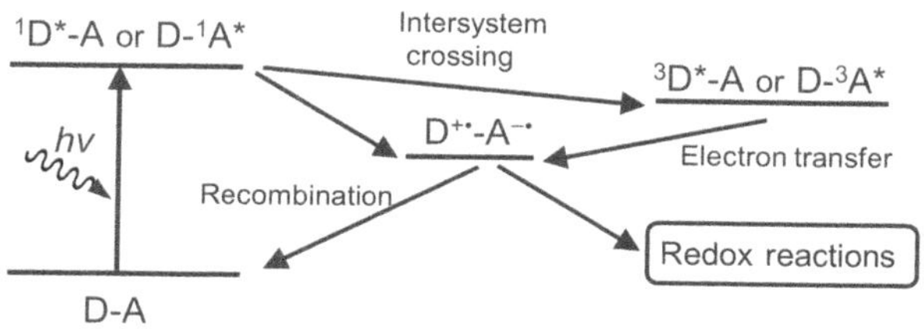

Fig. 3 Possible transient paths of a donor (D)–acceptor (A) paired system

D ($D^{\bullet+}$) and one-electron accepted A ($A^{\bullet-}$) [22] (Fig. 3). These photogenerated species are capable of permitting further intermolecular electron transfer reactions to the other reactive molecules [23]. In other words, $D^{\bullet+}$ and $A^{\bullet-}$ can lead to oxidation reactions or reduction reactions, respectively. In addition, promptly generated $D^{\bullet+}$ and $A^{\bullet-}$ can induce local electric field derived from the gap of the electric potential between positively charged $D^{\bullet+}$ and negatively charged $A^{\bullet-}$ [24]. Such an induced electric field may influence the surrounding environment in biological systems [25].

In the other molecules which have not been mentioned above, photoexcited molecules result in nonradiative deactivation, which release the light energy as a form of heat [26]. Not only molecules, but polymers or plasmonic metal nanoparticles are reported as excellent mediators for heat generation by light irradiation [27].

In summary, the following forms can be promising mediators to use light energy in cellular systems.

- A triplet excited state to permit energy transfer (a phenomenon called "photosensitization").
- Electron transferred donor (D)/acceptor (A) molecules.
- Heat, as a result of nonradiative deactivation.

2.2 *Light Control with Heat Generation*

Functionalities of cell membrane are highly responsive to temperature. One of the reasons is temperature-dependent nature of the fluidity of lipid molecules therein [28]. This may influence the structures and functionalities of the cell membrane [29].

Nontoxic plasmonic particles, such as gold nanoparticles, can be used for generating local heat in cells. Gold nanoparticles absorb light very efficiently at their plasmon resonance frequency [27]. P. Urban and coworkers investigated the effect of the local heating by gold nanoparticles on the permeability of lipid membranes [30] (Fig. 4). After putting gold nanoparticles on the outer membrane of HEK293 cells, they illuminated with 532 nm laser and recorded the change in the permeability of the cell membrane as electric currents by using patch-clamp. As a result, an increase of the ion permeability was verified when the membrane embedded by gold

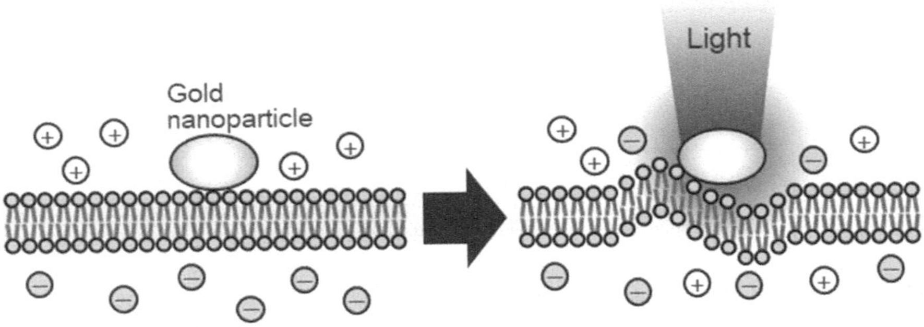

Fig. 4 Change in the permeability of ions in the cell membrane by local heating by the gold nanoparticles triggered by light irradiation [30]

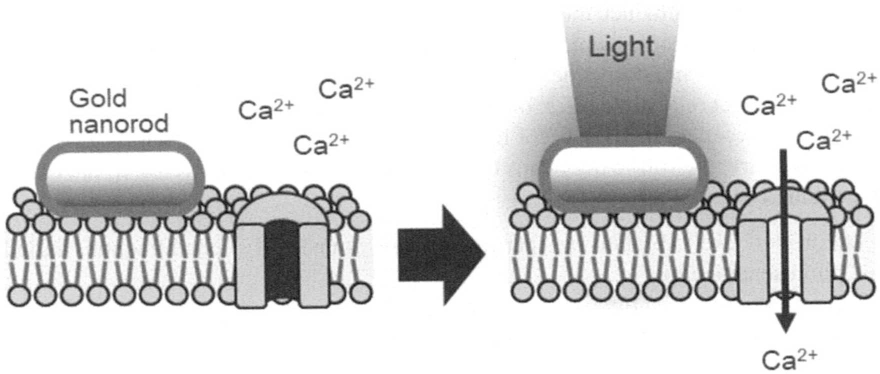

Fig. 5 A gold nanorod opening temperature-dependent ion channel, TRPV1, by illumination of NIR light resulting in local heating [32]

nanoparticles is illuminated. This was explained with the perturbation of cell membrane structure caused by the local heating by the gold nanoparticles.

Ion channels can also be a target for light control with heal generation. There are considerable amount of temperature-dependent ion channels in the cell membrane [31]. Transient receptor potential cation channel subfamily V member 1 (TRPV1) is one of such temperature-dependent proteins. TRPV1 is activated, initiating Ca^{2+} flux, at a temperature greater than 43 °C.

H. Nakatsuji and coworkers reported an artificial light control method of TRPV1 by using gold nanorods with illumination of 780 nm near-infrared (NIR) light [32] (Fig. 5). They successfully delivered the gold nanorods onto the cell membrane of primary cultured DRG neurons from wild type mice, which express TRPV1. Irradiation of the NIR light clearly triggered an influx of Ca^{2+} of the membrane, which leads to depolarization of the membrane. Since TRPV1 is a potential therapeutic target for nociceptive pain and cancer [33], this methodology provides useful information for novel phototherapeutic approaches controlling cell membrane by light followed by heat generation.

2.3 *Light Control by Photosensitization Generating $^1O_2$*

Membrane potentials can be perturbed by the destruction of membrane structure. Such a drastic change can be led by severe damage caused by oxidant [21]. Reactive oxygen species are well-known oxidants that form as a result of photosensitization of a molecule [20, 21]. Membrane photosensitization, which is a photo-oxidation of membrane molecules caused by reactive oxygen species generated as a result of photosensitization, has been shown to decrease the plasma membrane potential [34].

Because $^1O_2$ is highly oxidative, it can oxidize and degrade unsaturated lipid molecules readily [34]. Rose bengal, porphyrins, and phthalocyanines have been well studied as popular photosensitizers. Irene E. Lochevar and coworkers reported a photosensitized effect in the outer cell membrane using rose bengal as a sensitizer [35]. Rose bengal was incorporated into the outer membrane of P388D$_1$ cells, a mouse macrophage monocyte line, and photosensitizing effects were examined with the irradiation of 355 and 532 nm lights independently. Both conditions of irradiation in the presence of oxygen caused significant loss of membrane potentials, whereas photosensitized depolarization did not occur in the absence of oxygen. Photosensitized effect on the membrane permeability was also examined and it was found that the generation of $^1O_2$ causing oxidation of lipids molecules significantly increased the permeability of trypan blue.

Mitochondrial membrane and its potential can also be changed by photosensitization [36]. Porphyrin derivatives bearing triphenylphosphonium ion, which offers targeting ability to mitochondria [37], were demonstrated specific accumulation and induction of photooxidation reactions in mitochondria. The oxidation reactions would cause the destruction of mitochondrial membrane structure, resulting in cell death.

Meanwhile, it was reported by R.S. Ray and coworkers that an indirect approach using singlet oxygen could be useful and applicable to change mitochondrial membrane potential [38]. They used Ketoprofen, which is a widely used nonsteroidal anti-inflammatory drug as a photosensitizer. Under sunlight/UV exposure, significant ROS, mainly $^1O_2$, generation was observed in keratinocyte (HaCaT) cells. $^1O_2$ resulted in single and double-strand DNA breakage. It caused G2/M phase arrest, apoptosis, and depolarization of mitochondrial membrane potential.

Genetically encoded proteins are also used for photosensitization of cell membranes. Using light-activated proteins is an emerging and considerably active research field, which is called optogenetics today. KillerRed is a representative protein that permits photosensitization [39, 40]. An unique feature of this protein is, with using genetic modification, the protein can be expressed in a specific region in a cell. A reported main mechanism for generation of a reactive oxygen by illumination was through electron transfer to generate superoxide [41]. The outer membrane where KillerRed expressed demonstrated light-dependent change in the membrane structure and membrane potentials. When KillerRed was expressed in mitochondria, light stimulation on the protein caused depolarization of mitochondrial membrane and then led to apoptosis [42].

2.4 Light Control by Utilizing Electron Transfer Reactions

Photoinduced electron-transfer reactions between electron donors (D) and acceptors (A) involve the generation of donor radical cations and acceptor radical anions [22]. This is in sharp contrast with photoinduced energy transfer reactions where photoinduced energy transfer from energy donors to energy acceptors results in the formation of the excited-state of the energy acceptors. When, in the most general cases, the energy level of the acceptor triplet excited-state is typically higher than that of $^1O_2$ [20], photoinduced energy transfer reactions could be associated with cell damage [21, 34]. Using photoinduced electron-transfer reaction is beneficial for suppressing the effects of singlet oxygen during illumination because the triplet excited state lifetime is drastically shortened because of the transition to the energetically lower-lying electron transferred state.

The early study was performed by T.A. Moore and coworkers in 1997 [43]. They focused on the proton transfer in the respiratory chain and aimed to mimic this system by artificial molecules. First, they demonstrated successful light-driven proton transfer mediated by artificially synthesized carotenoid polyene (C)-porphyrin (P)-naphthoquinone (Q) linked molecule (Fig. 6). The C-P-Q triad and diphenyl benzoquinone (Qs) were incorporated in liposomes and efficient proton transfer was led by the coupling of protons with electrically reduced Qs ($Qs^{\bullet -}$) resulted by the intermolecular electron transfer from photoexcited C-P-Q which induced intramolecular charge separation ($C^{\bullet +}$-P-$Q^{\bullet -}$). Even though the quantum yield of the proton transfer is not high ($\Phi \sim 0.004$), this report provided a brilliant strategy for photocontrol of proton gradient around the cell membrane and thus control of membrane potential.

Using this light-driven proton pump system, they achieved the promotion of the conversion of ADP to ATP using CF_0F_1-ATP synthase [44] (Fig. 7). In proteoliposomes in which the C-P-Q triad, diphenyl benzoquinone (Qs) and CF_0F_1-ATP synthases were incorporated, laser light irradiation at a wavelength of 633 nm resulted in a light-mediated generation of ATP. This study demonstrated a possibility of controlling energy production in mammalian cells by light, which may be expressed as "artificial photosynthesis."

Charge separated state is a unique photoinduced excited state of D-A molecules as a result of an intramolecular electron transfer between the D and A molecules. To extract the full potential of this unique state and to minimize the side-effect caused by $^1O_2$, it is essential to produce a long-lived, charge-separated state efficiently. With

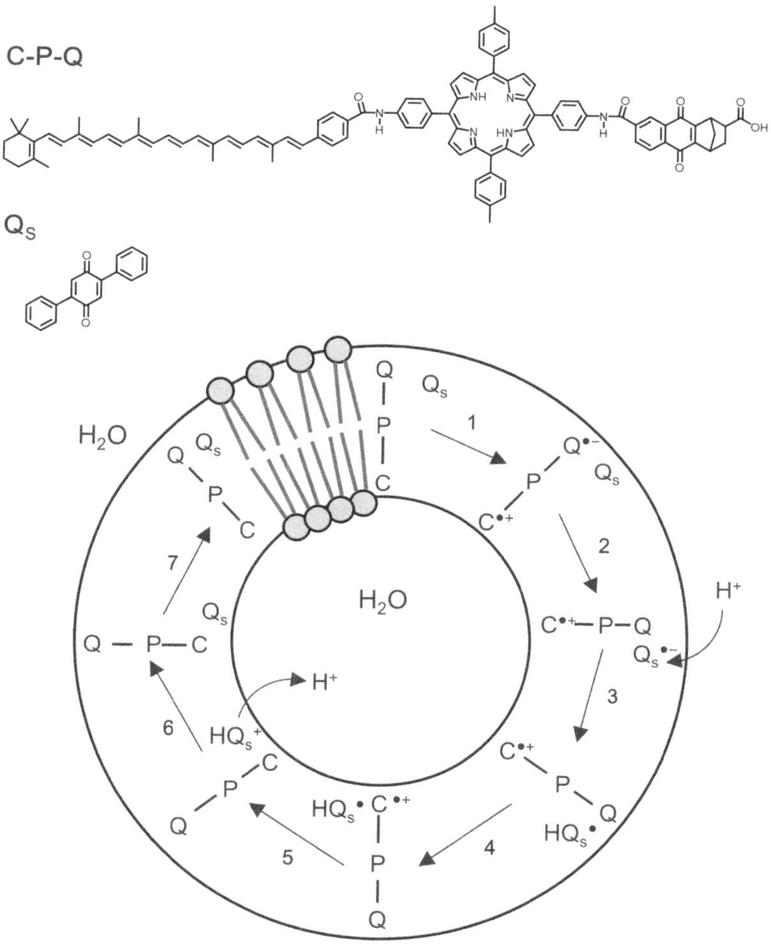

Fig. 6 (Upper) Molecular structures of the artificial light-driven proton pump consisting of an carotenoid polyene (C)–porphyrin (P)–naphthoquinone (Q) linked molecule and a quinone derivative (Q_S). (Bottom) Schematic drawing of the light driven proton pumping reported by T.A. Moore and coworkers [43]

using such efficient photoinduced charge separation, we can utilize unique properties of the photoinduced excited state.

As a pioneering work, in 2001, H. Imahori and colleagues reported an artificial molecular system based on fullerene (C_{60})–porphyrin (P)–ferrocene (Fc) triad that achieved high quantum yield up to 99% and a long-lived state of 7.7 ms of the charge separated state [45] (Fig. 8). In the molecule, C_{60} has excellent electron acceptability, and by combining this with a porphyrin skeleton excellent in light-capturing ability and a ferrocene moiety which has an excellent in electron donating ability, such high efficiency of charge separation was accomplished. It was a

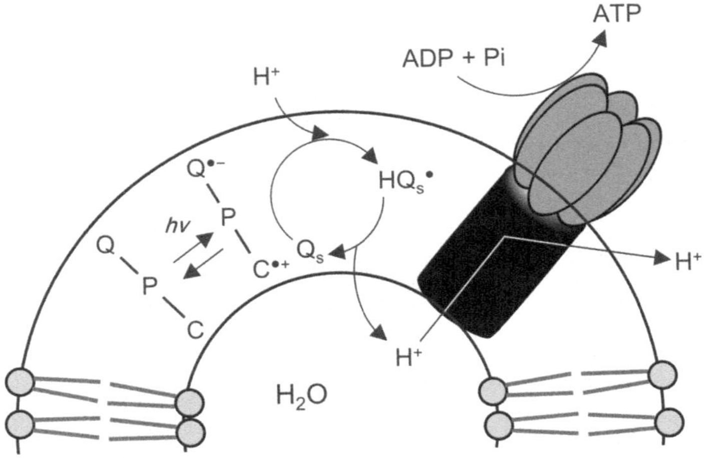

Fig. 7 Schematic drawing of an artificial system for promoting an activity of an ATP synthase (CF_0F_1-ATP synthase), by the light-driven proton pump [44]

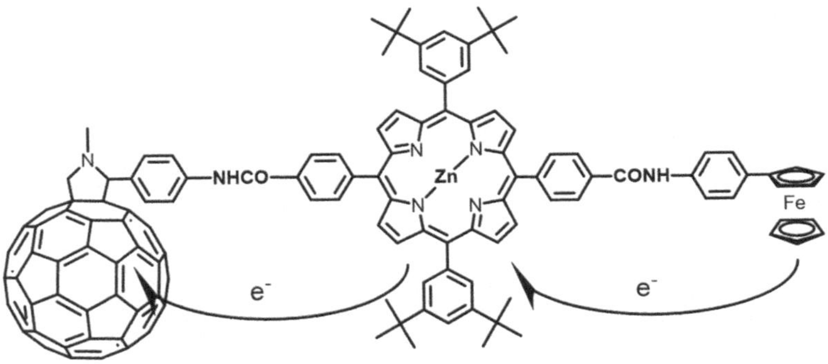

Fig. 8 Molecular structure of a fullerene (C_{60})–porphyrin (P)–ferrocene (Fc) triad that achieved excellent properties of photoinduced intramolecular electron transfer [45]

remarkable report that succeeded for the first time to realize an excellent artificial system to utilized photoinduced charge separation.

Recently, they reported such amphiphilic C_{60}-P-Fc triads for applying and developing cell functions by light control. In the charge separation state, since positive and negative charges are localized in the molecule, it is possible to induce a local electric field [24], and the photoexcited state has molecular polarity different from the ground state. On the basis of this hypothesis, it was anticipated that the properties unique to these charge separation states influence the cell membrane environment and can change the cell membrane potential [46] (Fig. 9). Changes in the cell membrane potential were recorded by patch-clamp method after introducing C_{60}-P-Fc triad imparted with hydrophilicity by the introduction of a quaternary ammonium moiety into the cell membrane of PC 12 cells, derived from a pheochromocytoma

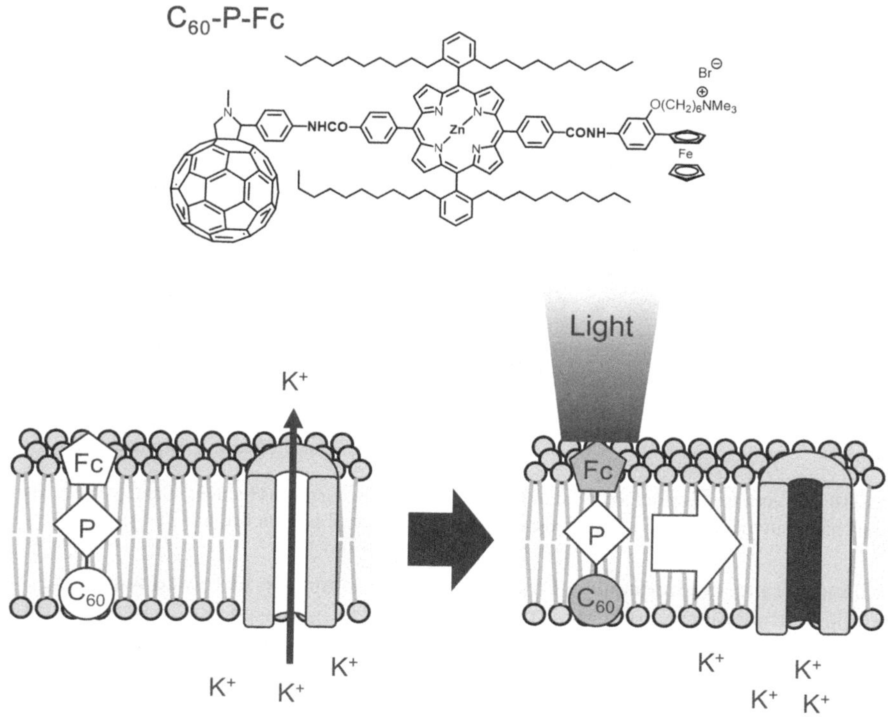

Fig. 9 (Upper) Molecular structure of an amphiphilic C_{60}-P-Fc triad. (Lower) Schematic drawing of the photocontrol of ion transportation by the C_{60}-P-Fc triad. Under illumination, the C_{60}-P-Fc triad is in the intramolecular charge separated state which influence ion channels

of the rat adrenal medulla. As a result, it was found that cell depolarization is caused by the generation of photoinduced charge separation state in cells, leading to the inhibition of outward K^+ channels.

In this study, fullerene and porphyrin were used as key molecules. Generally, however, these molecules show a high $^1O_2$ generation ability in the case of single molecules, not of D-A molecules [18, 19]. However, in the D-A molecule, the lifetime of the triplet excited state which permits the generation of $^1O_2$ is remarkably shortened due to the transition to the charge separation state occurring with high quantum yield, and as a result, Photocytotoxicity derived from the $^1O_2$ production did not appear. This is a great advantage by utilizing the charge separation state of the D-A molecule.

Because the C_{60}-P-Fc triad system could permit photoinduced change in the cell membrane potential without inducing photocytotoxicity, C_{60}-P-Fc molecules were introduced into primary cultured neurons derived from rats and membrane potential changes were observed while light irradiation [48] (Fig. 10). As a result, a significant change in membrane potential was observed only in the presence of D-A molecule. Furthermore, since this membrane potential change was depolarization of the

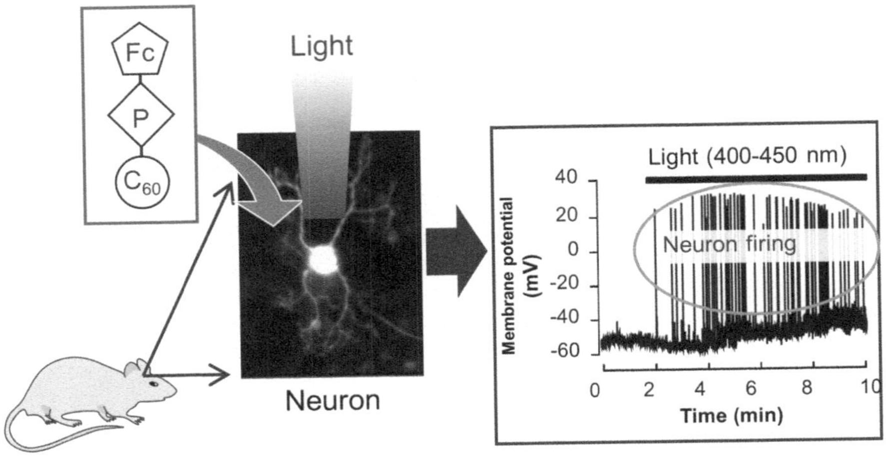

Fig. 10 Photoinduced depolarization of the cell membrane of neurons by using the C_{60}-P-Fc triad. The change in membrane potential modulates the frequency of neuron firing [47]

cell membrane, the frequency of the neuron firing, it may be increased with increasing the cell membrane potential, therewith was also increased. Since neuron firing is an important phenomenon triggering various signaling in nerve cells [7], the possibility of light control of various nerve activities was demonstrated by using this D-A molecule.

Although the D-A molecule is effective for light control of cell membrane potential, the quantum yield of the photoinduced charge separation in the reported system was only 27%. Even though this is a value sufficient to cause cell membrane depolarization by light irradiation, considering that molecules do not emit fluorescence, most of the light energy is lost via nonradiative inactivation. As a factor of the deactivation, it is considered that the intra- and/or intermolecular recombination of the charge separated state may be accelerated by aggregation of the D-A molecules.

In this regard, the authors investigated the effects of the number of quaternary ammonium moieties introduced in order to suppress the association of D-A molecules and improved the dispersibility in the cell membrane [48]. Molecules in which one to four quaternary ammonium moieties were introduced to the DA molecule were synthesized and studied. As a result, the molecule into which four quaternary ammonium were introduced, demonstrate homogeneous dispersion in the cell membrane and. Furthermore, when this molecule was introduced into the cell membrane and irradiated with light, large depolarization was observed, being consistent with the homogeneous dispersion which suppress undesired aggregation. The molecule succeeded in recording the champion record (86%) as the quantum yield of photoinduced charge separation reported in synthetic molecules in cells and biomimetic systems.

Robert H. Chow and coworkers took a unique approach utilizing change in the charge capacitively stored on the cell membrane for light-triggered changes in

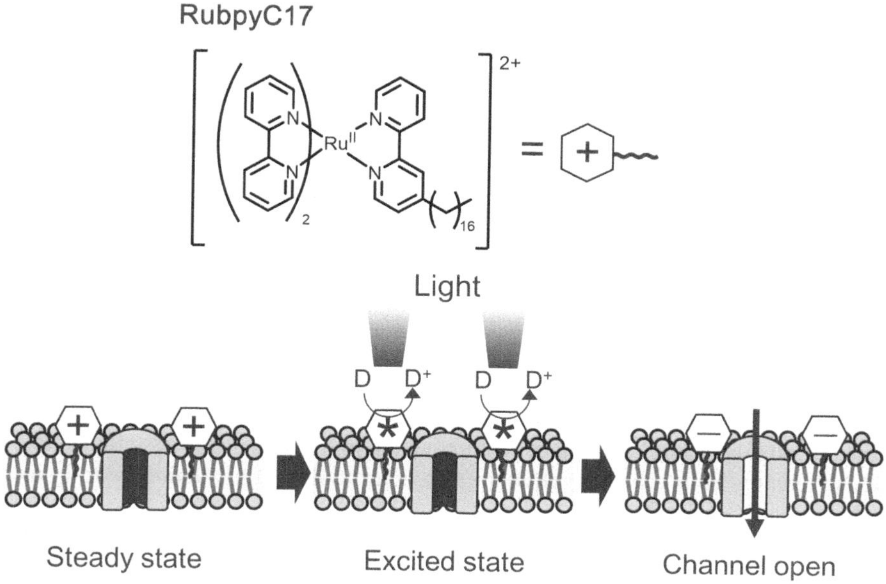

Fig. 11 (Upper) Molecular structure of a ruthenium-diimine complex tethered with septadecyl chain (RubpyC17). (Lower) Scheme of the light-induced electron flow mediated by RubpyC17 in the cell membrane. Resultant electronically reduced RubpyC17 changes the surface charge of the cell membrane

membrane potential [49] (Fig. 11). They used a ruthenium-diimine complex tethered with a septadecyl chain (RubpyC17) for anchoring to the outer cell membrane. Incorporation of RubpyC17 into plasma membranes of live cells, such as INS, HEK293T, and chromaffin cells, was first confirmed by fluorescence microscopic images. Cell membrane potentials of the cells were monitored by patch-clamp techniques and light-induced depolarizations were observed in the presence of RubpyC17. Light stimulation in the blue region was performed with a lamp equipped a 470/40 nm bandpass filter. They got rid out of the possibility of (1) light-induced direct interaction between RubpyC17 and ion channels and (2) light-induced pore formation in the plasma membrane, which might cause the change in cell membrane potential with carefully performed electrophysiological investigations. This strategy may allow us to control cell membrane potentials without changing the state of ion channels or lipid molecules.

An intermolecular electron transfer reaction following the photoinduced intramolecular charge separation in a D-A molecule was demonstrated to change the lipid structure in mitochondria. 9-Mesityl-10-methylacridinium and its derivative were used to induced mitochondria specific lipid photooxidation by using the intermolecular electron transfer reaction [50]. This molecular skeleton was reported in 2004 as an excellent D-A molecule permitting intramolecular charge separation with high quantum yield and long lifetime ($\Phi = 0.98$, lifetime >2 h, in acetonitrile)

[51]. Mitochondria-specific localization of the molecules was accomplished by their strong interaction with anionic lipids in mitochondria owing to its cationic and π-conjugated nature as reported for the mitochondria-specific fluorescent reagents [52, 53]. In HeLa cells, the molecules triggered mitochondrial lipid oxidation, which was followed by apoptotic cell death, under illumination within a few seconds. These results showed that the present molecular system is promising to utilize the electron transfer reaction for inducing oxidation reaction with a precise spatio-temporal manner in a cell by light.

3 Conclusions and Perspective

Noninvasive optical control of biological functions is an important ongoing challenge to establish novel approaches for understanding and controlling cellular functions. Seeking for much more efficient molecular tools and methodology to utilize light energy is desirable.

For instance, bringing the quantum yield of intramolecular and/or intermolecular electron transfer reactions close to 100% in the future will pave a way to develop light manipulated molecular probes that have high efficiency but reduced side reactions and side effects to the limit. For a viewpoint of in vivo study and clinical application, the use of NIR light instead of UV and visible light is favorable because NIR light is able to reach deep inside tissues. This feature derived from the transparent nature of most of the biomolecules to NIR light whereas visible and UV light are absorbed. On the other hand, from a viewpoint of the spatial resolution of a light control, light at short wavelength is more advantageous than NIR light. Therefore, the choice of a light source and of a way to use the full potential of it is important to construct efficient and beneficial compounds.

This research field will develop with the advancement of interdisciplinary research areas including chemistry, biology, photonics, and associated research efforts.

References

1. Zheng J, Trudeau MC (eds) (2015) Handbook of ion channels. CRC Press, Boca Raton, FL
2. Mannella CA (2006) Structure and dynamics of the mitochondrial inner membrane cristae. BBA-Mol Cell Res 1763(5–6):542–548. https://doi.org/10.1016/j.bbamcr.2006.04.006
3. Carafoli E (1991) Calcium pump of the plasma membrane. Physiol Rev 71(1):129–153
4. Palacin M, Estevez R, Bertran J, Zorzano A (1998) Molecular biology of mammalian plasma membrane amino acid transporters. Physiol Rev 78(4):969–1054
5. Fisher AB (2009) Redox signaling across cell membranes. Antioxid Redox Signal 11(6):1349–1356. https://doi.org/10.1089/ARS.2008.2378
6. Sherwood NT, Lesser SS, Lo DC (1997) Neurotrophin regulation of ionic currents and cell size depends on cell context. Proc Natl Acad Sci U S A 94(11):5917–5922

7. Fields RD, Eshete F, Stevens B, Itoh K (1997) Action potential-dependent regulation of gene expression: temporal specificity in Ca2+ cAMP-responsive element binding proteins, and mitogen-activated protein kinase signaling. J Neurosci 17(19):7252–7266
8. Bezanilla F (2000) The voltage sensor in voltage-dependent ion channels. Physiol Rev 80(2):555–592
9. Bagal S, Brown AD, Cox PJ, Omoto K, Owen RM, Pryde DC, Sidders B, Skerratt SE, Stevens EB, Storer RI, Swain NA (2013) Ion channels as therapeutic targets: a drug discovery perspective. J Med Chem 56(3):593–624. https://doi.org/10.1021/jm3011433
10. van Meer G, Voelker DR, Feigenson GW (2008) Membrane lipids: where they are and how they behave. Nat Rev Mol Cell Biol 9(2):112–124. https://doi.org/10.1038/nrm2330
11. Antonov VF, Petrov VV, Molnar AA, Predvoditelev DA, Ivanov AS (1980) Appearance of single-ion channels in unmodified lipid bilayer-membranes at the phase-transition temperature. Nature 283(5747):585–586. https://doi.org/10.1038/283585a0
12. Marsh D, Horvath LI (1998) Structure, dynamics and composition of the lipid-protein interface. Perspectives from spin-labelling. Biochim Biophys Acta 1376(3):267–296. https://doi.org/10.1016/S0304-4157(98)00009-4
13. Newmeyer DD, Ferguson-Miller S (2003) Mitochondria: releasing power for life and unleashing the machineries of death. Cell 112(4):481–490. https://doi.org/10.1016/S0092-8674(03)00116-8
14. Chan DC (2006) Mitochondria: dynamic organelles in disease, aging, and development. Cell 125(7):1241–1252. https://doi.org/10.1016/j.cell.2006.06.010
15. Klingenberg M (1980) The ADP-ATP translocation in mitochondria, a membrane potential controlled transport. J Membr Biol 56(2):97–105
16. Green DR, Reed JC (1998) Mitochondria and apoptosis. Science 281(5381):1309–1312. https://doi.org/10.1126/science.281.5381.1309
17. Kluck RM, BossyWetzel E, Green DR, Newmeyer DD (1997) The release of cytochrome c from mitochondria: a primary site for Bcl-2 regulation of apoptosis. Science 275(5303):1132–1136. https://doi.org/10.1126/science.275.5303.1132
18. Silva EFF, Serpa C, Dabrowski JM, Monteiro CJP, Formosinho SJ, Stochel G, Urbanska K, Simoes S, Pereira MM, Arnaut LG (2010) Mechanisms of singlet-oxygen and superoxide-ion generation by porphyrins and bacteriochlorins and their implications in photodynamic therapy. Chem Eur J 16(30):9273–9286. https://doi.org/10.1002/chem.201000111
19. Markovic Z, Trajkovic V (2008) Biomedical potential of the reactive oxygen species generation and quenching by fullerenes (C_{60}). Biomaterials 29(26):3561–3573. https://doi.org/10.1016/j.biomaterials.2008.05.005
20. Schweitzer C, Schmidt R (2003) Physical mechanisms of generation and deactivation of singlet oxygen. Chem Rev 103(5):1685–1757. https://doi.org/10.1021/cr010371d
21. Stark G (2005) Functional consequences of oxidative membrane damage. J Membr Biol 205(1):1–16. https://doi.org/10.1007/s00232-005-0753-8
22. Balzani V, Piotrowiak P, MAJ R, Mattay J, Astruc D, Gray HB, Winkler JR, Fukuzumi S, Mallouk TE, Haas Y, de Silva AP, Gould IR (2001) Electron transfer in chemistry, vol. 1, principles, theories, methods, and techniques. WILEY-VCH Verlag GmbH, Berlin. https://doi.org/10.1002/9783527618248
23. El-Khouly ME, Ito O, Smith PM, D'Souza F (2004) Intermolecular and supramolecular photoinduced electron transfer processes of fullerene-porphyrin/phthalocyanine systems. J Photochem Photobiol C 5(1):79–104. https://doi.org/10.1016/j.jphotochemrev.2004.01.003
24. Gosztola D, Yamada H, Wasielewski MR (1995) Electric-field effects of photogenerated ion-pairs on nearby molecules – a model for the carotenoid band shift in photosynthesis. J Am Chem Soc 117(7):2041–2048. https://doi.org/10.1021/Ja00112a018
25. Polizzi NF, Eibling MJ, Perez-Aguilar JM, Rawson J, Lanci CJ, Fry HC, Beratan DN, Saven JG, Therien MJ (2016) Photoinduced electron transfer elicits a change in the static dielectric constant of a de novo designed protein. J Am Chem Soc 138(7):2130–2133. https://doi.org/10.1021/jacs.5b13180

26. Yuan A, Wu JH, Tang XL, Zhao LL, Xu F, Hu YQ (2013) Application of near-infrared dyes for tumor imaging, photothermal, and photodynamic therapies. J Pharm Sci 102(1):6–28. https://doi.org/10.1002/jps.23356
27. Richardson HH, Carlson MT, Tandler PJ, Hernandez P, Govorov AO (2009) Experimental and theoretical studies of light-to-heat conversion and collective heating effects in metal nanoparticle solutions. Nano Lett 9(3):1139–1146. https://doi.org/10.1021/nl8036905
28. Janmey PA, Kinnunen PKJ (2006) Biophysical properties of lipids and dynamic membranes. Trends Cell Biol 16(10):538–546. https://doi.org/10.1016/j.tcb.2006.08.009
29. Urban AS, Fedoruk M, Horton MR, Radler J, Stefani FD, Feldmann J (2009) Controlled nanometric phase transitions of phospholipid membranes by plasmonic heating of single gold nanoparticles. Nano Lett 9(8):2903–2908. https://doi.org/10.1021/nl901201h
30. Urban P, Kirchner SR, Muhlbauer C, Lohmuller T, Feldmann J (2016) Reversible control of current across lipid membranes by local heating. Sci Rep 6:ARTN 22686. https://doi.org/10.1038/srep22686
31. Chowdhury S, Jarecki BW, Chanda B (2014) A molecular framework for temperature-dependent gating of ion channels. Cell 158(5):1148–1158. https://doi.org/10.1016/j.cell.2014.07.026
32. Nakatsuji H, Numata T, Morone N, Kaneko S, Mori Y, Imahori H, Murakami T (2015) Thermosensitive ion channel activation in single neuronal cells by using surface-engineered plasmonic nanoparticles. Angew Chem Int Ed 54(40):11725–11729. https://doi.org/10.1002/anie.201505534
33. Brederson JD, Kym PR, Szallasi A (2013) Targeting TRP channels for pain relief. Eur J Pharmacol 716(1–3):61–76. https://doi.org/10.1016/j.ejphar.2013.03.003
34. Specht KG, Rodgers MAJ (1990) Depolarization of mouse myeloma cell-membranes during photodynamic-action. Photochem Photobiol 51(3):319–324. https://doi.org/10.1111/j.1751-1097.1990.tb01717.x
35. Kochevar IE, Lambert CR, Lynch MC, Tedesco AC (1996) Comparison of photosensitized plasma membrane damage caused by singlet oxygen and free radicals. BBA-Biomembranes 1280(2):223–230. https://doi.org/10.1016/0005-2736(95)00297-9
36. Lei WH, Xie JF, Hou YJ, Jiang GY, Zhang HY, Wang PF, Wang XS, Zhang BW (2010) Mitochondria-targeting properties and photodynamic activities of porphyrin derivatives bearing cationic pendant. J Photochem Photobiol B 98(2):167–171. https://doi.org/10.1016/j.jphotobiol.2009.12.003
37. Ross MF, Kelso GF, Blaikie FH, James AM, Cocheme HM, Filipovska A, Da Ros T, Hurd TR, Smith RAJ, Murphy MP (2005) Lipophilic triphenylphosphonium cations as tools in mitochondrial bioenergetics and free radical biology. Biochemistry 70(2):222–230. https://doi.org/10.1007/s10541-005-0104-5
38. Ray RS, Mujtaba SF, Dwivedi A, Yadav N, Verma A, Kushwaha HN, Amar SK, Goel S, Chopra D (2013) Singlet oxygen mediated DNA damage induced phototoxicity by ketoprofen resulting in mitochondrial depolarization and lysosomal destabilization. Toxicology 314(2–3):229–237. https://doi.org/10.1016/j.tox.2013.10.002
39. Bulina ME, Chudakov DM, Britanova OV, Yanushevich YG, Staroverov DB, Chepurnykh TV, Merzlyak EM, Shkrob MA, Lukyanov S, Lukyanov KA (2006) A genetically encoded photosensitizer. Nat Biotechnol 24(1):95–99. https://doi.org/10.1038/nbt1175
40. Bulina ME, Lukyanov KA, Britanova OV, Onichtchouk D, Lukyanov S, Chudakov DM (2006) Chromophore-assisted light inactivation (CALI) using the phototoxic fluorescent protein KillerRed. Nat Protoc 1(2):947–953. https://doi.org/10.1038/nprot.2006.89
41. Vegh RB, Solntsev KM, Kuimova MK, Cho S, Liang Y, Loo BLW, Tolbert LM, Bommarius AS (2011) Reactive oxygen species in photochemistry of the red fluorescent protein "killer red". Chem Commun 47(17):4887–4889. https://doi.org/10.1039/c0cc05713d
42. Wang YQ, Nartiss Y, Steipe B, McQuibban GA, Kim PK (2012) ROS-induced mitochondrial depolarization initiates PARK2/PARKIN-dependent mitochondrial degradation by autophagy. Autophagy 8(10):1462–1476. https://doi.org/10.4161/auto.21211

43. SteinbergYfrach G, Liddell PA, Hung SC, Moore AL, Gust D, Moore TA (1997) Conversion of light energy to proton potential in liposomes by artificial photosynthetic reaction centres. Nature 385(6613):239–241. https://doi.org/10.1007/s10863-012-9435-2
44. Steinberg-Yfrach G, Rigaud JL, Durantini EN, Moore AL, Gust D, Moore TA (1998) Light-driven production of ATP catalysed by F_0F_1-ATP synthase in an artificial photosynthetic membrane. Nature 392(6675):479–482. https://doi.org/10.1038/33116
45. Imahori H, Tamaki K, Guldi DM, Luo CP, Fujitsuka M, Ito O, Sakata Y, Fukuzumi S (2001) Modulating charge separation and charge recombination dynamics in porphyrin fullerene linked dyads and triads: Marcus-normal versus inverted region. J Am Chem Soc 123(11):2607–2617. https://doi.org/10.1021/ja003346i
46. Numata T, Murakami T, Kawashima F, Morone N, Heuser JE, Takano Y, Ohkubo K, Fukuzumi S, Mori Y, Imahori H (2012) Utilization of photoinduced charge-separated state of donor-acceptor-linked molecules for regulation of cell membrane potential and ion transport. J Am Chem Soc 134(14):6092–6095. https://doi.org/10.1021/Ja3007275
47. Takano Y, Numata T, Fujishima K, Miyake K, Nakao K, Grove WD, Inoue R, Kengaku M, Sakaki S, Mori Y, Murakami T, Imahori H (2016) Optical control of neuronal firing via photoinduced electron transfer in donor-acceptor conjugates. Chem Sci 7(5):3331–3337. https://doi.org/10.1039/c5sc04135j
48. Cai N, Takano Y, Numata T, Inoue R, Mori Y, Murakami T, Imahori H (2017) Strategy to attain remarkably high photoinduced charge-separation yield of donor–acceptor linked molecules in biological environment via modulating their cationic moieties. J Phys Chem C 121(32):17457–17465. https://doi.org/10.1021/acs.jpcc.7b04466
49. Rohan JG, Citron YR, Durrell AC, Cheruzel LE, Gray HB, Grubbs RH, Humayun M, Engisch KL, Pikov V, Chow RH (2013) Light-triggered modulation of cellular electrical activity by ruthenium diimine nanoswitches. ACS Chem Neurosci 4(4):585–593. https://doi.org/10.1021/cn300213f
50. Takano Y, Hanai E, Imahori H (2017) Photoinduced electron transfer reaction in mitochondria for spatiotemporal selective photo-oxidation of lipids by donor/acceptor linked molecules. Nanoscale 9:17909
51. Fukuzumi S, Kotani H, Ohkubo K, Ogo S, Tkachenko NV, Lemmetyinen H (2004) Electron-transfer state of 9-mesityl-10-methylacridinium ion with a much longer lifetime and higher energy than that of the natural photosynthetic reaction center. J Am Chem Soc 126(6):1600–1601. https://doi.org/10.1021/Ja038656q
52. Erbrich U, Naujok A, Petschel K, Zimmermann HW (1982) The fluorescent staining of mitochondria in living Hela-cells and lm-cells with new acridine-dyes. Histochemistry 74(1):1–7. https://doi.org/10.1007/Bf00495046
53. Kobayashi H, Nagao S, Hirota S (2016) Characterization of the cytochromec membrane-binding site using cardiolipin-containing bicelles with NMR. Angew Chem Int Ed 55(45):14019–14022. https://doi.org/10.1002/anie.201607419

Physical Concepts Toward Cell–Material Integration

Motomu Tanaka and Akihisa Yamamoto

1 Interfaces: Where Materials Meet Cells

One of the most crucial steps in cell–material integration is the optimization of cell–material interactions. Namely, materials should not cause the protein denaturing by nonspecific adhesions or interfere with the endogenous functions of biological cells. Typical examples include stents and artificial heart valves, as the nonspecific adsorption and accumulation (fouling) of serum proteins might cause a serious consequence. On the other hand, cells should not interfere with the structural integrity of materials by irreversible chemical reactions or enzymatic degradation. The place where cells meet materials defines the "interface" between two different worlds.

M. Tanaka (✉)
Physical Chemistry of Biosystems, Institute of Physical Chemistry, Heidelberg University, Heidelberg, Germany

Center for Integrated Medicine and Physics, Institute for Advanced Study, Kyoto University, Kyoto, Japan
e-mail: tanaka@uni-heidelberg.de

A. Yamamoto
Center for Integrated Medicine and Physics, Institute for Advanced Study, Kyoto University, Kyoto, Japan

D. O. Wang, D. Packwood (eds.), *Cell-Inspired Materials and Engineering*, Fundamental Biomedical Technologies, https://doi.org/10.1007/978-3-030-55924-3_9

2 Roles of Interfaces in Biology: Why Are Many Functions Confined in 2D?

In biological cells, cell membranes define the boundary (interface) between extracellular and cytoplasmic spaces. When we look into cells, we can see a versatile of complex molecular machineries operating specific recognition, selective transport, and signaling are confined in cell membranes or in the close proximity of cell membranes. Why does nature confine so many key functions and processes in 2D interfaces? In late 1970s, Hardt extended the steady state of diffusion-limited reactions described by Smoluchowski, and demonstrated the "economy" of dimentionality with a very simple but straightforward calculation [1]. Here, the mean diffusion time τ for three-body collision in two- and three-dimensions can be given as

$$\tau_{2D} = \frac{x^2}{2D} \ln\left(\frac{x}{r}\right) \text{and } \tau_{3D} = \frac{x^3}{3Dr}, \quad (1)$$

where D is the diffusion coefficient, r the radius of diffusing particles, and x the separation distance between two particles. The dependence of mean diffusion time on the particle radius r is $\langle \tau_{2D} \rangle \propto - \ln(r)$ for two-dimensional systems, while $\langle \tau_{3D} \rangle \propto r^{-1}$ in three dimensions. A clear difference in the dependence of τ on r indicates the energetic and thus economic reason why many biochemical reactions are confined in 2D membranes [2].

3 Interplays of Interfacial Forces

To describe how various interfacial forces interplay at cell–cell, and cell–material interfaces, let us consider a very simple situation: two planes keeping a stable, finite separation distance via a thin "interlayer."

When a separation distance is large, the middle of the interlayer under equilibrium retains its intrinsic bulk properties (Fig. 1a). Thus, individual interfaces can be explained within the framework of the classical Gibbs capillary theory: a change in the interlayer thickness at a constant phase volume and contact area do not cost any change in the free energy of the system, and thus the pressure in the interlayer equals to zero. On the other hand, if the long-range force fields overlap (Fig. 1b), any change in the interlayer thickness will cost positive or negative work [3]. This work is originated from the interplay of attractive and repulsive forces in the interfacial region, which can range even more than several tens of nanometers.

Derjaguin introduced a simple measure to describe the analytical concept of the thermodynamics of interfaces, called a "disjoining pressure." The disjoining pressure is the sum of various interfacial forces acting per unit area [3], which can experimentally be determined by measuring the external pressures to keep the separation distance constant. The disjoining pressure can be defined in terms of the

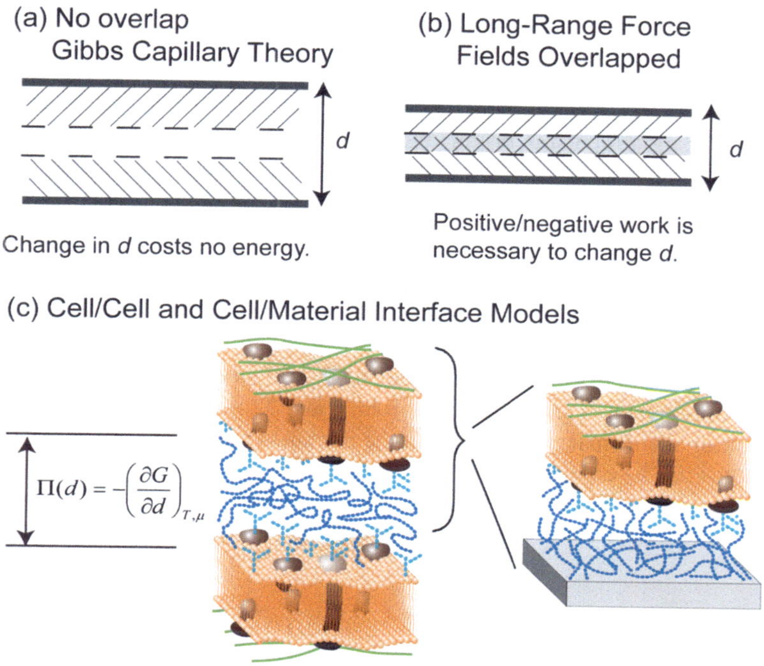

Fig. 1 Schematic illustrations showing the contact of two planes at (**a**) a large spacing (no overlap of long-range force fields) and (**b**) a short distance (the force field from each side partially overlaps). (**c**) The contact of two cell neighboring membranes keeping a finite separation distance d in the presence of polymer interlayer. A model called "supported membranes" [4, 5] can be taken as one half model of a cell–cell contact

lateral density of Gibbs free energy at constant temperature T:$\Pi(d) = -(\partial G/\partial d)_{\mu, T}$, where d is the interlayer thickness.

The minimum of the free energy that determines the equilibrium state corresponds to $\Pi = 0$ (Fig. 1c). One straightforward method to determine the disjoining pressure as a function of distance d is the surface force apparatus, which is based on the combination of a cantilever (spring), a piezoelectric element, and a microinterferometry [6, 7]. Compared to other instruments, such as atomic force microscopy, this technique is much more laborious but more sensitive to interactions acting over longer distance. Although the precise calculation of participating individual forces is nontrivial, simplified model systems enable one to dissect the significance of different interfacial forces.

We have utilized molecular constructs called "polymer supported membranes" [4, 8–11] that can be considered as "a half mode"" of the cell–cell contact mediated via biopolymers (Fig. 1c). Here, the force–distance relationship (note: the force is normalized by the area, that is, it is the pressure–distance relationship) can be obtained by the combination of several techniques [12, 13]. For example, the contribution of electrostatic interactions is negligible in the case of neutral (zwitterionic) phosphatidylcholine membranes deposited on cellulose supports in the presence of 100 mM NaCl, as the Debye screening length (<1 nm) is much thinner than the polymer supports (~10 nm). Here, the disjoining pressure can be classified into three major force contributors: (1) van der Waals force, (2) hydration repulsion, and (3) undulation repulsion originating from the membrane undulation. The first one, van der Waals force, is proportional to d^{-3}, which can be calculated using an asymmetric n-layer model, such as.

$$P_{\mathrm{vdW}} = \frac{1}{6\pi}\left[\frac{A_{234}}{d^3} - \frac{\sqrt{A_{121}A_{343}}}{(d+T_2)^3} - \frac{\sqrt{A_{545}A_{323}}}{(d+T_4)^3} + \frac{\sqrt{A_{545}A_{121}}}{(d+T_2+T_4)^3}\right] \tag{2}$$

for a 5-layer model [14]. A_{ijk} is the Hamaker constant corresponding to the interaction between media i and k across medium j, and T_{m} is the thickness of mth interlayer. Taking the Hamaker constant and the thickness of for each layer, the van der Waals force can be computed as a function of d. The second contributor, hydration repulsion originates from the work necessary to remove water from a hydrated layer to the infinitely thick, bulk liquid phase, which decays exponentially over distance [15, 16],

$$P_{\mathrm{Hyd.}} = P_0 \exp\left[-C\left(\frac{D}{D_0}\right)\right]. \tag{3}$$

Here, the force distance relationship can be obtained by measuring the membrane–substrate spacing at different osmotic pressures. The third force, the repulsive pressure caused by the thermal fluctuation of membranes, called Helfrich repulsion [17, 18].

$$P_{\mathrm{und}}(d) = \alpha_1 \frac{(k_{\mathrm{B}}T)^2}{\kappa d^3}, \tag{4}$$

where κ is the bending rigidity of the membrane (~ 10 $k_{\mathrm{B}}T$). The prefactor $\alpha_1 = (\pi^2/128)$ was taken from the previous analytical derivation, which was further confirmed by Monte Carlo simulations [18]. The red line in Fig. 2a represents the disjoining pressure $\Pi = P_{\mathrm{vdW}} + P_{\mathrm{Hyd}} + P_{\mathrm{und}}$ plotted as a function of d, which is calculated from the sum of three forces for a phosphatidylcholine membrane on a cellulose support (dry film thickness ≈4.1 nm). The calculated "point of zero force" suggests the equilibrium membrane–substrate distance of $d \approx 14$ nm, which shows good agreement with the experimental results of high energy X-ray reflectivity (17 nm) and neutron reflectivity (19 nm). As shown in the figure, the counterbalance between van der Waals attraction and hydration repulsion play dominant roles in confining the equilibrium membrane position, while the contribution of undulation forces turned out to be rather minor.

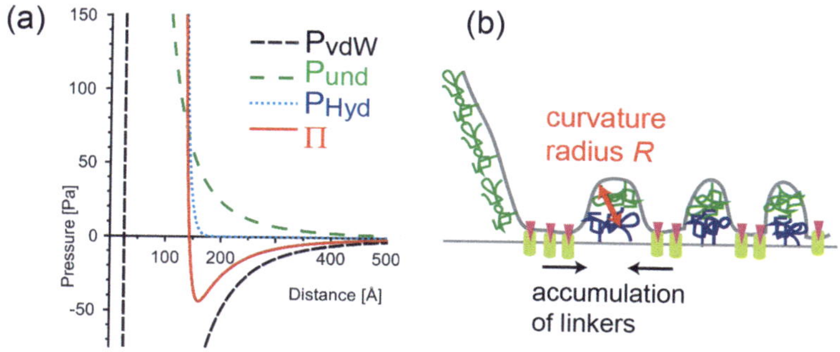

Fig. 2 (**a**) The pressure–distance relationships of a phosphatidylcholine membrane deposited on a cellulose support (dry thickness ≈4.1 nm). The point of zero force (Π = 0) predicted from the sum of three major forces (red) shows a good agreement with those measured by X-ray and neutron reflectivity experiments. (**b**) Schematic illustration of the adhesion-induced phase separation. The accumulation of small adhesion contacts reduces the elastic energy penalty by bending of cell membranes

To keep a finite equilibrium distance d between two planes, the disjoining pressure should fulfill the condition: $\partial\Pi(d)/\partial d < 0$. On the other hand, the interlayer becomes unstable when $\partial\Pi(d)/\partial d > 0$. Continuous thinning of the wetting layer below the free energy minimum leads to a negative disjoining pressure, resulting in dissipation of the film (also called as "dewetting"). A typical example is the rupture/collapse of films made out of complex fluids, such as polymers and surfactants, during thinning [19–21].

The "collapse" of interlayers is also found in biology. One of the most prominent examples is the formation of adhesion plaques, where two cell membranes are zipped by the accumulation of ligand–receptor pairs (linkers). Although this kind of lateral phase separation of linkers has been often described qualitatively in terms of specific protein–protein interactions, Sackmann and Bruinsma suggested that the concept of "wetting-dewetting transition" can be adopted to adhesion-induced phase separation of tight and weak adhesion contacts [22]. In dealing with biological adhesion, one has to consider entropic forces generated by polymer "brushes" grafted on cell surfaces, such as glycocalyx, in addition to the competing two dominant contributors, P_{vdW} and P_{Hyd}. Within the framework of the classical scaling concept, the entropic repulsive force generated by polymer chains can be simplified as [23, 24].

$$P_{\mathrm{poly}}(d) \approx \frac{k_B T}{d^3}\left[\left(\frac{d_{\mathrm{eq}}}{d}\right)^{\frac{9}{4}} - \left(\frac{d}{d_{\mathrm{eq}}}\right)^{\frac{3}{4}}\right], \tag{5}$$

where d_{eq} is the equilibrium thickness of the polymer chain.

As shown in Fig. 2b, small tight adhesion contacts formed between neighboring cells act as nucleation points, causing the growth of adhesion plaques in order to compensate the energy penalty due to bending cell membranes. As the bending energy described in the energy representation of Hook's law is proportional to the square of the curvature R^{-1}, $G_{ela} \propto R^{-2}$, the accumulation of loose contact zones with the radius R decreases the bending energy (Fig. 2b). In fact, a recent study suggested that the lateral mobility of ligand–receptor pairs significantly influences the apoptotic kinetics of cancer cells induced by the formation of death indicing clusters formed by CD95 [25].

4 Views from Surface Free Energy: Physics of Wetting

Stable contacts of cell–cell and cell–material interfaces can be established only in case the presence of any "additional" layer results in the gain of the total free energy of the system. This concept was well described in the classical framework of physics of wetting, where the establishment of stable interfaces is characterized by a positive "spreading power" S. S is defined as the difference between the surface density of free energy of the outermost system γ_0 and the sum of the free energies from all internal interfaces $\sum\gamma_{int}$ [26]. The system fulfills the prerequisite of "complete wetting" when the spreading power is positive:

$$S = \gamma_0 - \Sigma\gamma_{int} \geq 0. \tag{6}$$

If this condition is fulfilled, the film can be formed simultaneously from the bulk reservoir [27], heal local defects [28], and fill microstructured channels [29].

If one considers a biological cell adhering on a solid substrate (Fig. 3a), it shares a common feature with a droplet on a solid substrate establishing a finite contact angle at the three-phase contact (Fig. 3b). Taking the classical Young's equation describing the balance of surface tension γ, the free energy of adhesion per unit area can be given as: $W = \gamma_{SV} + \gamma_{LV} - \gamma_{SL} = \gamma_{LV}(1 + cos\theta)$, where S, L, and V in suffixes represent solid, liquid, and vapor phases, respectively.

Needless to say, the adhesion of cells includes interplays of specific and nonspecific (generic) forces acting on the molecular levels. Bell, Dembo, and Bongrand pioneered a theoretical model of cell adhesion that takes both static and dynamic aspects of molecular bonds [30]. For example, by exchanging "liquid (L)" by "membrane (M)" and "vapor (V)" by "liquid (L)," the Young's equation can be rewritten as $W = \gamma_{SL} + \gamma_{ML} - \gamma_{SM} = \gamma_{ML}(1 + \cos\theta)$. Namely, the increase in the membrane–substrate adhesion can be manifested by the decrease in contact angle θ and thus the increase in adhesion area by flattening of cells [31–33]. From the field of experimental physics, Sackmann and coworkers have pioneered a variety of synthetic model cells, such as giant lipid vesicles, in contact with well-defined surrogate substrates displaying specific ligand or receptor molecules [34, 35]. In addition to the molecular complexity, one major correction required to calculate the free

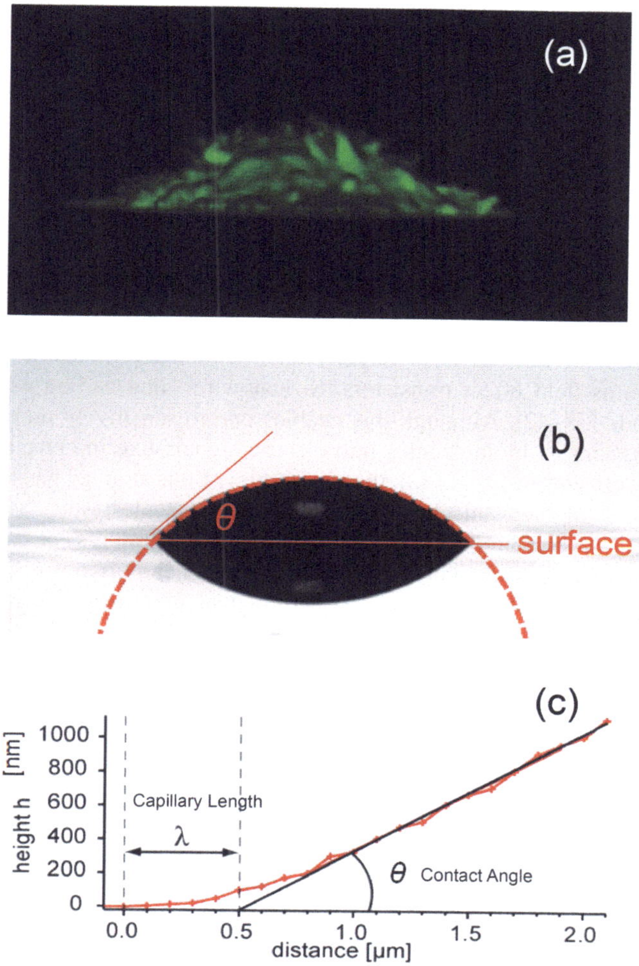

Fig. 3 (**a**) Side view of a pancreatic cancer cell adherent on a substrate displaying specific ligand molecules, reconstructed from confocal fluorescence image stacks. Actin cytoskeletons are stained with FITC-conjugated phalloidin. (**b**) Water droplet deposited on a silicon wafer. (**c**) Height profile of a giant vesicle in contact with a substrate. $\lambda = \sqrt{\kappa / \Gamma}$ is the capillary length, in which the elastic deformation dominates the shape

energy of cell adhesion is the fact that cells, similar to other "complex fluids," undergo both plastic and elastic deformations. This means that the Young's equation for classical Newtonian fluid (like liquid water) is not sufficient to describe the shape of cells on surfaces, which is also obvious from the side view of a biological cell (Fig. 3a). The quantitative analysis of the cell/vesicle shape can be carried out using the capillary length $\lambda = \sqrt{\kappa / \Gamma}$, which is given as a function of the bending rigidity κ and tension Γ of cell membranes [36, 37]. As shown in Fig. 3c, the capillary length is a characteristic distance from the contact point, in which the shape of cells/vesicles is not governed by the plastic deformation (tension) but by the elastic deformation. In the following sections, several examples are taken to demonstrate how the interface matching contributed to bridge two different worlds: biological systems and materials.

5 Connecting Two Worlds: Bio & Electronics

A most commonly taken strategy for the electronic coupling of biological matter and semiconductor devices is to directly anchor biomacromolecules to various devices, such as field effect transistors, Si nanowires, and carbon nanotubes by covalent bonds [38–42]. Although this enables one to sensitively translate highly selective functions of biomolecules into electric signals like the channel conductance of field effect transistors [43], the application of this strategy has been limited to chemically and mechanically robust biomacromolecules, such as DNA and enzymes. Actually, the direct coupling of transmembrane proteins, such as receptors and ion channels, onto semiconductors via covalent bonds is not possible. First, membrane proteins lose their natural structures and functions in the absence of membranes. Second, the direct contact to "hard" solids also causes the protein denaturation. Therefore, one of the key steps toward the integration of biological systems and solid based devices is to understand and manage the interface.

For example, the coupling of electrochemically active cells like neurons and semiconductor devices such as field effective transistors [44–46] is possible only when the gap distance and thus the interfacial interactions between cells and devices are fine adjusted to the optimal range. Fromherz and his coworkers pioneered the "synapse-on-a-chip" concept [47] and demonstrated that the gap distance influences not only the current flow around the junction and thus the potential between cell and semiconductor but also the voltage-dependent process in the cells [48]. As described in the previous sections, the establishment of a finite cell–chip distance could be explained by the interplay of the specific interactions (attraction between integrin and laminin) and generic interactions (van der Waals interaction, electrostatic interaction, hydration repulsion, and entropic repulsion).

The integration of biological cells and solid-state devices opens up a large potential for designing materials that can utilize biological cells to communicate with the surrounding environments. On the other hand, the interface between biological cells and semiconductors is often not stable or controllable. One possible solution is to reconstitute defined cell membrane models with the reduced number of components by the deposition of membrane-associated proteins into cell membrane models and deposit the protein–lipid composites onto semiconductors [28, 49–51]. In particular, the latter is a promising strategy to provide with a natural environment for the immobilization of membrane-associated proteins under nondenaturing conditions and in well-defined orientation [4, 8]. Moreover, in contrast to metal electrode systems [52, 53], the use of semiconductor allows for the detection of changes in membrane potentials as the change in flat band potentials, which enables one to reach the sensitivity close to detect the binding/unbinding of individual protein molecules [50, 54, 55]. The coupling of semiconductors with proteins and membranes is often impeded by irreversible decomposition of semiconductors under physiological conditions, that is, in the presence of 0.1 M salt at a neutral pH. For example, Si-based devices suffer from instability due to ion diffusion into SiO_2, and GaAs undergoes irreversible corrosion and releases toxic As in physiological buffers. Such problems can be bypassed by the protection with organic monolayers, which enables both stability and functionality to operate the materials under physiological environments [55–58].

More recently, GaN has been drawing increasing attentions because of its outstanding chemical/electrochemical stability and the capability to control the electronic properties by doping and band gap engineering by alloying with In or Al [59]. Oxidized GaN surfaces can be chemically functionalized with silane chemistry [38], which opens a large potential toward the functionalization with lipid membranes [51, 54] and peptides [60]. The former enables one to detect the change in membrane potentials sensitively as the change in semiconductor space charge capacitance and thus flat band potentials (Fig. 4a). On the other hand, the latter allows for the flexible modulation of electronic band structures of GaN semiconductors by utilizing macromolecular dipole moments of peptides (Fig. 4b), which offers an alternative to the classical inorganic doping and the deposition of additional inorganic layers.

6 Other Examples: Biominerals, Antifouling Materials

Another prominent example where the interface plays vital roles in connecting organic and inorganic materials is biominerals, such as bones, teeth, and nacres of shells. Biominerals consist of self-assembled, hierarchical composite structures across several orders of magnitude in length scales [61–63]. Inspired by the nature's design, material scientists have developed several synthetic approaches for the creation of such hybrid materials. For example, several studies have demonstrated that the characteristic size and structures of $CaCO_3$ can be modulated by the interfacial characteristics of (macro)molecular templates, such as their mechanical properties and polymer chain ordering [64–67]. For example, mollusk nacres consist of aragonite platelets formed in the matrix made out of chitin and silk-like proteins (enriched with alanine and glycine), taking periodic layer-by-layer structures [68]. A recent study demonstrated that the formation of stratified chitin–protein structures can be understood by the condensation of chitin at the interface, whose side chains are modified by hydrophobic amino acids [69]. The use of self-organized, oriented organic templates will bring about a paradigm shift from the top-down approach ("enforced" patterning by mean of micro/nanofabrication) to the bottom-up strategy (interfacial self-assembling) toward the creation of novel hybrid materials with

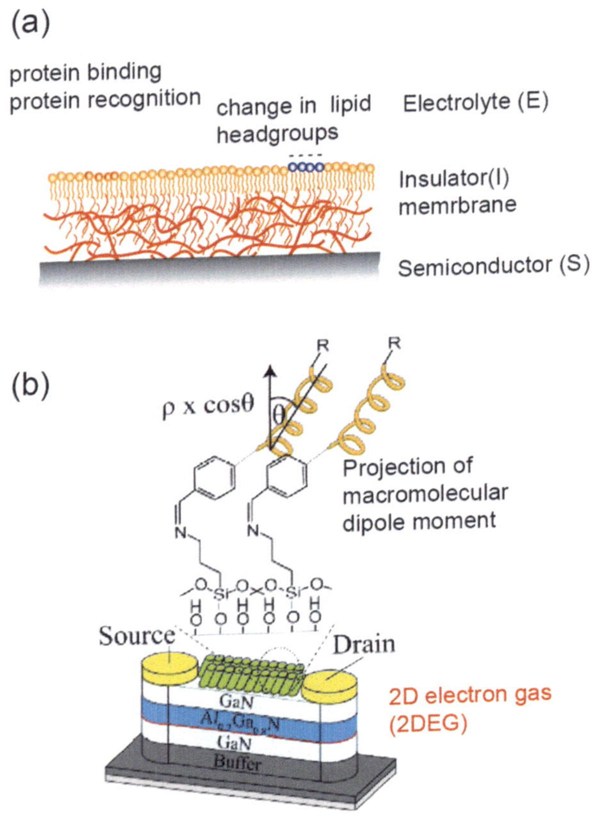

Fig. 4 (**a**) Example of bioelectronic coupling: "membrane voltmeter" based on electrolyte–insulator–semiconductor (EIS). Interface matching can be achieved by silane and/or soft polymer interlayers. The change in surface potential can be detected as the change in the semiconductor space charge capacitance and thus flat band potential. (**b**) Flexible modulation of electronic band structures of GaN semiconductors by molecular dipole moments. For example, the carrier mobility in two-dimensional electron gas layers can be fine adjusted by the sign and magnitude of molecular dipole moments without any additional doping or layer deposition

outstanding mechanical toughness inspired by the hierarchical structural design of nature.

Spontaneous fouling of proteins and cells happens once diagnostic devices or implants are exposed to tissues and body fluids like blood. This is a major concern in a number of applications, ranging from the nonspecific accumulation of proteins on contact lenses, the biofilm formation in indwelling catheters, to the occlusion of cardiovascular implants [70, 71]. A widely taken strategy to prevent biofouling is the grafting of self-assembled monolayers and polymers [72, 73]. Many of antifouling polymers consist of hydrophilic monomers that carry neutral or weakly negative charges, and one of the most widely used ones is polyethylene glycol (PEG). In

recent years, zwitterionic polymers have been drawing attentions of material scientists as a new alternative [74–76]. In fact, zwitterionic molecules are widely found in biological systems: cell membranes mainly consist of zwitterionic phospholipids that prevents nonspecific adsorption of proteins to cell surfaces. Phosphatidylcholine and sphingomyelin rendered with zwitterionic phosphatidylcholine moieties, making up >50% of total lipids in animal cells [77]. Previous vibrational spectroscopy [78] and quasi-elastic neutron scattering [79] suggested that water molecules in the vicinity of zwitterionic surfaces are characterized with the loosely bound (freezing bound) state. Although these techniques do not possess spatial resolution to obtain structures and electronic states of water molecules in the close proximity of interfaces, ample evidence suggests that water is not only a space-filling medium but also plays key roles in modulating interactions [80].

For example, one can calculate the effective interaction potential $V(\Delta h)$ by monitoring the height fluctuation of cells or particles Δh undergoing vertical Brownian motion (Fig. 5a):

$$V(\Delta h) = -k_{\mathrm{B}}T \ln\left(P(\Delta h)\right) + \text{const.}, \tag{7}$$

where $P(\Delta h)$ is the probability function of fluctuation amplitude of the particle (Fig. 5b). Within the framework of Derjaguin's approximation [3], the potential near the first minimum can be approximated as a harmonic potential. Here, the second derivative of the potential $V''(\Delta h)$ is nothing but the "spring constant" characterizing the curvature (sharpness) of potential. Within the fluctuation–dissipation theorem [81], one can also gain the hydrodynamic friction γ from the height–height autocorrelation function:

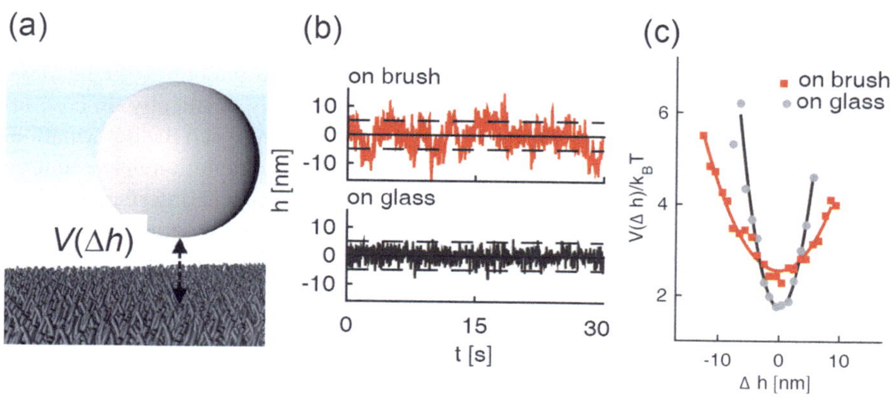

Fig. 5 (**a**) Schematic illustration of cell/particle undergoing vertical Brownian motion. (**b**) An osmotically tense, human erythrocyte shows a pronounced height fluctuation on zwitterionic polymer brushes compared to that on passivated glass substrates. (**c**) Effective interfacial potentials corresponding to the data presented in panel **b**, implying a significant softening of potential confinement by the presence of polymer brushes

$$\Delta h(\delta t)\Delta h(0) \approx \Delta h^2(0)e^{-\frac{\delta t}{\tau}}, \text{where} \tau = \frac{\gamma}{V''} = \frac{6\pi\eta_{\text{eff}}R^2}{hV''}. \tag{8}$$

Recently, we found that the curvature of interaction potential between zwitterionic polymer brushes and human erythrocytes and the hydrodynamic friction are very low (Fig. 5c), which seems to explain the excellent blood repellency of zwitterionic polymer brushes [82]. The combination with spectroscopic studies would enable one to quantitatively unravel the physical mechanism of antifouling.

7 Concluding Remarks and Perspectives

In this section, several physical key concepts toward cell–material integration have been introduced. Within the fundamental framework of thermodynamics of interfaces (disjoining pressure), one can understand how the interplay of different forces determines the finite distance between neighboring cells and avoids undesired, non-specific adhesion. Although it is practically almost impossible to dissect individual force contributors, the combination of several experimental techniques and theoretical calculation enables one to understand the relative significance of interfacial forces as a function of spacing distance. For example, by taking polymer-supported membranes as a half model of cell–cell interactions, the combination of experiments and theoretical calculation demonstrated that the equilibrium substrate–membrane spacing distance is predominantly determined by the counterbalance between van der Waals attraction and hydration repulsion. If one sheds light on the surface free energy, the establishment of stable contacts can be achieved only if the complete wetting conditions are fulfilled. In fact, cell–cell contacts and cell–matrix adhesion can be treated as a problem of wetting physics. One major correction to be taken into consideration is the fact that cell membranes and many biomacromolecules undergo both plastic and elastic deformation, which is in contrast to the classical Newtonian liquid (like liquid water) that deforms only plastically. In case of wetting of cell membranes, the characteristic length scale in which the deformation is governed by elasticity (capillary length) is determined by the ration between tension and bending modulus. Though interactions at biological interface further includes specific molecular interactions that lead to the phase separation of "linkers" and "repellers," the understanding of physical principles helps us a lot fine-adjust the contact between cells and materials. In fact, the precise control of the spacing distance between cells and underlying substrate materials become extremely important in electric coupling of cells (such as neurons) and electric materials (such as field effect transistors). Fine-tuning of interfacial forces and wetting properties (including optimal surface chemistry) allows for the design of bioelectronic materials that can translate highly sensitive and specific functions of biological materials into electric signals.

In recent years, an increasing number of studies have postulated the importance of water in the context of "biocompatibility" of materials, including antifouling proteins and blood repellency. In contrast to the classical consensus treating water as a space-filling medium, many studies suggest that water molecules actively control the structure and function of biomacromolecules and membranes [80]. For example, modern spectroscopic and dynamic studies implied the coexistence of tightly bound water molecules and loosely bound water molecules near zwitterionic surfaces [78, 79], but little is understood how they interact with charged moieties and ions. One open question that is medically relevant is the mechanism how zwitterionic phosphatidylcholines protect the intestinal mucus layer against pathological bacteria [83]. The effective interaction potentials between the artificial model of intestinal mucus and silica particles coated with differently charged lipids implied that zwitterionic phosphatidylcholines interact most strongly with heavily sialylated and hence negatively charged mucin proteins [84, 85].

It is well established that electrostatics of biological interfaces is nontrivial both experimentally and theoretically. Except for some specialized techniques, such as grazing incidence X-ray fluorescence [86–88], the localization of ions (elements) at molecular level resolution is not possible. Second, the charge distribution in such interfaces is never homogeneous or static, and multivalent ions (such as Ca^{2+}) bridge the moieties carrying opposite charges. This makes the classical Derjaguin–Landau–Verwey–Overbeek theory inadequate, which requires a creation of minimal coarse-grained model [89, 90]. Thus, the further understanding and controlling of hydrating water in the vicinity of cell–material contacts by the combination of new experimental techniques and complementary theoretical models will bring about a paradigm shift in cell–material integrations.

Acknowledgments M.T. is thankful to all the past and current lab members for their enormous efforts and scientific inputs. Especially the following people contributed to the works presented in this chapter: S. Kaufmann, F.F. Rossetti, E. Schneck, W. Abuillan, T. Kaindl, A. Burk, T. Schubert, R.G. Oliveira, H.Y. Yoshikawa, S. Mehlhose, N. Frenkel, B. Fröhlich, and F. Amadei. M.T. is grateful to A.D. Ho, W. Stremmel, J. Sleeman, M. Lanzer (Heidelberg University), O.V. Konovalov (ESRF), B. Demé and G. Fragneto (ILL), S. Kimura (Kyoto University), M. Eickhoff (Bremen), K. Arinaga (Fujitsu), A. Martin-Villanba (DKFZ), D.A. Pink (St. Francis Xavier Univ.), K. Brandenburg (Research Center Borstel), Y. Higaki and A. Takahara (Kyushu Univ.) for long-lasting, highly interdisciplinary collaboration. These works have been supported by the DFG (SPP2171, Germany's Excellence Strategy, 2082/1-390761711) and JSPS (WPI Program and 19H05719). M.T. thanks Nakatani Foundation for support.

References

1. Hardt SL (1979) Rates of diffusion controlled reactions in one, two and three dimensions. Biophys Chem 10:239–243
2. Tanaka M (2013) Physics of interactions at biological and biomaterial interfaces. Curr Opin Colloid Interface Sci 18:432–439
3. Derjaguin BV, Churaev NV (1987) Surface forces. Consultants Bureau, New York, NY

4. Tanaka M, Sackmann E (2005) Polymer-supported membranes as models of the cell surface. Nature 437:656–663
5. Sackmann E (1996) Supported membranes: scientific and practical applications. Science 271:43–48
6. Israelachvili JN, Adams GE (1978) Measurement of forces between two mica surfaces in aqueous electrolyte solutions in the range 0–100 nm. J Chem Soc Faraday Trans 74:975–1001
7. Parker JL, Christenson HK, Ninham BW (1989) Device for measuring the force and separation between two surfaces down to molecular separations. Rev Sci Instrum 60:3135–3138
8. Tanaka M (2006) Polymer-supported membranes: physical models of cell surfaces. MRS Bull 31:513–520
9. Sackmann E, Tanaka M (2000) Supported membranes on soft polymer cushions: fabrication, characterization and applications. Trends Biotechnol 18:58–64
10. Tanaka M, Wong AP, Rehfeldt F, Tutus M, Kaufmann S (2004) Selective deposition of native cell membranes on biocompatible micropatterns. J Am Chem Soc 126:3257–3260
11. Tanaka M, Kaufmann S, Nissen J, Hochrein M (2001) Orientation selective immobilization of human erythrocyte membranes on ultrathin cellulose films. Phys Chem Chem Phys 3:4091–4095
12. Seitz PC, Reif MD, Konovalov OV, Jordan R, Tanaka M (2009) Modulation of substrate-membrane interactions by linear poly(2-methyl-2-oxazoline) spacers revealed by X-ray reflectivity and ellipsometry. ChemPhysChem 10:2876–2883
13. Rossetti FF, Schneck E, Fragneto G, Konovalov OV, Tanaka M (2015) Generic role of polymer supports in the fine adjustment of interfacial interactions between solid substrates and model cell membranes. Langmuir 31:4473–4480
14. Israelachvili JN (1992) Intermolecular and surface forces, 2nd edn. Academic, Cambridge, MA
15. Rau DC, Parsegian VA (1990) Direct measurement of forces between linear polysaccharides xanthan and schizophyllan. Science 249:1278–1281
16. Leikin S, Parsegian VA, Rau DC, Rand RP (1993) Hydration forces. Annu Rev Phys Chem 44:369–395
17. Helfrich W (1978) Steric interaction of fluid membranes in multilayer systems. Zeitschrift für Naturforschung Teil A 33:305–315
18. Bachmann M, Kleinert H, Pelster A (2001) Fluctuation pressure of a stack of membranes. Phys Rev E 63:051709
19. Vrij A (1966) Possible mechanism for spontaneous rupture of thin free liquid films. Discuss Faraday Soc 42:23
20. Langevin D (1998) Dynamics of surfactant layers. Curr Opin Colloid Interface Sci 3:600–607
21. Reiter G et al (1999) Thin film instability induced by long-range forces. Langmuir 15:2551–2558
22. Sackmann E, Bruinsma RF (2002) Cell adhesion as wetting transition? Chem Phys Chem 3:262–269
23. De Gennes PG (1976) Scaling theory of polymer adsorption. J Phys France 37:1445–1452
24. Alexander S (1977) Polymer adsorption on small spheres. A scaling approach. J Phys France 38:977–981
25. Gülcüler Balta GS et al (2019) 3D cellular architecture modulates tyrosine kinase activity, thereby switching CD95-mediated apoptosis to survival. Cell Rep 29:2295–2306.e2296
26. Brochard-Wyart F, de Gennes PG (1992) Dynamics of partial wetting. Adv Colloid Interf Sci 39:1–11
27. Nissen J, Gritsch S, Wiegand G, Rädler J (1999) Wetting of phospholipid membranes on hydrophilic surfaces-concepts towards self-healing membranes. Eur Phys J B 10:335–344
28. Hillebrandt H, Wiegand G, Tanaka M, Sackmann E (1999) High electric resistance polymer/lipid composite films on indium-tin-oxide electrodes. Langmuir 15:8451–8459
29. Furukawa K, Nakashima H, Kashimura Y, Torimitsu K (2006) Microchannel device using self-spreading lipid bilayer as molecule carrier. Lab Chip 6:1001–1006

30. Bell GI, Dembo M, Bongrand P (1984) Cell adhesion. Competition between nonspecific repulsion and specific bonding. Biophys J 45:1051–1064
31. Rieger H et al (2015) Cytoadhesion of Plasmodium falciparum-infected erythrocytes to chondroitin-4-sulfate is cooperative and shear enhanced. Blood 125:383–391
32. Kaindl T et al (2012) Spatio-temporal patterns of pancreatic cancer cells expressing CD44 isoforms on supported membranes displaying hyaluronic acid oligomers arrays. PLoS One 7:e42991
33. Burk AS et al (2015) Quantifying adhesion mechanisms and dynamics of human hematopoietic stem and progenitor cells. Sci Rep 5:9370
34. Albersdörfer A, Feder T, Sackmann E (1997) Adhesion-induced domain formation by interplay of long-range repulsion and short-range attraction force: a model membrane study. Biophys J 73:245–257
35. Goennenwein S, Tanaka M, Hu B, Moroder L, Sackmann E (2003) Functional incorporation of Integrins into solid supported membranes on ultrathin films of cellulose: impact on adhesion. Biophys J 85:846–855
36. Bruinsma R. In: Proc. NATO Adv. Inst. Phys. Biomater. NATO ASI Ser. pp 61–75
37. Purrucker O et al (2007) Polymer-tethered membranes as quantitative models for the study of integrin-mediated cell adhesion. Soft Matter 3:333–336
38. Baur B et al (2005) Chemical functionalization of GaN and AlN surfaces. Appl Phys Lett 87:263901
39. Cui Y, Wei Q, Park H, Lieber CM (2001) Nanowire nanosensors for highly sensitive and selective detection of biological and chemical species. Science 293:1289–1292
40. Fritz J, Cooper EB, Gaudet S, Sorger PK, Manalis SR (2002) Electronic detection of DNA by its intrinsic molecular charge. Proc Natl Acad Sci 99:14142–14146
41. van der Schoot BH, Bergveld P (1988) ISFET based enzyme sensors. Biosensors 3:161–186
42. Katz E, Willner I (2004) Biomolecule-functionalized carbon nanotubes: applications in nanobioelectronics. ChemPhysChem 5:1084–1104
43. Bergveld P (2003) Thirty years of ISFETOLOGY – what happened in the past 30 years and what may happen in the next 30 years. Sensors Actuators B Chem 88:1–20
44. Fromherz P, Offenhausser A (1991) A neuron-silicon junction: a Retzius cell of the leech on an insulated-gate field-effect transistor. Science 252:1290
45. Hess LH et al (2011) Graphene transistor arrays for recording action potentials from electrogenic cells. Adv Mater 23:5045–5049
46. Steinhoff G et al (2005) Recording of cell action potentials with Al Ga N/ Ga N field-effect transistors. Appl Phys Lett 86:033901
47. Zeck G, Fromherz P (2001) Noninvasive neuroelectronic interfacing with synaptically connected snail neurons immobilized on a semiconductor chip. Proc Natl Acad Sci 98:10457–10462
48. Zeck G, Fromherz P (2003) Repulsion and attraction by extracellular matrix protein in cell adhesion studied with nerve cells and lipid vesicles on silicon chips. Langmuir 19:1580–1585
49. Atanasov V et al (2005) Membrane on a chip: a functional tethered lipid bilayer membrane on silicon oxide surfaces. Biophys J 89:1780–1788
50. Hillebrandt H, Tanaka M, Sackmann E (2002) A novel membrane charge sensor: sensitive detection of surface charge at polymer/lipid composite films on indium tin oxide electrodes. J Phys Chem B 106:477–486
51. Schubert T et al (2009) Gallium nitride electrodes for membrane-based electrochemical biosensors. Eur Phys J E 30:233–238
52. Cornell BA, Braach-Maksvytis V, King L, Osman P (1997) A biosensor that uses ion-channel switches. Nature 387:580
53. Ross M, Steinem C, Galla H-J, Janshoff A (2001) Visualization of chemical and physical properties of calcium-induced domains in DPPC/DPPS Langmuir–Blodgett layers. Langmuir 17:2437–2445

54. Frenkel N et al (2014) High precision, electrochemical detection of reversible binding of recombinant proteins on wide bandgap GaN electrodes functionalized with biomembrane models. Adv Funct Mater 24:4927–4934
55. Gassull D, Ulman A, Grunze M, Tanaka M (2008) Electrochemical sensing of membrane potential and enzyme function using gallium arsenide electrodes functionalized with supported membranes. J Phys Chem B 112:5736–5741
56. Tutus M, Purrucker O, Adlkofer A, Eickhoff M, Tanaka M (2005) Electrochemical stabilization of crystalline silicon with aromatic self-assembled monolayers in aqueous electrolytes. Phys Stat Sol A 242:2838–2845
57. Adlkofer K, Tanaka M (2001) Stable surface coating of gallium arsenide with octadecylthiol monolayers. Langmuir 17:4267–4273
58. Kaindl T et al (2010) Modulation of band bending of gallium arsenide with oriented helical peptide monolayers. J Phys Chem C 114:22677–22683
59. Steinhoff G, Purrucker O, Tanaka M, Stutzmann M, Eickhoff M (2003) AlxGa1−xN—a new material system for biosensors. Adv Funct Mater 13:841–846
60. Mehlhose S et al (2018) Flexible modulation of electronic band structures of wide band gap GaN semiconductors using bioinspired, nonbiological helical peptides. Adv Funct Mater 28:1704034
61. Addadi L, Weiner S (1992) Control and design principles in biological mineralization. Angew Chem Int Ed 31:153–169
62. Cölfen H, Mann S (2003) Higher-order organization by mesoscale self-assembly and transformation of hybrid nanostructures. Angew Chem Int Ed 42:2350–2365
63. Aizenberg J et al (2005) Skeleton of Euplectella sp.: structural hierarchy from the nanoscale to the macroscale. Science 309:275–278
64. Kato T, Sakamoto T, Nishimura T (2010) Macromolecular templating for the formation of inorganic-organic hybrid structures. MRS Bull 35:127–132
65. Sugawara A, Ishii T, Kato T (2003) Self-organized calcium carbonate with regular surface-relief structures. Angew Chem 115:5457–5461
66. Nishimura T, Ito T, Yamamoto Y, Yoshio M, Kato T (2008) Macroscopically ordered polymer/CaCO3 hybrids prepared by using a liquid-crystalline template. Angew Chem Int Ed 47:2800–2803
67. Han YJ, Wysocki LM, Thanawala MS, Siegrist T, Aizenberg J (2005) Template-dependent morphogenesis of oriented calcite crystals in the presence of magnesium ions. Angew Chem Int Ed 44:2386–2390
68. Levi-Kalisman Y, Falini G, Addadi L, Weiner S (2001) Structure of the nacreous organic matrix of a bivalve mollusk Shell examined in the hydrated state using Cryo-TEM. J Struct Biol 135:8–17
69. Weiss IM, Kaufmann S, Heiland B, Tanaka M (2009) Covalent modification of chitin with silk-derivatives acts as an amphiphilic self-organizing template in nacre biomineralisation. J Struct Biol 167:68–75
70. Banerjee I, Pangule RC, Kane RS (2011) Antifouling coatings: recent developments in the design of surfaces that prevent fouling by proteins, bacteria, and marine organisms. Adv Mater 23:690–718
71. Krishnan S, Weinman CJ, Ober CK (2008) Advances in polymers for anti-biofouling surfaces. J Mater Chem 18:3405–3413
72. Mrksich M, Whitesides GM (1996) Using self-assembled monolayers to understand the interactions of man-made surfaces with proteins and cells. Annu Rev Biophys Biomol Struct 25:55–78
73. Nath N, Hyun J, Ma H, Chilkoti A (2004) Surface engineering strategies for control of protein and cell interactions. Surf Sci 570:98–110
74. Iwata R et al (2004) Control of nanobiointerfaces generated from well-defined biomimetic polymer brushes for protein and cell manipulations. Biomacromolecules 5:2308–2314

75. Schlenoff J, Zwitteration B (2014) Coating surfaces with Zwitterionic functionality to reduce nonspecific adsorption. Langmuir 30:9625–9636
76. Shao Q, Jiang S (2015) Molecular understanding and design of zwitterionic materials. Adv Mater 27:15–26
77. Van Deenen L (1966) Some structural and dynamic aspects of lipids in biological membranes. Ann N Y Acad Sci 137:717–730
78. Murakami D et al (2013) Spreading and structuring of water on superhydrophilic polyelectrolyte brush surfaces. Langmuir 29:1148–1151
79. Yamada T, Takahashi N, Tominaga T, Takata, S.-i. & Seto, H. (2017) Dynamical behavior of hydration water molecules between phospholipid membranes. J Phys Chem B 121:8322–8329
80. Ball P (2017) Water is an active matrix of life for cell and molecular biology. Proc Natl Acad Sci 114:13327–13335
81. Monzel C et al (2015) Fine adjustment of interfacial potential between pH-responsive hydrogels and cell-sized particles. Langmuir 31:8689–8696
82. Higaki Y et al (2017) Ion-specific modulation of interfacial interaction potentials between solid substrates and cell-sized particles mediated via Zwitterionic, super-hydrophilic poly (sulfobetaine) brushes. J Phys Chem B 121:1396–1404
83. Stremmel W et al (2012) Mucosal protection by phosphatidylcholine. Dig Dis 30(suppl 3):85–91
84. Amadei F, Fröhlich B, Stremmel W, Tanaka M (2018) Nonclassical interactions of phosphatidylcholine with mucin protect intestinal surfaces: a microinterferometry study. Langmuir 34:14046–14057
85. Korytowski A et al (2017) Accumulation of phosphatidylcholine on gut mucosal surface is not dominated by electrostatic interactions. Biochim Biophys Acta Biomembr 1859:959–965
86. Novikova NN et al (2003) Total reflection X-ray fluorescence study of Langmuir monolayers on water surface. J Appl Crystallogr 36:727–731
87. Schneck E et al (2010) Quantitative determination of ion distributions in bacterial lipopolysaccharide membranes by grazing-incidence X-ray fluorescence. Proc Natl Acad Sci 107:9147–9151
88. Abuillan W et al (2013) Physical interactions of fish protamine and antisepsis peptide drugs with bacterial membranes revealed by combination of specular x-ray reflectivity and grazing-incidence x-ray fluorescence. Phys Rev E 88:012705
89. Netz RR (1999) Debye-Huckel theory for interfacial geometries. Phys Rev E 60:3174–3182
90. Oliveira RG et al (2009) Physical mechanism of bacterial survival revealed by combined grazing-incidence X-ray scattering and Monte Carlo simulation. Comptes Rendus French Acad Sci 12:209–217

Part V
Artificial Environments for Cell Control

Using Stem Cells and Synthetic Scaffolds to Model Ethically Sensitive Human Placental Tissue

Georgia R. Kafer

1 Introduction

The placenta is the only transient human organ and arguably the least well understood tissue in the body (or not in the body if you are male or a nonpregnant female). The placenta serves as a life support machine for the developing fetus. The placenta can do the work of the hepatic, endocrine, immune, gastrointestinal, renal, and respiratory systems for the embryo while the organs necessary for these functions develop. If placental function is compromised, then the health of the fetus is in jeopardy. Given the importance of the placenta, it is somewhat surprising that most of what we know about the human placenta has not changed for many decades. In fact, despite ongoing research efforts, our knowledge of human placental development has been sparingly built upon over the last 20 years. A primary reason for this is that studying the human placenta is ethically sensitive due to the rights and terms of consent from an unborn fetus. In addition, the most meaningful studies of human placental development need to first be performed on normal, disease free tissues. While there are some research groups that have access to placental tissue from all stages of gestation, most labs are forced to rely on placental samples collected from term placenta. At term, the placenta's life support job is finished, and it is typically discarded. Term placenta are therefore more readily attainable as sampling the tissue after the baby is born does not pose any danger to the life of mother or baby. However, at term the placenta has finished developing and will begin to degrade, making it difficult to study both placental development and healthy placental function. To study these aspects of placentation, it would be ideal to sample placental tissue across gestation, however this could jeopardize the wellbeing of a healthy

G. R. Kafer (✉)
Genome Integrity Unit, Children's Medical Research Institute, Westmead, NSW, Australia
e-mail: gkafer@cmri.org.au

D. O. Wang, D. Packwood (eds.), *Cell-Inspired Materials and Engineering*, Fundamental Biomedical Technologies, https://doi.org/10.1007/978-3-030-55924-3_10

fetus. Because of this dilemma we are forced to study placental development using model systems that are not entirely ideal. This chapter will begin with a brief overview of our current knowledge of human placentation. The models routinely used to study human placental biology will be reviewed before a discussion of new 3D based modeling methods and the benefits that such methods may have for future placental research avenues.

2 The Human Placenta

The human placenta begins as a handful of trophectoderm cells that form about 5 days following fertilization. Compared to other organs, we currently understand very little about human placental tissue, particularly during the first trimester (0–12 weeks gestation). This is because accessing human placental tissue, especially during the initial stages of pregnancy is technically and ethically challenging. What we do know has largely been derived from histological studies of placental tissue, some of which date back to the 1940s. Histological studies of the human placenta typically consist of sampling placental tissue along a developmental timeline. From such work, we know that the trophectoderm cells, which are situated on the outer edges of the developing embryo and are responsible for contacting the uterine wall (Fig. 1). The trophectoderm matures into several different trophoblast cell lineages. Through a series of events that involve the embryo secreting enzymes that digest part of the uterine lining, the embryo attaches to the uterine wall at around 7 days after fertilization. Invasion into the uterine wall occurs between 8 and 9 days later and is initiated by the invasive trophoblast cells (Fig. 1). At approximately 12 days two major types of trophoblast populate the early human placenta, the cytotrophoblast cells and syncytiotrophoblast (Fig. 1). The cytotrophoblast cells begin to invade further into the uterine wall to generate primary, secondary and then tertiary villi. The syncytiotrophoblast are terminally differentiated and serve as the barrier between maternal and fetal blood. Syncytiotrophoblast cells fuse together to form a layer of multinucleated cells. It is through the syncytiotrophoblast cell layer that the exchange of nutrients and waste occurs. During pregnancy, the syncytiotrophoblast cells will naturally die off in waves and be shed into the maternal circulation [1, 2]. The cytotrophoblast cells serve as the progenitor population for the syncytiotrophoblast. Cytotrophoblast have a single nucleus and will proliferate during the lifetime of the placenta and to repopulate the overlaying syncytiotrophoblast. From approximately 15 days following fertilization the cytotrophoblast and syncytiotrophoblast begin to self-organize into villus-like structures (Fig. 1). At this stage cytotrophoblast are also differentiating into other types of cells that are known as extra villous syncytiotrophoblast which are responsible for further invasion of the placenta into the uterus. These cells also remodel the maternal arteries and help the placenta to gain access to the maternal blood. By 18 days, the basic structure of the human placenta has been established. The placenta now will consist of many villi which are predominantly made of cells which have differentiated from the mesoderm

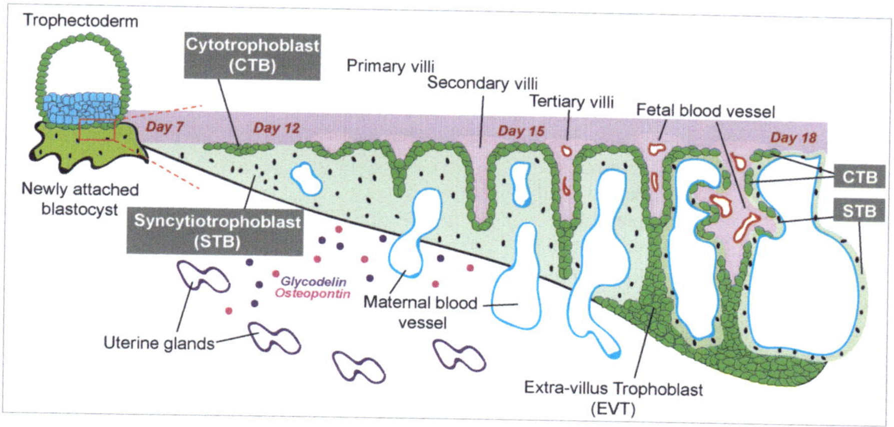

Fig. 1 Schematic overview of early trophoblast growth. The early embryo contacts the uterine epithelium approximately 5 days postfertilization. The outer trophectoderm mediates the connection with the uterus and a population of invasive trophoblast cells will digest part of the uterine lining to facilitate invasion. Between 5 and 12 days after fertilization the trophoblast cells have begun to mature and a population of syncytiotrophoblast cells has formed. The two major types of trophoblast are proliferative cytotrophoblast and fused, nonproliferative syncytiotrophoblast. The cytotrophoblast initiates the formation of primary villi which will continue to grow into secondary and tertiary villi. At this stage, the placental cells must rely on glycodelin and osteopontin secreted from maternal uterine glands for energy. Between 15 and 18 days of gestation the basic structure of the placenta is established. Cytotrophoblast gives rise to extra villous trophoblast which colonize and remodel the maternal blood vessels and anchor the growing placenta to the uterine wall. From around day 15 the fetal blood vessels develop and are initially separated from maternal blood spaces by a now thin layer or syncytiotrophoblast which lines the outer edges of the immature placental villi

of the embryo proper. The cytotrophoblast cells sit as a layer of cells lining the mesoderm-derived villi and the with the syncytiotrophoblast sitting on top as a big, seamless, multinucleated cell layer (for an excellent review on the development of early human placental structures see [3]). While histology can reveal intricate details pertaining to cell structure such techniques are unable to reveal detailed information on function and cell mechanics, and because of this, our knowledge gaps pertain primarily to how the placenta develops.

3 Current Model Systems for the Study of Human Placentation

Several model systems have been developed to study the functional aspects of placental development (see Table 1 for a summary). A common approach is to use cell lines derived from human cancerous "placenta-like" tissue known as choriocarcinoma. BeWo, Jeg-3 and Jar cells are three common choriocarcinoma cell lines

Table 1 Current non-3D models and approaches used by researchers to study the human placenta

Model/approach	Advantages	Disadvantages
Animal models (mice, rats, and sheep)	Less ethically restricted than human studies Tissue can be collected at all stages of placental development	No animal has an identical placental structure to that of a human
Use of "placental like" cell lines (BeWo, jar, and JEG-3)	Allows for the in vitro manipulation of cells	Cell lines are generated from tumors of cells that were "placental like" (such as choriocarcinomas) and so are likely to not fully recapitulate placental cell biology
Use of donated "term" placenta	Allows for the study of human placental structure	Due to the transient nature of the placenta, term placentas have finished developing and are largely necrotic tissue, thus preventing the study of any "developmental" features
Use of donated placental from unsuccessful pregnancies		Although placental tissue may be from early trimesters, the tissue is likely to be abnormal and so is unable to reveal information about normal placental biology
Use of donated placental tissue from terminated pregnancies		Although tissue does come from early trimester pregnancies, the tissue may not be well preserved due to the procedures required to extract the placental tissue from the uterus

which have been extensively used to model human placental biology over the years [4]. Choriocarcinoma cell lines are advantageous because the cells proliferate in vitro, express most trophoblast markers from both syncytiotrophoblast and cytotrophoblast type cells and secrete hormones and growth factors associated with the placenta. The major disadvantage of using these lines is that because these cells are derived from tumors it is likely that their biology will not reflect normal placental cells [5]. In addition, although these lines contain a heterogenous mixture of both cytotrophoblast and syncytiotrophoblast, it is difficult to discern exactly which stage of development these cells might correspond to. Further, as these cells grow as a monolayer it is near-impossible to study critical structural aspects of placental biology such as villi formation and the phenomenon whereby an underlying layer of cytotrophoblast gives rise to syncytiotrophoblast.

Explant cultures are another common model system for human placental biology. For explant cultures, bona fide human placental tissue is used. This tissue is typically from term placentas and is donated after the baby has been delivered. However, because the placenta is a transient organ that is programmed to perish following birth, the tissue from term placentas, which have reached the end of their lifespan, has limited usefulness in the study of placental development. Some developments have been made where trophoblast "stem" cells can be isolated as "side" populations from term placentae. Both cytotrophoblast and syncytiotrophoblast can be differentiated from these side-population trophoblasts, suggesting that there is a potential for these cells to be used to study human placentation [6]. In rarer cases, placental tissue from the second trimester (13–26 weeks of gestation) or even first trimester (0–12 weeks of gestation) can be sourced from unsuccessful pregnancies. Unfortunately, it is possible that the procedures involved in harvesting this tissue can damage the cells within the explants. Further, explants do not typically grow well in culture and cannot be maintained in culture for extended periods of time [7, 8]. The exception to this could be cells that are isolated from the chorion in very early (7-week-old) placentas. Cells isolated from this tissue have recently been shown mimic a cytotrophoblast population and differentiate into syncytiotrophoblast like cells in culture [9]. However, the generation of these cells is still difficult due to the scarcity and ethical sensitivity of harvesting early human fetal tissues.

Small laboratory animals have also been heavily relied upon to investigate the events that occur during early placentation. Laboratory animals are however not ideal models of human pregnancy due to significant differences in placental structure and as such it is difficult to interpret what we learn from animals into a human context. As an example, women have a single uterus (except in rare examples) but a mouse has two uterine horns and embryos will grow in both. In addition, mice have a much shorter window of development (implantation occurs 4 days postfertilization and gestation are 19–20 days) relative to humans. Further, the cell types found in the mouse uterus and the mouse placenta differ vastly to those found in human reproductive organs (for a recent review see [10]).

4 Modeling Human Placentation with Human Embryonic Stem Cells

The generation of human embryonic stem cells (hESC) in 1998 marked the beginning of a new era in the study of development [11]. hESC were subsequently used in countless studies to understand more about pluripotency and the development of the embryonic lineages (endoderm, mesoderm, and ectoderm). The original study also revealed that hESC could differentiate into trophoblast cells, yet this feature of hESC biology has not been exploited to the same extent. Four years after the discovery of hESCs, Xu et al. reported that treating hESCs with the growth factor BMP4 could enhance trophoblast differentiation in vitro [12]. Following this initial report, many other studies reported the induction of trophoblast differentiation from hESCs using a combination of growth factors such as BMP4 [13–15] or BMP4 treatment in combination with FGF and Activin/Nodal signaling inhibition [16–19]. However, generating trophoblast like cells from hESC has been met with some criticism [20]. This criticism came from researchers who attested that hESC are pluripotent and not totipotent, and as such should not be able to differentiate into extraembryonic tissues. Instead, it was argued that the trophoblast cells which Xu et al., Thomson et al., and others had described were representative of mesodermal lineages [21]. Since then, many groups have made significant progress towards establishing that hESC can generate legitimate human trophoblast-like cells. Confirmation of a trophoblast-like lineage has been provided from studies that directly compare the transcriptome and proteome of hESC generated trophoblasts to bona fide human placental tissue from different trimesters. These studies have found that hESC derived trophoblast like cells match the methylome of isolated trophoblast cells from the placenta [22] and exhibit very similar gene expression profiles to that of primary first trimester human placental cells [23]. A comprehensive analysis of hESC differentiation into multiple lineages also revealed that hESC derived trophoblast like cells have a gene expression profile that is more like profiles of human placental tissue and primary cultured cytotrophoblast cells, than to cells that are of mesodermal lineage and are derived from the stroma or amnion. Also, these trophoblast-like cells appear to differentiate into cells which have a very similar gene expression profile to both syncytiotrophoblast and extra villous trophoblast [24]. In addition, it has recently been shown that syncytiotrophoblast differentiated in vitro from hESC derived trophoblast-like cells exhibit gene expression profiles similar to very early human syncytium [25]. Proteomic analysis of hESC derived trophoblast like cells show that the proteomic profile of epigenetic proteins is very consistent with that of first trimester placenta derived trophoblasts [26].

One of the difficulties in conclusively determining whether hESC generated trophoblast like cells are bona fide examples of early human trophoblast cells arises from the fact that we simply do not know enough about what an early human trophoblast cell is. In animal models such as mice, trophoblast stem (TS) cells have been isolated from blastocyst outgrowths [27]. These TS cells are like embryonic stem (ES) cells in that they are self-renewing and can be induced to differentiate into trophoblast lineages. However, not only have we failed so far at generating human TS cells, but we do not know whether a human TS cell population even exists (for an excellent review on "true" trophoblasts see [28]). Recent work has begun to optimize culture conditions to promote the maintenance of both placental and blastocyst derived trophoblast cells which appear to be "stem" like and capable of differentiating into mature trophoblast lineages [29]. However, it is still difficult for us to ascertain whether these cells are true human TS cells as we do not have a complete picture of what a human TS cell should look like. It is also likely that the TS cell markers which we use to assess populations of mouse TS cells may not be applicable to human TS cells [20]. Instead, the most reliable marker of human TS like cells is the ability of these cells to generate mature trophoblast lineage cells such as proliferative cytotrophoblasts and ultimately nonproliferative, fused syncytiotrophoblast. Despite some limitations, for many researchers, hESCs are currently the best option to study the "birth" of human TS cells and their subsequent differentiation into mature trophoblast lineages. The remaining parts of this chapter will explore the approaches currently being employed to create the best possible model of early human placentation using hESC derived trophoblast cells.

5 The 3D Modeling Era

The limitations of the aforementioned model systems reveal the need to have access to a more robust and legitimate model of human placenta biology. This is particularly true for initial stages of placental development around the window of implantation and initial villus formation, which we currently know very little about. There is a need for new models of human placentation which could negate the ethical sensitivities involved in performing research on human placentae while improving our knowledge of human placental biology.

In the body, cells typically grow in a 3D environment. Despite this, most areas of scientific research which use cell culture grow cells in two-dimensional (2D) monolayers. This is largely due to the ease of growing cells in monolayers rather than 3D structures. Three-dimensional cell culture can improve our ability to model true cell biology. This is because a 3D structure can facilitate the growth of cells in a manner that is similar to their natural microenvironment. A 3D culture system is more likely to recapitulate environmental cues which are highly likely to impact upon many aspects of the cell's biology, such as mechanical stresses, cell-to-cell contact and cell crosstalk which relies on mesoscopic levels of organization. The beneficial impact of such 3D culture systems is likely to be especially critical for cells which are typically found within highly organized organs. Several types of cells including cells from liver, prostate and breast are now often grown in 3D structures. Data from multiple sources indicates that if 2D and 3D cultured cells have striking differences in key biological features such as morphology, gene and protein expression, motility, mobility, and even differentiation state [30].

A large driving factor for our new ability to culture cells in 3D is the invention of multiple "artificial" extracellular matrix compounds. Currently, the most commonly used compound is "Matrigel," a mixture of basement membrane proteins from mouse sarcoma tissue that are soluble and sterile [31]. When cold, Matrigel exists as a liquid, yet when warmed to 37 °C, Matrigel will solidify to form a gelatinous material. In vitro cultured cells can be cultured underneath, on top of, or even inside solidified Matrigel material. If cells are grown inside Matrigel then grown in the typical 2D monolayer format can be avoided as cells are free to associate with other cells in an orientation of their choosing. In addition to facilitating 3D organization, Matrigel is also rich in growth factors as well as laminin and collagen fragments which together can provide in vitro cultured cells with signals that are otherwise only found in living organs or tissue. Indeed, many different formulations of Matrigel are now sold commercially, with each formulated and tested to enhance the growth of different cell types (for a review on how different cells respond differently to Matrigel see [32]).

The 3D growth of cells grown in vitro can also be facilitated by altering the shape or orientation of a culture vessel. If cultured cells are provided with inert scaffolds onto which cells can attach, then growth can be encouraged into almost any shape. The use of scaffolds in cell culture has been catapulted due to the invention and now widespread commercialization of 3D printing technology. Printers can either directly print scaffold material or print negative molds into which inert materials can be made into any desired shape (for a review see [33]). In the following sections, the role of both Matrigel and 3D scaffolding will be discussed in the context of modeling the human uterus and the developing placenta.

6 Modeling Human Placentation in 3D

There have been two recent reports of 3D placental organoids or "placentoid" culture using human placental cells. Research from Haider et al. and Turco et al. has shown that 3D placentoid structures can be made from first trimester placental isolates and that these placentoids appear to self-renew [34, 35]. However, these models are limited as (1) cells must still be sourced from ethically sensitive placental tissue and (2) the placental structure does not resemble true placental villus structure as cytotrophoblasts are located on the outer edge of the structure and syncytiotrophoblast are located internally. This structural flaw makes it difficult to study placental function (i.e., exchange across the syncytiotrophoblast layer) and does not allow researchers to adequately study "normal" placental development.

The first evidence that hESCs and 3D culture could be utilized as a model of early placentation came from a paper showing that embryoid bodies made with hESCs developed structures reminiscent of human placental tissue [36]. In these studies, hESCs were induced to form embryoid bodies (EBs) through growth on a nonadherent surface following FGF2 withdrawal. Eight-day old EBs were transferred into small, 100 μl volumes of solidified droplets and grown for up to 8 weeks. Makers for both cytotrophoblast and syncytiotrophoblast cells were found in the cultures and placental hormones were secreted from the structures. From approximately 2 weeks the structures displayed intricate cell outgrowths, which morphologically appeared similar to early villous trophoblast structures. Despite encouraging results, generating trophoblast lineages through this method was not efficient. However, this work had demonstrated that encouraging cells to grow in a 3D organization facilitated the differentiation of trophoblast cells and the formation of placental like structures not otherwise seen in 2D culture models.

Following the work of Gerami-Naini et al., several other groups also showed that human trophoblast like cells could be generated using an EB model. One study used a semisolid medium containing 0.6% methylcellulose to facilitate EB growth in 3D [37]. In addition to generating trophoblast like cells, the method used by Peiffer et al. also resulted in the differentiation of endothelial like cells in culture, but only when the semisolid medium was used. Given that the placenta is a highly vascularized organ, the codifferentiation of trophoblast cells and endothelial cells, which in vivo would initiate early vasculature, is a particularly promising model for studying early human placentation. Another study showed that hESCs could generate cells that very closely resembled cytotrophoblasts and spontaneously formed spheroid structures. This was not triggered through 3D culture but by treatment with fibroblast growth factor 4 (FGF4) and heparin [38], which are typically used to maintain cultures of mouse TS cells [27].

To further facilitate trophoblast differentiation from hESC generated EBs several groups have experimented with culture substrates following EB formation. Sivasubramaiyan et al. used agarose, gelatin, and fibronectin to create a "biomimetic" platform [39]. This platform appeared to enhance trophoblast differentiation, measured by heightened secretion of placental hormones hCG and progesterone.

Other groups have attempted to refine the trophoblast cell populations obtained through EB formation using successive rounds of selection for cells secreting high levels of hCG [40]. These studies appear to specifically facilitate the generation of cytotrophoblasts which can be maintained in culture and spontaneously give rise to syncytiotrophoblast populations. Although the induced cytotrophoblast populations were grown in monolayers, Udayashankar et al. also performed coculture experiments where trophoblast cells which grew in vessel like structures were grown with human uterine stromal cells. When grown in a coculture system the hESC derived trophoblast vessels became highly invasive, and the invasiveness of these cells was influenced by different oxygen levels. These studies suggest that (1) culture environment is critical for directed hESC differentiation and (2) that hESC derived trophoblast can mimic both the invasive nature and the oxygen sensitivity of early trophoblast which is essential for successful implantation and placentation [41].

An alternative to growing hESC derived trophoblast cells from EBs is to grow monolayer differentiated trophoblast cells in 3D. hESCs differentiating into trophoblast like cells using BMP4 and inhibition of FGF2 and activin/nodal tend to self-aggregate into subpopulations of "like" cells (cytotrophoblast and syncytiotrophoblast) and grow into 3D structures even when grown as monolayers (Kafer, G.R., Kamei, K. and Carlton P.M., unpublished) (Fig. 2). This observation suggested it may be possible to facilitate bona fide placental structures in vitro by providing a means through which the self-aggregation of trophoblast like cells could be enhanced. 3D-printing technology has the advantage of allowing the creation of almost any structure imaginable. Using this technology, a variety of surfaces can be cast from printed "molds" and cells can be grown on different surfaces in effort to move away from standard flat monolayer culture systems. Virtually any cell culture safe, nontoxic material can be used to cast these molds. One such material which has been extensively used is polydimethylsiloxane (PDMS). PDMS is optically clear, inert, and nontoxic organosilicon. PDMS can be made in any lab by mixing two liquid components, which solidify within several hours. As a liquid, PDMS has a very low flow time, meaning that the material will cover any surface and faithfully mold to any small surface imperfections. Using 3D-printed resin molds, PDMS castings of "synthetic" placental villi of varying shapes and sizes can be generated (Kafer, G.R., Kamei, K., and Carlton P.M., unpublished) (Fig. 3a–c). hESC differentiating into trophoblast cells can be grown on PDMS villi that are first overlaid with Matrigel (Fig. 3d, e). Due to the optically clear nature of PDMS, cells grown on these synthetic villi can be directly processed for microscopic analysis (Fig. 3f, g). hESC generated trophoblast cells growing on PDMS villi are maintainable in culture for longer periods than cells grown in monolayers (Fig. 4a–h).

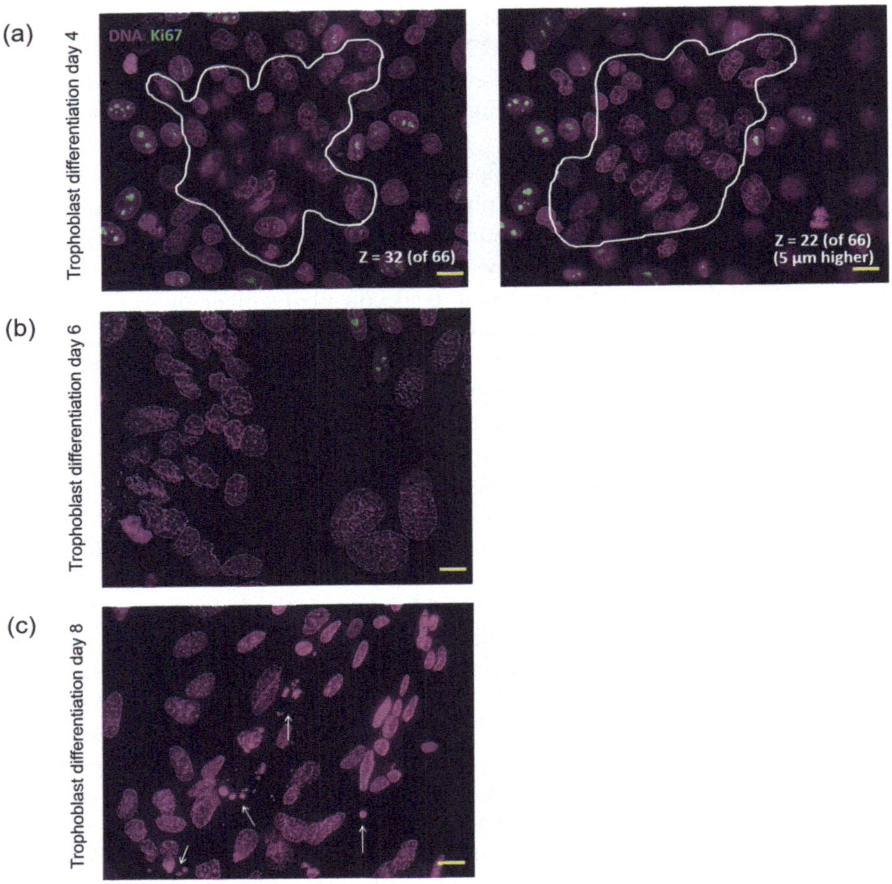

Fig. 2 Immunofluorescent images of hESC which have been induced to differentiate for 4, 6 or 8 days. Cells were grown on glass and stained with Ki67 to mark proliferative cells. Cells that are proliferating are hypothesized to be cytotrophoblast like while cells which are not proliferating and Ki67 negative are hypothesized to be syncytiotrophoblast like cells. (**a**) Cytotrophoblast and syncytiotrophoblast like cells appear to cluster into like populations after 4 days of differentiation. Cytotrophoblast like cells appear to populate the flat surface, while syncytiotrophoblast cells grow into structures that appear to be higher than typical monolayer cell cultures (outlined in white). (**b**) As differentiation proceeds more cells become syncytiotrophoblast like and loose proliferative markers. (**c**) By 8 days in culture no Ki67 positive cells are found and cells undergo apoptosis (white arrows) and detach from the glass surface they were growing on. Yellow scale bars = 10 µm

Interestingly, if the PDMS molds are inverted and cells are allowed to grow upon the synthetic villi in a hanging culture system then self-aggregation into cytotrophoblast and syncytiotrophoblast populations is encouraged (Kafer, G.R., Kamei, K., and Carlton P.M., unpublished). More specifically, proliferating cytotrophoblast cells become located at the base of the villi and nonproliferating syncytiotrophoblast cells appear to extend from the ends of the villi (Kafer, G.R., Kamei, K., and

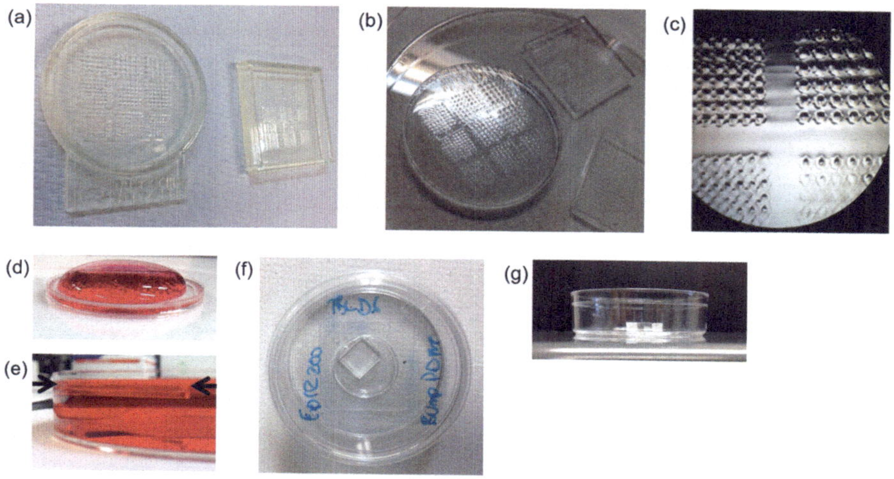

Fig. 3 Polydimethylsiloxane (PDMS) villi molds. (**a**) Resin molds created using 3D-printer technology. (**b**) PDMS castings of villi of different sizes and shapes. (**c**) Magnified illustration of artificial PDMS villi. (**d**) Illustration of Matrigel coating of the PDMS villi. (**e**) Illustration of the inverted floating PDMS disk. Black arrows indicate the floating PDMS disk. (**f**, **g**) Illustration of PDMS which has been used as a surface to culture cells. Cells cultured on PDMS can be fixed and processed for immunofluorescence analysis. PDMS can be easily cut and mounted onto a glass bottom dish for microscopy

Carlton P.M., unpublished) (Fig. 4f). In this system, endothelial-like cell structures also appear to develop (Kafer, G.R., Kamei, K., and Carlton P.M., unpublished) (Fig. 4h), like what has been previously described [37]. In addition, extended culture of hESC derived trophoblast cells can result in the accumulation of free-floating accumulations of cells (Kafer, G.R., Kamei, K., and Carlton P.M., unpublished) (Fig. 4i). These cell accumulations were similar in size to structures referred to as syncytial nuclear aggregates (SNA) or "macrovesicles" [1, 2]. SNAs are generated as a result of normal human placental function and consist of groups syncytiotrophoblast cells which have detached from the placenta and are shed into the maternal blood, These structures have a defined size of between 25 and 200 μm and increased SNA shedding has been implicated in the development of the life-threatening pregnancy associated condition preeclampsia (reviewed in [42]. Collectively, it appears that the facilitated growth of hESC derived trophoblast on synthetic villi by hanging culture may provide a model through which both villi growth and SNA formation could be studied.

In addition to in vitro models of early villi growth and cytotrophoblast to syncytiotrophoblast formation, several groups are also using 3D culture to study very early trophoblast outgrowth. To study initial blastocyst-stage trophoblast outgrowth, a recent study has described the generation of spheroids from hESC derived trophoblast cells by facilitating 3D culture [43]. Culture vessels known as "Aggrewells" were used to achieve 3D structures. Aggrewells are essentially cell culture plates

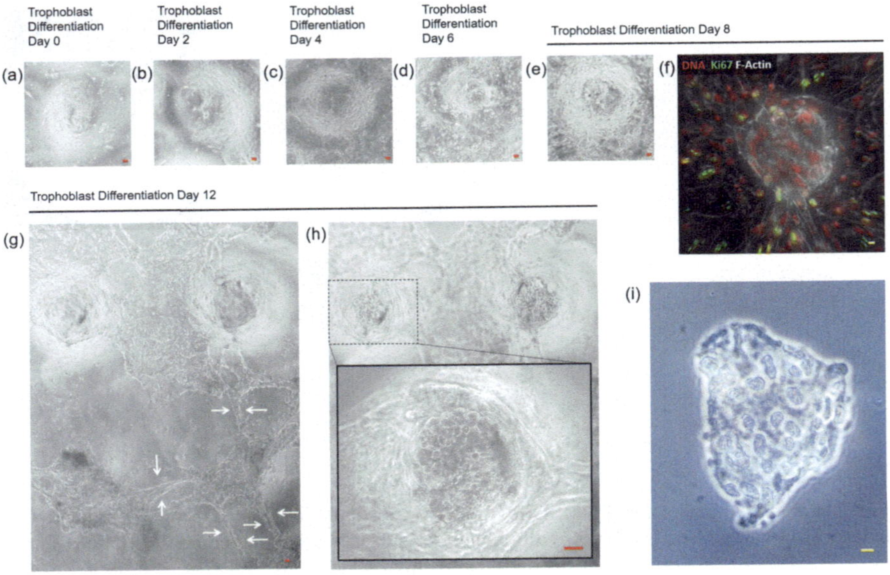

Fig. 4 Morphology of trophoblast cells derived from hESC and grown on Matrigel coated PDMS villi structures. (**a–e**) Bright field images of differentiating cells growing on PDMS molds. Cells were grown on inverted, free-floating PDMS disks to simulate a hanging culture system. Cells adhered well to the Matrigel coated PDMS and at 8 days of differentiation did not appear to undergo apoptosis. (**f**) Day 8 trophoblast cells were stained using immunofluorescence. Cell nuclei are shown using DAPI stain (red), proliferative cells are shown using Ki67 immunolabeling and F-actin was used to stain cytoplasm. Unlike cells grown as monolayers on glass, cells grown on PDMS villi molds still have populations of proliferative cytotrophoblast cells 8 days following the induction of trophoblast differentiation. Cells growing at the ends of villi molds appeared to be of nonproliferative syncytiotrophoblast lineage. (**g**, **h**) Bright field images of day 12 trophoblast cells taken at the bottom of the focal plane (bottom of the PDMS villi) and (**h**) the same cells imaged from the top of the focal plane (top of the PDMS villi). Higher magnification of cells growing on a PDMS villus shown in the boxed insert. White arrows indicate cylindrical, endothelial-like structures which spontaneously form in culture and appear to connect cells growing on PDMS villi. (**i**) A bright field image of an aggregate of cells found floating in the media at day 12 of differentiation. Yellow scale bars = 10 μm. Red scale bars = 20 μm

with an inner surface made of nonadhesive plastic that is shaped into small funnels. The unique shape and material of the Aggrewells effectively direct cells to group together while preventing the formation of a monolayer. Using this system, Lee et al. have shown that adhesiveness and invasiveness of hESC-derived trophoblast spheroids is heightened, which further supports the notion that to generate good in vitro placentation models a 3D culture system must be utilized. This notion has recently been strengthened by work showing that human embryos grown inside a 3D "embedded" system composed of laminin-rich extracellular matrix (lrECM) can facilitate the growth of postimplantation-like trophoblast cells [44]. The use of an embedded culture system to facilitate embryo growth recently reached new heights with the description of a coculture system that appeared to facilitate embryo formation using only stem cells. Harrison et al. have shown that mouse ES cells cocultured with mouse TS cells in a 3D environment made of Matrigel results in the self-assembly of cells to form structures that very closely mimic the early mouse embryo [45]. The success of this innovative approach suggests that in the future we might be able to utilize 3D culture systems to create human-pseudo embryos by coculturing hESCs with hESC derived trophoblast cells. Such work would not only enable the study of the very early events of implantation and placentation for the first time but also provide the means through which early human placentation can be experimentally manipulated in a manner that is potentially free of ethical concern.

7 The Future and Implications of Placental-Organoid Culture

Aside from providing a model to study how the human placenta develops, new in vitro models of human trophoblast outgrowth and placentation would foreseeably benefit other areas of human health. Models could be used in preclinical trials of drugs and their ability to cross the fetoplacental barrier and the study of how infections, such as HIV or Zika virus, are transmitted to the fetus and the ways in which transmission can be prevented (for a review see [46]). The importance of developing new in vitro models of human placentation to address our knowledge gaps has recently been recognized by leading government bodies. The American National Institute of Health has just recently announced a targeted initiative named the "Human Placenta Project" to study the human placenta by developing new and innovative models of human placentation [47]. Given our growing appreciation for how 3D culture environments can alter cell biology, it is highly likely that any future models of human placentation will benefit from the incorporation of 3D culture systems and the use of synthetic scaffolds. The advent of such systems will not only improve our knowledge of human placentation and our understanding of placenta related medicine, but could also negate the ethical troubles that face current placental research including the use of animals and difficult to obtain human samples.

References

1. Askelund KJ, Chamley LW (2011) Trophoblast deportation part I: review of the evidence demonstrating trophoblast shedding and deportation during human pregnancy. Placenta 32(10):716–723
2. Calvert SJ, Longtine MS, Cotter S, Jones CJP, Sibley CP, Aplin JD, Nelson DM, Heazell AEP (2016) Studies of the dynamics of nuclear clustering in human syncytiotrophoblast. Reproduction 151(6):657–671
3. James JL, Carter AM, Chamley LW (2012) Human placentation from Nidation to 5 weeks of gestation. Part I: what do we know about formative placental development following implantation? Placenta 33(5):327–334

4. Rothbauer M, Patel N, Gondola H, Siwetz M, Huppertz B, Ertl P (2017) A comparative study of five physiological key parameters between four different human trophoblast-derived cell lines. Sci Rep 7(1):5892
5. Horii M, Touma O, Bui T, Parast MM (2020) Modeling human trophoblast, the placental epithelium at the maternal fetal Interface. Reproduction 160(1):R1–R11
6. Gamage TKJB, Perry JJ, Fan V, Groom K, Chamley LW, James JL (2020) Side-population trophoblasts exhibit the differentiation potential of a trophoblast stem cell population, persist to term, and are reduced in fetal growth restriction. Stem Cell Rev Rep. https://doi.org/10.1007/s12015-020-09991-8
7. Di Santo S, Malek A, Sager R, Andres A-C, Schneider H (2003) Trophoblast viability in perfused term placental tissue and explant cultures limited to 7–24 hours. Placenta 24(8–9):882–894
8. Newby D, Marks L, Cousins F, Duffie E, Lyall F (2005) Villous explant culture: characterization and evaluation of a model to study Trophoblast invasion. Hypertens Pregn 24(1):75–91
9. Genbacev O, Donne M, Kapidzic M, Gormley M, Lamb J, Gilmore J, Larocque N et al (2011) Establishment of human trophoblast progenitor cell lines from the chorion. Stem Cells 29(9):1427–1436
10. Hafez S (2017) Comparative placental anatomy: divergent structures serving a common purpose. In: Huckle WR (ed) Progress in molecular biology and translational science, vol 145. Academic, Cambridge, MA, pp 1–28
11. Thomson JA, Itskovitz-Eldor J, Shapiro SS, Waknitz MA, Swiergiel JJ, Marshall VS, Jones JM (1998) Embryonic stem cell lines derived from human blastocysts. Science 282(5391):1145–1147
12. Xu R-H, Chen X, Li DS, Li R, Addicks GC, Glennon C, Zwaka TP, Thomson JA (2002) BMP4 initiates human embryonic stem cell differentiation to trophoblast. Nat Biotechnol 20(12):1261–1264
13. Das P, Ezashi T, Schulz LC, Westfall SD, Livingston KA, Michael Roberts R (2007) Effects of fgf2 and oxygen in the bmp4-driven differentiation of trophoblast from human embryonic stem cells. Stem Cell Res 1(1):61–74
14. Schulz LC, Ezashi T, Das P, Westfall SD, Livingston KA, Roberts RM (2008) Human embryonic stem cells as models for trophoblast differentiation. Placenta 29(Suppl A):S10–S16
15. Xu R-H (2006) *In vitro* induction of trophoblast from human embryonic stem cells. Methods Mol Med 121:189–202
16. Amita M, Adachi K, Alexenko AP, Sinha S, Schust DJ, Schulz LC, Michael Roberts R, Ezashi T (2013) Complete and unidirectional conversion of human embryonic stem cells to trophoblast by BMP4. Proc Natl Acad Sci U S A 110(13):E1212–E1221
17. Sarkar P, Randall SM, Collier TS, Nero A, Russell TA, Muddiman DC, Rao BM (2015) Activin/nodal signaling switches the terminal fate of human embryonic stem cell-derived trophoblasts. J Biol Chem 290(14):8834–8848
18. Sudheer S, Bhushan R, Fauler B, Lehrach H, Adjaye J (2012) FGF inhibition directs BMP4-mediated differentiation of human embryonic stem cells to syncytiotrophoblast. Promotion of trophoblast stem cell proliferation by FGF4. Stem Cells Dev 21(16):2987–3000
19. Yang Y, Adachi K, Sheridan MA, Alexenko AP, Schust DJ, Schulz LC, Ezashi T, Michael Roberts R (2015) Heightened potency of human pluripotent stem cell lines created by transient BMP4 exposure. Proc Natl Acad Sci U S A 112(18):E2337–E2346
20. Roberts RM, Loh KM, Amita M, Bernardo AS, Adachi K, Alexenko AP, Schust DJ et al (2014) Differentiation of trophoblast cells from human embryonic stem cells: to be or not to be? Reproduction 147(5):D1–D12
21. Bernardo AS, Faial T, Gardner L, Niakan KK, Ortmann D, Senner CE, Callery EM et al (2011) BRACHYURY and CDX2 mediate BMP-induced differentiation of human and mouse pluripotent stem cells into embryonic and extraembryonic lineages. Cell Stem Cell 9(2):144–155
22. Xie W, Schultz MD, Lister R, Hou Z, Rajagopal N, Ray P, Whitaker JW et al (2013) Epigenomic analysis of multilineage differentiation of human embryonic stem cells. Cell 153(5):1134–1148

23. Telugu BP, Adachi K, Schlitt JM, Ezashi T, Schust DJ, Roberts RM, Schulz LC (2013) Comparison of extravillous trophoblast cells derived from human embryonic stem cells and from first trimester human placentas. Placenta 34(7):536–543
24. Li Y, Moretto-Zita M, Soncin F, Wakeland A, Wolfe L, Leon-Garcia S, Pandian R et al (2013) BMP4-directed trophoblast differentiation of human embryonic stem cells is mediated through a ΔNp63+ cytotrophoblast stem cell state. Development 140(19):3965–3976
25. Yabe S, Alexenko AP, Amita M, Yang Y, Schust DJ, Sadovsky Y, Ezashi T, Michael Roberts R (2016) Comparison of syncytiotrophoblast generated from human embryonic stem cells and from term placentas. Proc Natl Acad Sci U S A 113:E2598
26. Sarkar P, Mischler A, Randall SM, Collier TS, Dorman KF, Boggess KA, Muddiman DC, Rao BM (2016) Identification of epigenetic factor proteins expressed in human embryonic stem cell-derived trophoblasts and in human placental trophoblasts. J Proteome Res 15(8):2433–2444
27. Tanaka S, Kunath T, Hadjantonakis AK, Nagy A, Rossant J (1998) Promotion of trophoblast stem cell proliferation by FGF4. Science 282(5396):2072–2075
28. Gamage TK, Chamley LW, James JL (2016) Stem cell insights into human trophoblast lineage differentiation. Hum Reprod Update 23(1):77–103
29. Okae H, Toh H, Sato T, Hiura H, Takahashi S, Shirane K, Kabayama Y, Suyama M, Sasaki H, Arima T (2018) Derivation of human trophoblast stem cells. Cell Stem Cell 22(1):50–63.e6
30. Picollet-D'hahan N, Dolega ME, Liguori L, Marquette C, Le Gac S, Gidrol X, Martin DK (2016) A 3D toolbox to enhance physiological relevance of human tissue models. Trends Biotechnol 34(9):757–769
31. Kleinman HK, Martin GR (2005) Matrigel: basement membrane matrix with biological activity. Semin Cancer Biol 15(5):378–386
32. Benton G, George J, Kleinman HK, Arnaoutova IP (2009) Advancing science and technology via 3D culture on basement membrane matrix. J Cell Physiol 221(1):18–25
33. Do A-V, Khorsand B, Geary SM, Salem AK (2015) 3D printing of scaffolds for tissue regeneration applications. Adv Healthc Mater 4(12):1742–1762
34. Haider S, Meinhardt G, Saleh L, Kunihs V, Gamperl M, Kaindl U, Ellinger A et al (2018) Self-renewing trophoblast organoids recapitulate the developmental program of the early human placenta. Stem Cell Rep 11(2):537–551
35. Turco MY, Gardner L, Kay RG, Hamilton RS, Prater M, Hollinshead MS, McWhinnie A et al (2018) Trophoblast organoids as a model for maternal-fetal interactions during human placentation. Nature 564(7735):263–267
36. Gerami-Naini B, Dovzhenko OV, Durning M, Wegner FH, Thomson JA, Golos TG (2004) Trophoblast differentiation in Embryoid bodies derived from human embryonic stem cells. Endocrinology 145(4):1517–1524
37. Peiffer I, Belhomme D, Barbet R, Haydont V, Zhou Y-P, Fortunel NO, Li M, Hatzfeld A, Fabiani J-N, Hatzfeld JA (2007) Simultaneous differentiation of endothelial and trophoblastic cells derived from human embryonic stem cells. Stem Cells Dev 16(3):393–402
38. Harun R, Ruban L, Matin M, Draper J, Jenkins NM, Liew GC, Andrews PW, Li TC, Laird SM, Moore HDM (2006) Cytotrophoblast stem cell lines derived from human embryonic stem cells and their capacity to mimic invasive implantation events. Hum Reprod 21(6):1349–1358
39. Sivasubramaiyan K, Totey S, Bhat V, Totey SM, Deb K (2009) Y-27632 enhances differentiation of blastocyst like cystic human embryoid bodies to endocrinologically active trophoblast cells on a biomimetic platform. J Biomed Sci 16:88
40. Udayashankar R, Baker D, Tuckerman E, Laird S, Li TC, Moore HD (2011) Characterization of invasive trophoblasts generated from human embryonic stem cells. Hum Reprod 26(2):398–406
41. Huppertz B, Gauster M, Orendi K, König J, Moser G (2009) Oxygen as modulator of trophoblast invasion. J Anat 215(1):14–20
42. Roland CS, Hu J, Ren C-E, Chen H, Li J, Varvoutis MS, Leaphart LW, Byck DB, Zhu X, Jiang S-W (2016) Morphological changes of placental syncytium and their implications for the pathogenesis of preeclampsia. Cell Mol Life Sci 73(2):365–376

43. Lee Y-L, Fong S-W, Chen ACH, Li T, Yue C, Lee C-L, Ng EHY, Yeung WSB, Lee K-F (2015) Establishment of a novel human embryonic stem cell-derived trophoblastic spheroid implantation model. Hum Reprod 30(11):2614–2626
44. Shahbazi MN, Jedrusik A, Vuoristo S, Recher G, Hupalowska A, Bolton V, Fogarty NNM et al (2016) Self-organization of the human embryo in the absence of maternal tissues. Nat Cell Biol 18(6):700–708
45. Harrison SE, Sozen B, Christodoulou N, Kyprianou C, Zernicka-Goetz M (2017) Assembly of embryonic and extraembryonic stem cells to mimic embryogenesis in vitro. Science 356(6334):eaal1810
46. Arora N, Sadovsky Y, Dermody TS, Coyne CB (2017) Microbial vertical transmission during human pregnancy. Cell Host Microbe 21(5):561–567
47. Guttmacher AE, Maddox YT, Spong CY (2014) The human placenta project: placental structure, development, and function in real time. Placenta 35(5):303–304

Nanofiber Extracellular Matrices in Regenerative Medicine

Ken-ichiro Kamei

1 Introduction

Stem cells hold great promise in regenerative medicine [1–3]. More specifically, mesenchymal stem cells (MSCs) and pluripotent stem cells (PSCs), including embryonic (ESCs) [4]) and induced pluripotent (iPSCs) [5] stem cells, have unique characteristics compared with other tissue-specific stem cells. These characteristics include unlimited self-renewal in vitro and differentiation capability. For practical application of stem cells in regenerative medicine, they need to be cultured in defined conditions with respect to their requirements, and per Good Manufacture Practices (GMP) [6] and Good Cell Culture Practices (GCCP) [7]. In the past decade, numerous studies developed new cell culture media to maintain or differentiate stem cells using a combination of growth factors and small compounds [8–10]. Although such culture media have been developed, the extracellular matrix (ECM) is less discussed because ECMs are difficult to control using conventional cell-culture wares.

In living organisms, ECMs exist in all organs and tissues. ECMs are components of cellular microenvironments. ECMs not only fill extracellular spaces but also form tissue/organ structures, provide chemical/physical cues, act as cell scaffolds, and contain growth factors [11–13]. Each tissue has unique ECM composition and structure; however, ECMs are not stable, but rather dynamic in their chemical [14–22] and physical [23–25] attributes with respect to tissue development and diseases [26–28]. Therefore, to regulate cellular phenotypes and functions, we need to consider what the ECMs are and how they should be utilized. Although adhesive molecules, coated on conventional culture dishes, have been widely used for two decades, the structure of ECM has not been sufficiently utilized, leaving room for improvement in growth efficiency.

K.-i. Kamei (✉)
Institute for Integrated Cell-Material Sciences (WPI-iCeMS), Kyoto University, Kyoto, Japan
e-mail: kamei.kenichiro.7r@kyoto-u.ac.jp

D. O. Wang, D. Packwood (eds.), *Cell-Inspired Materials and Engineering*, Fundamental Biomedical Technologies, https://doi.org/10.1007/978-3-030-55924-3_11

In this chapter, I will discuss how we use ECM technology for applications in regenerative medicine.

2 ECM Sources

2.1 Natural or Recombinant ECM Proteins

ECMs consist of water, proteins, and sugars, and represent the unique structures and properties used to regulate cellular functions. The variety of ECM proteins includes collagens [29, 30], fibronectin [31–33], vitronectin [34], and laminin [31, 35–38]. These proteins are expressed in different locations and show particular functionality in these locations. Several recombinant proteins have been used for cell culture. However, recombinant proteins are difficult in terms of quality control and are costly. Most ECM proteins possess large molecular weights, often result in lowered production yields, and require intensive purification. To tackle this issue, protein engineering helps to use only the functional domains of ECM proteins. For example, laminin-511 can maintain the self-renewal properties of hPSCs [35], but possesses a high molecular weight (780 kDa) [38]. Miyazaki et al. generated a shorter fragment of laminin-511, named the Laminin E8 fragment; this fragment can serve as ECM and can maintain the self-renewal properties of hPSCs [36].

Recombinant proteins are generally produced using bacterial, insect, yeast, or mammalian cells. Each type of host cells has a different protein-modification process; thus, the best approach is to use human cells, such as human embryonic kidney (HEK293) cells, for protein production. However, these cells often result in low production yields, compared with those of other cells.

Conversely, gelatin, which is used for medical capsules, has also been extensively used as cell-culture ECM. Gelatin is a mixture of partially hydrolyzed forms of collagen. Bovine gelatin is used for medical capsules; however, concerns remain about possible immune rejection, as well as contamination by prion proteins, which can cause bovine spongiform encephalopathy in cases of transplantation. Additionally, because of hydrolysis, gelatin does not possess the specific structure of ECMs and its activity, as ECM, is limited.

2.2 Synthetic ECM Polymers

In addition to recombinant proteins, synthetic polymers (e.g., poly-L-lysine) have been used for cell culture as coating on cell culture plates [39] or hydrogels for 3D cell culture [40]. Compared with the requirements for the production of recombinant proteins, these polymers are relatively straightforward to produce, and thus, can be generated in large amounts and with high purity. Most conventional polymers are not designed to target specific cell-adhesion proteins. Block-co-polymers [41, 42], or composites of peptides and proteins [43–46], have been generated to encompass this functionality. However, polymer design and manufacture have not yet been perfected, and there is much room for improvement with respect to their production and application.

2.3 Decellularized Matrix

Finally, there are ECMs obtained from decellularized tissues. In this case, cells are removed using detergents, but the structure of the tissues is preserved. Therefore, most ECM proteins and their structure are preserved, which is useful for whole-tissue reconstruction [47–50]. However, it is challenging to obtain original tissues, for decellularization, from human donors. Although there are trials assessing the use of animal tissues, there are concerns about immune rejection with respect to their use.

3 ECM Nanoengineering

3.1 Nanotechnology for ECM Engineering

Previous studies have used recombinant ECM proteins (e.g., vitronectin [34] and laminin [35–37]) and synthetic polymers [45, 46]; however, this does not ensure the presence of ECM topographical cues, which are crucial for regulating stem-cell phenotypes [23–25, 51]. Therefore, both the materials and topological cues must be taken into consideration. For this, the optimal approach is using nanofiber technology, because it enables the fabrication of nanostructured matrices featuring various materials that can act as artificial ECMs [52–66].

Nanolithography

Nanolithography, such as optical, electron-beam, nanoimprint, multiple-photon, and scanning-probe lithography, has been used to fabricate nanometer scale structure (1–100 nm), and is now applied for fabricating the unique patterns of extracellular matrix structure (Fig. 1) [24, 67, 68]. Lithographical technology allows for the production of the nanotopological structures of ECM at the nanometer scale, which is advantageous for investigating the roles of ECM in cell behavior [67]. Previously, Dalby and colleagues reported that nanopore substrates facilitate the self-renewal properties of mesenchymal stem cells [24, 68]. Conversely, nanolithography is also challenging in practical application. These technologies require silicon, glass, and

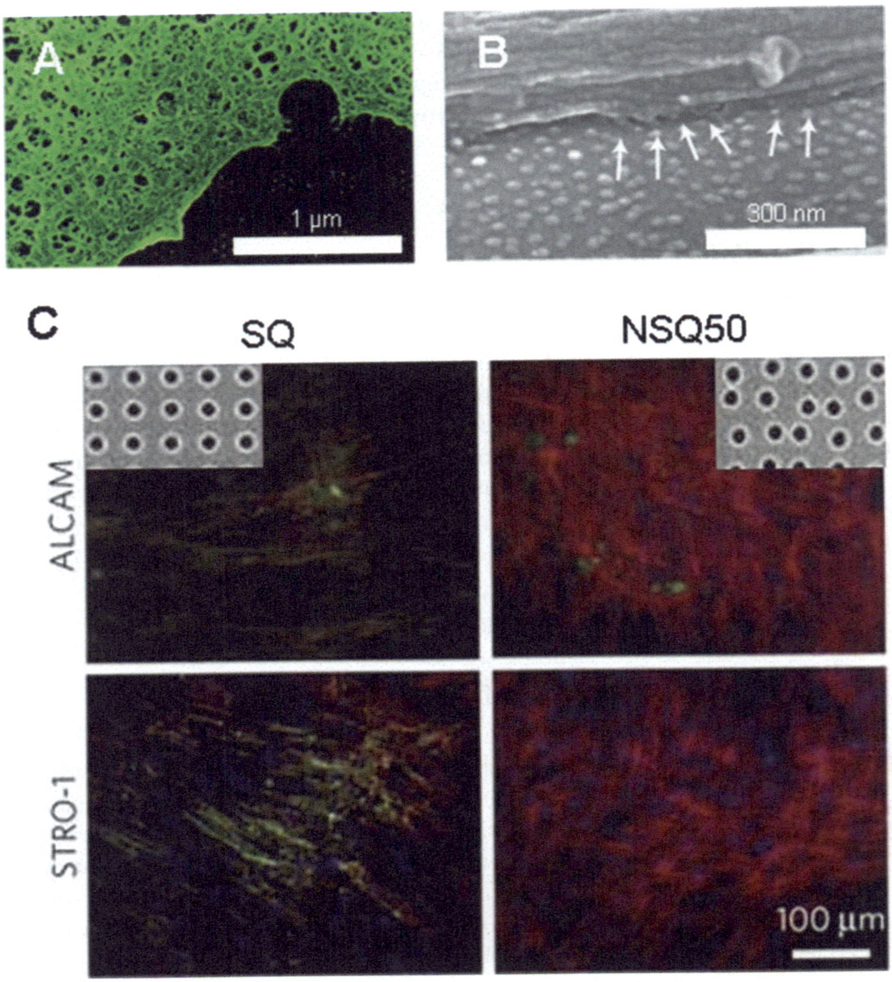

Fig. 1 Nanolithography used to pattern ECM peptides and structures. (**a**) Clusters of RGD peptide nanopatterns in a hexagonal arrangement, and each cluster were separated with 200-nm spaces. A cell was spread on the RGD patterns (A figure reproduced with permission with Ref. [67] by Nature Publishing Group). (**B**) A cell was able to adhere on the topographical cues below 10 nm (A figure reproduced with permission with Ref. [67] by Nature Publishing Group). (**C**) Mesenchymal stem cells (MSCs) cultured on SQ (absolute square lattice symmetry with 120-nm pits) and NSQ50 (a square arrangement with 300-nm center-to-center spacing, but with ±50 nm offset in pit placement in *x–y* axes) stained with MSC markers, such as STRO-1 and ALCAM (activated leukocyte cell adhesion molecule) after 8 weeks. Higher levels of STRO-1 and ALCAM were observed in MSCs cultured on SQ than those of cells cultured on the NSQ50. Since ALCAM is expressed by not only MSCs but also progenitor cells, STRO-1 is a more stringent marker for MSCs. The cells on NSQ50 at 8 weeks expressed ALCAM, but not STRO-1 expressed, and this result suggested that osteoprogenitor cells were still present, but the actual MSCs were reduced. Green, red, and blue indicate phenotypic markers (ALCAM or STRO-1), actin and nucleus, respectively. Scanning electron microscope (SEM) images of the SQ and NSQ50 surfaces are shown inset (A figure reproduced with permission with Ref. [24] by Nature Publishing Group)

even plastics as base materials. This affects flexibility, rendering it difficult to apply the nanofabricated substrate for cell/tissue transplantation. Moreover, these nanolithographical technologies require costly instruments, which can only be operated by trained personnel, resulting in high costs of fabrication. Although this technology allows for the fabrication of nanometer structures, it has not yet been optimized for large-scale production, and the time, required for fabrication, is still too long, such as over a week for 1 cm^2.

Nanofiber Strategy

In addition to nanolithography, electrospinning technology has widely been used for producing nanofibers (diameter: 3–5000 nm) from such diverse materials as synthetic polymers [52, 53, 57, 69–71], biomaterials [61, 62, 72], and combinations thereof [73–75] (Fig. 2). Nanofibers present advantages over conventional ECMs, such as increased cell–ECM interaction and mimicking of in vivo ECM-like structures; this is useful for stem cell research and application. Several research groups, including ours, were able to use simple and inexpensive materials to develop nanofiber matrices that maintain the self-renewal properties of hPSCs [73, 76]. Because this technology allows for the fabrication of relatively large amounts of nanostructured substrates, numerous studies have used it for tissue engineering.

3.2 *Functionalization*

In addition to the ECM nanotopology, the chemical properties of ECMs are also important for regulating cellular function. The surface chemistry of nanofiber ECMs is important for creating peptide and protein composites, stimulating cell signaling

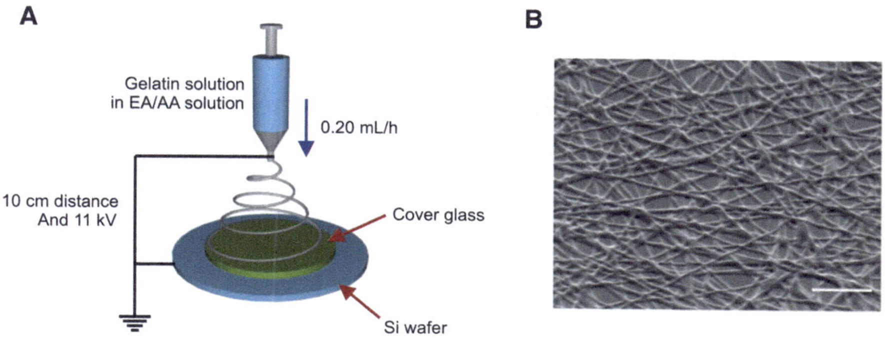

Fig. 2 Electrospinning method to fabricate nanofiber matrix. (**a**) The solution containing materials (here, gelatin) was electrospun on a glass slide placed in the center of a silicon wafer by applying a high voltage between the injection needle and the grounded silicon wafer. (**b**) SEM image of cross-linked gelatin nanofiber matrix. Scale bar, 5 μm (Figures reproduced with permission with Ref. [72] by Elsevier)

pathways, and determining cell fate [77]. These peptides and proteins can be modified into nanofiber ECMs with concentration gradients [78–80]. Such nanofiber ECMs, with gradient modification, allow us to investigate the mechanisms of stem cell differentiation and migration.

In addition to peptides and proteins, nanomaterials can also be integrated into nanofiber ECMs to provide additional functionality. For example, nanofiber ECMs embedded with gold nanoparticles facilitate cardiogenic differentiation of MSCs [81, 82]. Here, gold nanoparticles with a diameter range between 15 and 30 nm are used to adjust the stiffness of ECMs, mimicking that of the myocardium. Additionally, integrating bioactive glass nanoparticles with collagen nanofiber ECMs exerts synergetic effects on osteogenic and angiogenic differentiation of MSCs [83].

3.3 ECM Screening

As we have shown, tremendous efforts have been devoted to mimicking ECM in vitro. Some studies have developed platforms for screening ECMs, in order to identify optimal conditions and investigate the underlying mechanisms [21, 84, 85]. These methods, however, consider only the materials for the fabrication of ECMs. Only a few research groups, including ours, have reported on screening platforms for nanostructured ECMs [86, 87]. Recently, we reported on a robust screening platform used to analyze the structure of nanofiber ECM that can maintain self-renewing hPSCs [87] (Fig. 3). The platform consisted with a microfluidic structure and a nanofiber array, to test not only nanofiber ECMs but also cell–cell interactions. We named this platform the multiplexed artificial cellular microenvironment (MACME) array. Then, image-based cytometry, followed by statistical analysis, is used to systematically analyze how hPSC self-renewal is altered by cellular environments. This enables quantitative interpretation of individual cellular phenotypic responses to environmental cues.

4 Nanofiber ECMs for Regenerative Medicine and Tissue Engineering

4.1 Scaled-Up Culture

Establishing efficient and robust 3D culture systems is one of the main challenges in large-scale cell production for regenerative medicine and drug discovery [1, 3]. Most of the conventional 3D culture systems, such as suspension [88–92], microcarriers [93, 94], hydrogels [95], and microencapsulation [96, 97], have limitations,

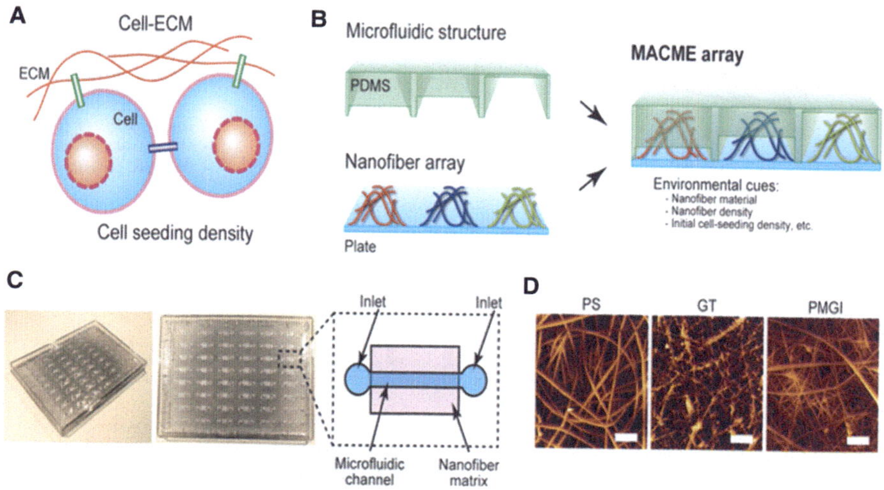

Fig. 3 Screening strategy to identify optimal cellular microenvironments for regulating cellular functions. (**a**) Systematic analysis of cellular microenvironmental cues that regulate cellular functions, by collecting a variety of interactions between cells and extracellular matrices (cell–ECMs) as well as initial cell seeding densities within a single platform. (**b**) Design of the multiplexed artificial cellular microenvironment (MACME) array. The array consists of two parts: a polydimethylsiloxane (PDMS)-based microfluidic structure and a nanofiber array. The microfluidic structure features three microfluidic-channel heights (250, 500, and 1000 μm) to regulate initial cell-seeding densities. (**c**) Photographs of bird and top views of the MACME array and illustration of a microfluidic channel (light blue) with a nanofiber matrix (pink). Each microfluidic channel has two inlets at the ends. (**d**) Atomic force microscopy images of nanofiber matrices made of polystyrene (PS), gelatin (GT), and polymethylglutarimide (PMGI). Scale bars: 6 μm (Figures reproduced with permission with Ref. [87] by John Wiley and Sons)

which cause insufficient cell growth, undesirable differentiation, difficulty in controlling cell-aggregate size, scalability, and hydrodynamic shear stress. Therefore, we established the "fiber-on-fiber (FF)" matrix, which consists of layered nano- and microfibers, for a new scaled-up culture method (Fig. 4). Previously, we have shown that gelatin nanofiber matrix can maintain self-renewing hPSCs long-term [72]; however, this matrix was too fragile to apply for scaled-up cell culture. Although microfiber matrices have a diameter that is too wide to interact with cells, combining nanofiber and microfiber matrices offers improved mechanical stability and flexibility compared with those of nanofiber matrices [76]. Moreover, because of the mesh-like structure, soluble factors in the culture medium are able to penetrate an FF matrix and deliver nutrients without stirring. Thus, compared with conventional 3D culture systems, our method allows us to minimize the mechanical stress on cells during culture. We also demonstrated the many advantages of nanofiber ECMs even for scaled-up culture of hPSCs.

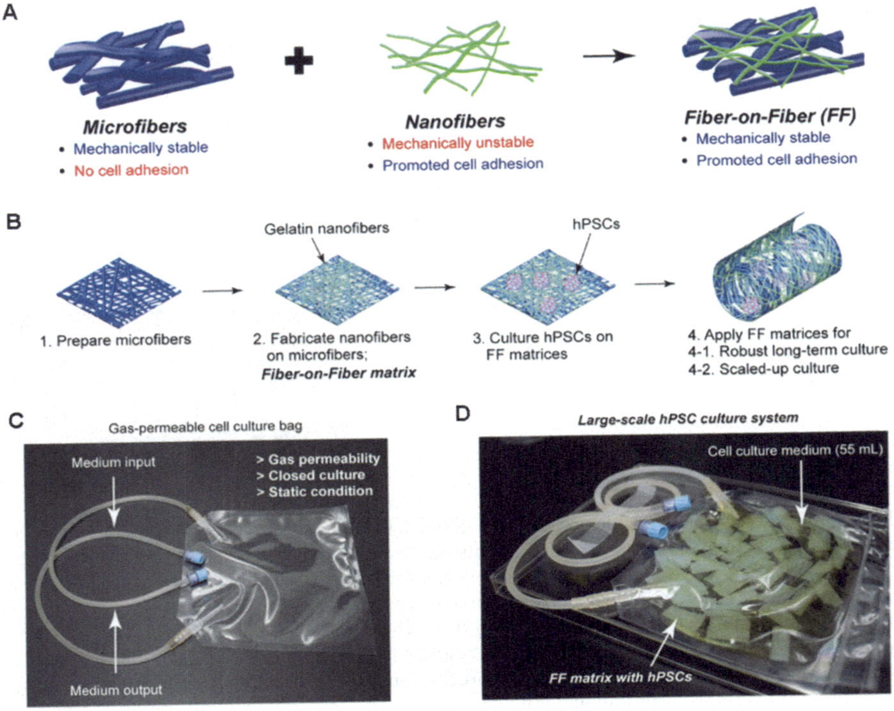

Fig. 4 Fiber-on-fiber (FF) matrix for scaled-up culture of hPSCs. (**a**) Conceptual illustration of an FF matrix. This matrix has a layered structure of a nanofiber mesh on a microfiber sheet. Although the nanofiber matrix allows for facilitating adhesion and growth of hPSCs with maintenance of pluripotency, it is mechanically unstable during culture. On the other hand, while microfibers do not support hPSC adhesion and growth, they have better mechanical stability. Therefore, FF matrix combines the advantages of both nano- and microfibers. (**b**) The procedure of FF matrix fabrication and hPSC culture. An FF matrix allows for folding, even with cultured cells. (**c**) A gas-permeable cell culture bag used for scaling up the hPSC culture. The bag occupies minimal space in an incubator, and maximizes the cell number/culture medium volume ratio. The tubings of bags allow for exchanging culture medium without contamination. (**d**) Photographs of the culture system. Sixty FF matrices cultured with hPSCs were placed in a gas-permeable culture bag. Figures reproduced with permission with Ref. [76] by Elsevier

4.2 *Targeted Differentiation*

Nanofiber ECMs, which facilitate cellular signaling pathways, play important roles in direct differentiation of stem cells to targeted tissue sites. Yamazoe et al. indicated that synthetic nanofiber ECMs facilitate in vitro differentiation of PSCs into hepatocyte-like cells via activation of Rac1 [98]. Several groups have reported that nanofiber ECMs facilitate cardiac differentiation of PSCs via activation of the Wnt/β-catenin signaling pathway [99]. Another advantage of nanofiber ECMs involves our ability to fabricate aligned nanofibers. Because some tissues, such as muscle and heart, possess aligned cells, it is very useful to mimic this tissue construction. Indeed, aligned nanofiber ECMs facilitate the differentiation process of PSCs towards cardiomyocytes [69, 100]. Nanofibers also show great promise for neural regeneration because neurons have fiber-like axon structures, and nanofiber ECMs facilitate axon guidance [101]. Hyssalo et al. showed that nanofiber ECMs facilitate the differentiation of hPSCs into neurons, astrocytes, and oligodendrocyte precursor cells in vitro [102]. Bone microenvironments require rigid conditions to differentiate stem cells for bone regeneration. Carbon nanotubes can be incorporated into nanofiber ECMs to increase the mechanical properties of the ECMs, enhancing bone regeneration in vivo and in vitro [103]. Because the chemical and physical properties of nanofiber ECMs can be optimized, these ECMs are useful for stem cell differentiation into targeted tissues, which is difficult to achieve with conventional ECMs.

4.3 Transplantation

One of the current challenges in cell transplantation is widespread cell death in patients or host animals. The transplanted cells, or tissues, show low engraftment efficiency caused by the lack of interaction between transplanted cells and host tissues, as well as low survival rates due to immune reaction from the host. Nanofiber ECMs may improve the efficiency of cell/tissue transplantation. Wang et al. demonstrated that functionalized nanofiber ECM with glial derived neurotrophic factor (GDNF), which enhances cell survival, improves engraftments of dopaminergic progenitor cells in parkinsonian mice [104]. Moreover, nanofiber ECMs, which consist of graphene oxide and biopolymer (such as chitosan and polyvinylpyrrolidone), improve wound healing in rat models [105]. Self-assembled, peptide-based, nanofiber ECMs improve the efficiency of bone regeneration using human iPSCs [106]. Thus, nanofiber ECMs have great potential in cell transplantation.

5 Future Perspectives

As we have discussed, nanofiber ECMs are likely to play important roles in applying stem cells for tissue engineering and regenerative medicine. To optimize nanofiber ECM for these applications, we need to integrate nanoengineering, material science, and cell biology. Although numerous nanofiber ECMs are mentioned in this chapter, the mechanisms underlying regulation of cell behavior, by nanofiber ECMs, remain unclear. To further develop nanofiber ECMs, it is critical to delineate these mechanisms; cellular biological analyses, such as transcriptomics, proteomics, and phenomics, combined with material informatics [107, 108], will help us achieve this understanding.

It is also important to investigate how natural ECMs work to form well-structured tissues/organs and regulate cellular functions. To generate advanced nanofiber ECMs, it is important to understand the natural ECMs.

Finally, the advantages of artificial nanofiber ECMs include novel functionalities that do not exist in natural biological systems. Examples of such recent progress are wearable devices to monitor physiological changes; nanofiber ECMs will contribute to advancement of these devices [109]. This approach allows for drug delivery or treatment that is controlled from the outside of the body. Thus, nanofiber ECMs can contribute to improving the quality of life.

Acknowledgments Funding was generously provided by the Japan Society for the Promotion of Science (JSPS; 17H02083 and 16K14660). WPI-iCeMS is supported by the World Premier International Research Centre Initiative (WPI), MEXT, Japan.

References

1. Serra M, Brito C, Correia C, Alves PM (2012) Process engineering of human pluripotent stem cells for clinical application. Trends Biotechnol 30(6):350–359. https://doi.org/10.1016/j.tibtech.2012.03.003
2. Pouton CW, Haynes JM (2007) Embryonic stem cells as a source of models for drug discovery. Nat Rev Drug Discov 6(8):605–616. https://doi.org/10.1038/nrd2194
3. Chen KG, Mallon BS, McKay RD, Robey PG (2014) Human pluripotent stem cell culture: considerations for maintenance, expansion, and therapeutics. Cell Stem Cell 14(1):13–26. https://doi.org/10.1016/j.stem.2013.12.005
4. Thomson JA, Itskovitz-Eldor J, Shapiro SS, Waknitz MA, Swiergiel JJ, Marshall VS, Jones JM (1998) Embryonic stem cell lines derived from human blastocysts. Science 282(5391):1145–1147
5. Takahashi K, Tanabe K, Ohnuki M, Narita M, Ichisaka T, Tomoda K, Yamanaka S (2007) Induction of pluripotent stem cells from adult human fibroblasts by defined factors. Cell 131(5):861–872. https://doi.org/10.1016/j.cell.2007.11.019
6. Sensebé L, Gadelorge M, Fleury-Cappellesso S (2013) Production of mesenchymal stromal/stem cells according to good manufacturing practices: a review. Stem Cell Res Ther 4(3):66. https://doi.org/10.1186/scrt217
7. Pamies D (2016) Good cell culture practice for stem cells and stem-cell-derived models. Altex. https://doi.org/10.14573/altex.1607121
8. Usta SN, Scharer CD, Xu J, Frey TK, Nash RJ (2014) Chemically defined serum-free and xeno-free media for multiple cell lineages. Ann Transl Med 2(10):97. https://doi.org/10.3978/j.issn.2305-5839.2014.09.05
9. Chen G, Gulbranson DR, Hou Z, Bolin JM, Ruotti V, Probasco MD, Smuga-Otto K, Howden SE, Diol NR, Propson NE, Wagner R, Lee GO, Antosiewicz-Bourget J, Teng JMC, Thomson JA (2011) Chemically defined conditions for human iPSC derivation and culture. Nat Methods 8(5):424–429. https://doi.org/10.1038/nmeth.1593
10. Burridge PW, Holmström A, Wu JC (2015) Chemically defined culture and cardiomyocyte differentiation of human pluripotent stem cells. Curr Protoc Hum Genet 87:21.3.1–21.3.15. https://doi.org/10.1002/0471142905.hg2103s87
11. Frantz C, Stewart KM, Weaver VM (2010) The extracellular matrix at a glance. J Cell Sci 123(Pt 24):4195–4200. https://doi.org/10.1242/jcs.023820

12. Discher DE, Mooney DJ, Zandstra PW (2009) Growth factors, matrices, and forces combine and control stem cells. Science 324(5935):1673–1677. https://doi.org/10.1126/science.1171643
13. Joyce JA, Pollard JW (2009) Microenvironmental regulation of metastasis. Nat Rev Cancer 9(4):239–252. https://doi.org/10.1038/nrc2618
14. Danhier F, Feron O, Preat V (2010) To exploit the tumor microenvironment: passive and active tumor targeting of nanocarriers for anti-cancer drug delivery. J Control Release 148(2):135–146. https://doi.org/10.1016/j.jconrel.2010.08.027
15. Murphy WL, McDevitt TC, Engler AJ (2014) Materials as stem cell regulators. Nat Mater 13(6):547–557. https://doi.org/10.1038/Nmat3937
16. Patel AK, Celiz AD, Rajamohan D, Anderson DG, Langer R, Davies MC, Alexander MR, Denning C (2015) A defined synthetic substrate for serum-free culture of human stem cell derived cardiomyocytes with improved functional maturity identified using combinatorial materials microarrays. Biomaterials 61:257–265. https://doi.org/10.1016/j.biomaterials.2015.05.019
17. Zhang R, Mjoseng HK, Hoeve MA, Bauer NG, Pells S, Besseling R, Velugotla S, Tourniaire G, Kishen REB, Tsenkina Y, Armit C, Duffy CRE, Helfen M, Edenhofer F, de Sousa PA, Bradley M (2013) A thermoresponsive and chemically defined hydrogel for long-term culture of human embryonic stem cells. Nat Commun 4:ARTN 1335. https://doi.org/10.1038/ncomms2341
18. Anderson DG, Putnam D, Lavik EB, Mahmood TA, Langer R (2005) Biomaterial microarrays: rapid, microscale screening of polymer-cell interaction. Biomaterials 26(23):4892–4897. https://doi.org/10.1016/j.biomaterials.2004.11.052
19. Celiz AD, Smith JGW, Patel AK, Hook AL, Rajamohan D, George VT, Flatt L, Patel MJ, Epa VC, Singh T, Langer R, Anderson DG, Allen ND, Hay DC, Winkler DA, Barrett DA, Davies MC, Young LE, Denning C, Alexander MR (2015) Discovery of a novel polymer for human pluripotent stem cell expansion and multilineage differentiation. Adv Mater 27(27):4006–4012. https://doi.org/10.1002/adma.201501351
20. Hansen A, Mjoseng HK, Zhang R, Kalloudis M, Koutsos V, de Sousa PA, Bradley M (2014) High-density polymer microarrays: identifying synthetic polymers that control human embryonic stem cell growth. Adv Healthc Mater 3(6):848–853. https://doi.org/10.1002/adhm.201300489
21. Mei Y, Hollister-Lock J, Bogatyrev SR, Cho SW, Weir GC, Langer R, Anderson DG (2010) A high throughput micro-array system of polymer surfaces for the manipulation of primary pancreatic islet cells. Biomaterials 31(34):8989–8995. https://doi.org/10.1016/j.biomaterials.2010.08.029
22. Saha K, Mei Y, Reisterer CM, Pyzocha NK, Yang J, Muffat J, Davies MC, Alexander MR, Langer R, Anderson DG, Jaenisch R (2011) Surface-engineered substrates for improved human pluripotent stem cell culture under fully defined conditions. Proc Natl Acad Sci U S A 108(46):18714–18719. https://doi.org/10.1073/pnas.1114854108
23. Bettinger CJ, Langer R, Borenstein JT (2009) Engineering substrate topography at the micro- and nanoscale to control cell function. Angew Chem Int Ed Engl 48(30):5406–5415. https://doi.org/10.1002/anie.200805179
24. McMurray RJ, Gadegaard N, Tsimbouri PM, Burgess KV, McNamara LE, Tare R, Murawski K, Kingham E, Oreffo RO, Dalby MJ (2011) Nanoscale surfaces for the long-term maintenance of mesenchymal stem cell phenotype and multipotency. Nat Mater 10(8):637–644. https://doi.org/10.1038/nmat3058
25. Sun Y, Jallerat Q, Szymanski JM, Feinberg AW (2015) Conformal nanopatterning of extracellular matrix proteins onto topographically complex surfaces. Nat Methods 12(2):134–136. https://doi.org/10.1038/nmeth.3210
26. Bonnans C, Chou J, Werb Z (2014) Remodelling the extracellular matrix in development and disease. Nat Rev Mol Cell Biol 15(12):786–801. https://doi.org/10.1038/nrm3904

27. Trappmann B, Gautrot JE, Connelly JT, Strange DG, Li Y, Oyen ML, Cohen Stuart MA, Boehm H, Li B, Vogel V, Spatz JP, Watt FM, Huck WT (2012) Extracellular-matrix tethering regulates stem-cell fate. Nat Mater 11(7):642–649. https://doi.org/10.1038/nmat3339
28. Brizzi MF, Tarone G, Defilippi P (2012) Extracellular matrix, integrins, and growth factors as tailors of the stem cell niche. Curr Opin Cell Biol 24:645. https://doi.org/10.1016/j.ceb.2012.07.001
29. Pot MW, de Kroon LMG, van der Kraan PM, van Kuppevelt TH, Daamen WF (2017) Unidirectional BMP2-loaded collagen scaffolds induce chondrogenic differentiation. Biomed Mater 13(1):015007. https://doi.org/10.1088/1748-605X/aa8960
30. Ishikawa M, Ohnishi H, Skerleva D, Sakamoto T, Yamamoto N, Hotta A, Ito J, Nakagawa T (2017) Transplantation of neurons derived from human iPS cells cultured on collagen matrix into Guinea-pig cochleae. J Tissue Eng Regen Med 11(6):1766–1778. https://doi.org/10.1002/term.2072
31. Hirata M, Yamaoka T (2017) Effect of stem cell niche elasticity/ECM protein on the self-beating cardiomyocyte differentiation of induced pluripotent stem (iPS) cells at different stages. Acta Biomater 65:44. https://doi.org/10.1016/j.actbio.2017.10.032
32. Cimino M, Goncalves RM, Bauman E, Barroso-Vilares M, Logarinho E, Barrias CC, Martins MCL (2017) Optimization of the use of a pharmaceutical grade xeno-free medium for in vitro expansion of human mesenchymal stem/stromal cells. J Tissue Eng Regen Med 12. https://doi.org/10.1002/term.2588
33. Miroshnikova YA, Rozenberg GI, Cassereau L, Pickup M, Mouw JK, Ou G, Templeman KL, Hannachi EI, Gooch KJ, Sarang-Sieminski AL, Garcia AJ, Weaver VM (2017) alpha5beta1-integrin promotes tension-dependent mammary epithelial cell invasion by engaging the fibronectin synergy site. Mol Biol Cell 28(22):2958–2977. https://doi.org/10.1091/mbc.E17-02-0126
34. Braam SR, Zeinstra L, Litjens S, Ward-van Oostwaard D, van den Brink S, van Laake L, Lebrin F, Kats P, Hochstenbach R, Passier R, Sonnenberg A, Mummery CL (2008) Recombinant vitronectin is a functionally defined substrate that supports human embryonic stem cell self-renewal via alpha V beta 5 integrin. Stem Cells 26(9):2257–2265. https://doi.org/10.1634/stemcells.2008-0291
35. Miyazaki T, Futaki S, Hasegawa K, Kawasaki M, Sanzen N, Hayashi M, Kawase E, Sekiguchi K, Nakatsuji N, Suemori H (2008) Recombinant human laminin isoforms can support the undifferentiated growth of human embryonic stem cells. Biochem Biophys Res Commun 375(1):27–32. https://doi.org/10.1016/j.bbrc.2008.07.111
36. Miyazaki T, Futaki S, Suemori H, Taniguchi Y, Yamada M, Kawasaki M, Hayashi M, Kumagai H, Nakatsuji N, Sekiguchi K, Kawase E (2012) Laminin E8 fragments support efficient adhesion and expansion of dissociated human pluripotent stem cells. Nat Commun 3:10. https://doi.org/10.1038/ncomms2231
37. Rodin S, Domogatskaya A, Strom S, Hansson EM, Chien KR, Inzunza J, Hovatta O, Tryggvason K (2010) Long-term self-renewal of human pluripotent stem cells on human recombinant laminin-511. Nat Biotechnol 28(6):611–U102. https://doi.org/10.1038/nbt.1620
38. Sroka IC, Chen ML, Cress AE (2008) Simplified purification procedure of laminin-332 and laminin-511 from human cell lines. Biochem Biophys Res Commun 375(3):410–413. https://doi.org/10.1016/j.bbrc.2008.08.029
39. Shimizu T, Yamato M, Kikuchi A, Okano T (2001) Two-dimensional manipulation of cardiac myocyte sheets utilizing temperature-responsive culture dishes augments the pulsatile amplitude. Tissue Eng 7(2):141–151. https://doi.org/10.1089/107632701300062732
40. Kamei KI, Koyama Y, Tokunaga Y, Mashimo Y, Yoshioka M, Fockenberg C, Mosbergen R, Korn O, Wells C, Chen Y (2016) Characterization of phenotypic and transcriptional differences in human pluripotent stem cells under 2D and 3D culture conditions. Adv Healthc Mater 5(22):2951–2958. https://doi.org/10.1002/adhm.201600893

41. Nagase K, Hatakeyama Y, Shimizu T, Matsuura K, Yamato M, Takeda N, Okano T (2015) Thermoresponsive cationic copolymer brushes for mesenchymal stem cell separation. Biomacromolecules 16(2):532–540. https://doi.org/10.1021/bm501591s
42. Nagase K, Hatakeyama Y, Shimizu T, Matsuura K, Yamato M, Takeda N, Okano T (2013) Hydrophobized thermoresponsive copolymer brushes for cell separation by multistep temperature change. Biomacromolecules 14(10):3423–3433. https://doi.org/10.1021/bm4006722
43. Xu Y, Luong D, Walker JM, Dean D, Becker ML (2017) Modification of poly(propylene fumarate)-bioglass composites with peptide conjugates to enhance bioactivity. Biomacromolecules 18(10):3168–3177. https://doi.org/10.1021/acs.biomac.7b00828
44. Wang Y, You C, Wei R, Zu J, Song C, Li J, Yan J (2017) Modification of human umbilical cord blood stem cells using polyethylenimine combined with modified TAT peptide to enhance BMP-2 production. Biomed Res Int 2017:2971413. https://doi.org/10.1155/2017/2971413
45. Villa-Diaz LG, Nandivada H, Ding J, Nogueira-De-Souza NC, Krebsbach PH, O'Shea KS, Lahann J, Smith GD (2010) Synthetic polymer coatings for long-term growth of human embryonic stem cells. Nat Biotechnol 28(6):581–583. https://doi.org/10.1038/Nbt.1631
46. Melkoumian Z, Weber JL, Weber DM, Fadeev AG, Zhou YE, Dolley-Sonneville P, Yang JW, Qiu LQ, Priest CA, Shogbon C, Martin AW, Nelson J, West P, Beltzer JP, Pal S, Brandenberger R (2010) Synthetic peptide-acrylate surfaces for long-term self-renewal and cardiomyocyte differentiation of human embryonic stem cells. Nat Biotechnol 28(6):606–695. https://doi.org/10.1038/Nbt.1629
47. Zhang J, Hu ZQ, Turner NJ, Teng SF, Cheng WY, Zhou HY, Zhang L, Hu HW, Wang Q, Badylak SF (2016) Perfusion-decellularized skeletal muscle as a three-dimensional scaffold with a vascular network template. Biomaterials 89:114–126. https://doi.org/10.1016/j.biomaterials.2016.02.040
48. Figliuzzi M, Bonandrini B, Remuzzi A (2017) Decellularized kidney matrix as functional material for whole organ tissue engineering. J Appl Biomater Funct Mater 15(4):0. https://doi.org/10.5301/jabfm.5000393
49. Ohno M, Fuchimoto Y, Hsu HC, Higuchi M, Komura M, Yamaoka T, Umezawa A, Enosawa S, Kuroda T (2017) Airway reconstruction using decellularized tracheal allografts in a porcine model. Pediatr Surg Int 33(10):1065–1071. https://doi.org/10.1007/s00383-017-4138-8
50. Bhumiratana S, Bernhard JC, Alfi DM, Yeager K, Eton RE, Bova J, Shah F, Gimble JM, Lopez MJ, Eisig SB, Vunjak-Novakovic G (2016) Tissue-engineered autologous grafts for facial bone reconstruction. Sci Transl Med 8(343):343ra383. https://doi.org/10.1126/scitranslmed.aad5904
51. McNamara LE, McMurray RJ, Biggs MJP, Kantawong F, Oreffo ROC, Dalby MJ, Salih V (2010) Nanotopographical control of stem cell differentiation. J Tissue Eng 1(1):120623. https://doi.org/10.4061/2010/120623
52. Lee S, Leach MK, Redmond SA, Chong SY, Mellon SH, Tuck SJ, Feng ZQ, Corey JM, Chan JR (2012) A culture system to study oligodendrocyte myelination processes using engineered nanofibers. Nat Methods 9(9):917–922. https://doi.org/10.1038/nmeth.2105
53. Liu L, Yuan Q, Shi J, Li X, Jung D, Wang L, Yamauchi K, Nakatsuji N, Kamei K, Chen Y (2012) Chemically-defined scaffolds created with electrospun synthetic nanofibers to maintain mouse embryonic stem cell culture under feeder-free conditions. Biotechnol Lett 34(10):1951–1957. https://doi.org/10.1007/s10529-012-0973-9
54. Mahairaki V, Lim SH, Christopherson GT, Xu L, Nasonkin I, Yu C, Mao HQ, Koliatsos VE (2011) Nanofiber matrices promote the neuronal differentiation of human embryonic stem cell-derived neural precursors in vitro. Tissue Eng Part A 17(5–6):855–863. https://doi.org/10.1089/ten.TEA.2010.0377
55. Nandakumar A, Fernandes H, de Boer J, Moroni L, Habibovic P, van Blitterswijk CA (2010) Fabrication of bioactive composite scaffolds by electrospinning for bone regeneration. Macromol Biosci 10(11):1365–1373. https://doi.org/10.1002/mabi.201000145

56. Nandakumar A, Truckenmuller R, Ahmed M, Damanik F, Santos DR, Auffermann N, de Boer J, Habibovic P, van Blitterswijk C, Moroni L (2013) A fast process for imprinting micro and nano patterns on electrospun fiber meshes at physiological temperatures. Small 9(20):3405–3409. https://doi.org/10.1002/smll.201300220
57. Nitanan T, Opanasopit P, Akkaramongkolporn P, Rojanarata T, Ngawhirunpat T, Supaphol P (2012) Effects of processing parameters on morphology of electrospun polystyrene nanofibers. Korean J Chem Eng 29(2):173–181. https://doi.org/10.1007/s11814-011-0167-5
58. Nur-E-Kamal A, Ahmed I, Kamal J, Schindler M, Meiners S (2006) Three-dimensional nanofibrillar surfaces promote self-renewal in mouse embryonic stem cells. Stem Cells 24(2):426–433. https://doi.org/10.1634/stemcells.2005-0170
59. Orlova Y, Magome N, Liu L, Chen Y, Agladze K (2011) Electrospun nanofibers as a tool for architecture control in engineered cardiac tissue. Biomaterials 32(24):5615–5624. https://doi.org/10.1016/j.biomaterials.2011.04.042
60. Siddappa R, Fernandes H, Liu J, van Blitterswijk C, de Boer J (2007) The response of human mesenchymal stem cells to osteogenic signals and its impact on bone tissue engineering. Curr Stem Cell Res Ther 2(3):209–220
61. Song JH, Kim HE, Kim HW (2008) Production of electrospun gelatin nanofiber by water-based co-solvent approach. J Mater Sci Mater Med 19(1):95–102. https://doi.org/10.1007/s10856-007-3169-4
62. Truong YB, Glattauer V, Briggs KL, Zappe S, Ramshaw JA (2012) Collagen-based layer-by-layer coating on electrospun polymer scaffolds. Biomaterials 33(36):9198–9204. https://doi.org/10.1016/j.biomaterials.2012.09.012
63. Xie JW, MacEwan MR, Li XR, Sakiyama-Elbert SE, Xia YN (2009) Neurite outgrowth on nanofiber scaffolds with different orders, structures, and surface properties. ACS Nano 3(5):1151–1159. https://doi.org/10.1021/Nn900070z
64. Zhao S, Zhou Q, Long Y-Z, Sun G-H, Zhang Y (2013) Nanofibrous patterns by direct electrospinning of nanofibers onto topographically structured non-conductive substrates. Nanoscale 5(11):4993–5000. https://doi.org/10.1039/c3nr00676j
65. Dvir T, Timko BP, Kohane DS, Langer R (2010) Nanotechnological strategies for engineering complex tissues. Nat Nanotechnol 6(1):13–22. https://doi.org/10.1038/nnano.2010.246
66. Mendes PM (2013) Cellular nanotechnology: making biological interfaces smarter. Chem Soc Rev 42(24):9207. https://doi.org/10.1039/c3cs60198f
67. Dalby MJ, Gadegaard N, Oreffo ROC (2014) Harnessing nanotopography and integrin–matrix interactions to influence stem cell fate. Nat Mater 13(6):558–569. https://doi.org/10.1038/nmat3980
68. Dalby MJ, Gadegaard N, Tare R, Andar A, Riehle MO, Herzyk P, Wilkinson CDW, Oreffo ROC (2007) The control of human mesenchymal cell differentiation using nanoscale symmetry and disorder. Nat Mater 6(12):997–1003. https://doi.org/10.1038/nmat2013
69. Khan M, Xu Y, Hua S, Johnson J, Belevych A, Janssen PM, Gyorke S, Guan J, Angelos MG (2015) Evaluation of changes in morphology and function of human induced pluripotent stem cell derived cardiomyocytes (HiPSC-CMs) cultured on an aligned-nanofiber cardiac patch. PLoS One 10(5):e0126338. https://doi.org/10.1371/journal.pone.0126338
70. Feng ZQ, Wang T, Zhao B, Li J, Jin L (2015) Soft graphene nanofibers designed for the acceleration of nerve growth and development. Adv Mater 27(41):6462–6468. https://doi.org/10.1002/adma.201503319
71. Alamein MA, Wolvetang EJ, Ovchinnikov DA, Stephens S, Sanders K, Warnke PH (2015) Polymeric nanofibrous substrates stimulate pluripotent stem cells to form three-dimensional multilayered patty-like spheroids in feeder-free culture and maintain their pluripotency. J Tissue Eng Regen Med 9(9):1078–1083. https://doi.org/10.1002/term.1960
72. Liu L, Yoshioka M, Nakajima M, Ogasawara A, Liu J, Hasegawa K, Li SS, Zou JL, Nakatsuji N, Kamei K, Chen Y (2014) Nanofibrous gelatin substrates for long-term expansion of human pluripotent stem cells. Biomaterials 35(24):6259–6267. https://doi.org/10.1016/j.biomaterials.2014.04.024

73. Gauthaman K, Venugopal JR, Yee FC, Peh GS, Ramakrishna S, Bongso A (2009) Nanofibrous substrates support colony formation and maintain stemness of human embryonic stem cells. J Cell Mol Med 13(9B):3475–3484. https://doi.org/10.1111/j.1582-4934.2009.00699.x
74. Huang CC, Lu CT, Chao CY, Huang TC, Lou CW, Lin JH (2010) Evaluation of electrospun poly (vinyl alcohol) (PVA)/gelatin nanofiber membrane and biocompatibility. JMSE 1–5:97–101. https://doi.org/10.4028/www.scientific.net/AMR.97-101.2249
75. Huang C, Fu X, Liu J, Qi Y, Li S, Wang H (2012) The involvement of integrin beta1 signaling in the migration and myofibroblastic differentiation of skin fibroblasts on anisotropic collagen-containing nanofibers. Biomaterials 33(6):1791–1800. https://doi.org/10.1016/j.biomaterials.2011.11.025
76. Liu L, Kamei K, Yoshioka M, Nakajima M, Liu J, Fujimoto N, Terada S, Tokunaga Y, Koyama Y, Sato H, Hasegawa K, Nakatsuji N, Chen Y (2017) Nano-on-micro fibrous extracellular matrices for scalable expansion of human ES/iPS cells. Biomaterials 124:47–54. https://doi.org/10.1016/j.biomaterials.2017.01.039
77. Jordan AM, Viswanath V, Kim S-E, Pokorski JK, Korley LTJ (2016) Processing and surface modification of polymer nanofibers for biological scaffolds: a review. J Mater Chem B 4(36):5958–5974. https://doi.org/10.1039/c6tb01303a
78. Lancuški A, Fort S, Bossard F (2012) Electrospun azido-PCL nanofibers for enhanced surface functionalization by click chemistry. ACS Appl Mater Interfaces 4(12):6499–6504. https://doi.org/10.1021/am301458y
79. Zheng J, Hua G, Yu J, Lin F, Wade MB, Reneker DH, Becker ML (2015) Post-electrospinning "Triclick" functionalization of degradable polymer nanofibers. ACS Macro Lett 4(2):207–213. https://doi.org/10.1021/mz500759n
80. Lipner J, Liu W, Liu Y, Boyle J, Genin GM, Xia Y, Thomopoulos S (2014) The mechanics of PLGA nanofiber scaffolds with biomimetic gradients in mineral for tendon-to-bone repair. J Mech Behav Biomed Mater 40:59–68. https://doi.org/10.1016/j.jmbbm.2014.08.002
81. Sridhar S, Venugopal JR, Sridhar R, Ramakrishna S (2015) Cardiogenic differentiation of mesenchymal stem cells with gold nanoparticle loaded functionalized nanofibers. Colloids Surf B Biointerfaces 134:346–354. https://doi.org/10.1016/j.colsurfb.2015.07.019
82. Ravichandran R, Sridhar R, Venugopal JR, Sundarrajan S, Mukherjee S, Ramakrishna S (2014) Gold nanoparticle loaded hybrid nanofibers for cardiogenic differentiation of stem cells for infarcted myocardium regeneration. Macromol Biosci 14(4):515–525. https://doi.org/10.1002/mabi.201300407
83. Kim J-J, El-Fiqi A, Kim H-W (2017) Synergetic cues of bioactive nanoparticles and nanofibrous structure in bone scaffolds to stimulate osteogenesis and angiogenesis. ACS Appl Mater Interfaces 9(3):2059–2073. https://doi.org/10.1021/acsami.6b12089
84. Mei Y, Saha K, Bogatyrev SR, Yang J, Hook AL, Kalcioglu ZI, Cho S-W, Mitalipova M, Pyzocha N, Rojas F, Van Vliet KJ, Davies MC, Alexander MR, Langer R, Jaenisch R, Anderson DG (2010) Combinatorial development of biomaterials for clonal growth of human pluripotent stem cells. Nat Mater 9(9):768–778. https://doi.org/10.1038/nmat2812
85. Kraehenbuehl TP, Langer R, Ferreira LS (2011) Three-dimensional biomaterials for the study of human pluripotent stem cells. Nat Methods 8(9):731–736. https://doi.org/10.1038/nmeth.1671
86. Cabezas MD, Eichelsdoerfer DJ, Brown KA, Mrksich M, Mirkin CA (2014) Combinatorial screening of mesenchymal stem cell adhesion and differentiation using polymer pen lithography. Methods Cell Biol 119:261–276. https://doi.org/10.1016/b978-0-12-416742-1.00013-5
87. K-i K, Mashimo Y, Yoshioka M, Tokunaga Y, Fockenberg C, Terada S, Koyama Y, Nakajima M, Shibata-Seki T, Liu L, Akaike T, Kobatake E, How S-E, Uesugi M, Chen Y (2017) Microfluidic-nanofiber hybrid array for screening of cellular microenvironments. Small 13(18). https://doi.org/10.1002/smll.201603104
88. Krawetz R, Taiani JT, Liu S, Meng G, Li X, Kallos MS, Rancourt DE (2010) Large-scale expansion of pluripotent human embryonic stem cells in stirred-suspension bioreactors. Tissue Eng Part C Methods 16(4):573–582. https://doi.org/10.1089/ten.TEC.2009.0228

89. Olmer R, Haase A, Merkert S, Cui W, Palecek J, Ran C, Kirschning A, Scheper T, Glage S, Miller K, Curnow EC, Hayes ES, Martin U (2010) Long term expansion of undifferentiated human iPS and ES cells in suspension culture using a defined medium. Stem Cell Res 5(1):51–64. https://doi.org/10.1016/j.scr.2010.03.005
90. Zweigerdt R, Olmer R, Singh H, Haverich A, Martin U (2011) Scalable expansion of human pluripotent stem cells in suspension culture. Nat Protoc 6(5):689–700. https://doi.org/10.1038/nprot.2011.318
91. Amit M, Laevsky I, Miropolsky Y, Shariki K, Peri M, Itskovitz-Eldor J (2011) Dynamic suspension culture for scalable expansion of undifferentiated human pluripotent stem cells. Nat Protoc 6(5):572–579. https://doi.org/10.1038/nprot.2011.325
92. Otsuji TG, Bin J, Yoshimura A, Tomura M, Tateyama D, Minami I, Yoshikawa Y, Aiba K, Heuser JE, Nishino T, Hasegawa K, Nakatsuji N (2014) A 3D sphere culture system containing functional polymers for large-scale human pluripotent stem cell production. Stem Cell Rep 2(5):734–745. https://doi.org/10.1016/j.stemcr.2014.03.012
93. Chen AK, Chen X, Choo AB, Reuveny S, Oh SK (2010) Expansion of human embryonic stem cells on cellulose microcarriers. Curr Protoc Stem Cell Biol Chapter 1:Unit 1C 11. https://doi.org/10.1002/9780470151808.sc01c11s14
94. Lecina M, Ting S, Choo A, Reuveny S, Oh S (2010) Scalable platform for human embryonic stem cell differentiation to cardiomyocytes in suspended microcarrier cultures. Tissue Eng Part C Methods 16(6):1609–1619. https://doi.org/10.1089/ten.TEC.2010.0104
95. Lei Y, Schaffer DV (2013) A fully defined and scalable 3D culture system for human pluripotent stem cell expansion and differentiation. Proc Natl Acad Sci U S A 110(52):E5039–E5048. https://doi.org/10.1073/pnas.1309408110
96. Serra M, Correia C, Malpique R, Brito C, Jensen J, Bjorquist P, Carrondo MJ, Alves PM (2011) Microencapsulation technology: a powerful tool for integrating expansion and cryopreservation of human embryonic stem cells. PLoS One 6(8):e23212. https://doi.org/10.1371/journal.pone.0023212
97. Chayosumrit M, Tuch B, Sidhu K (2010) Alginate microcapsule for propagation and directed differentiation of hESCs to definitive endoderm. Biomaterials 31(3):505–514. https://doi.org/10.1016/j.biomaterials.2009.09.071
98. Yamazoe T, Shiraki N, Toyoda M, Kiyokawa N, Okita H, Miyagawa Y, Akutsu H, Umezawa A, Sasaki Y, Kume K, Kume S (2013) A synthetic nanofibrillar matrix promotes in vitro hepatic differentiation of embryonic stem cells and induced pluripotent stem cells. J Cell Sci 126(Pt 23):5391–5399. https://doi.org/10.1242/jcs.129767
99. Chen Y, Zeng D, Ding L, Li XL, Liu XT, Li WJ, Wei T, Yan S, Xie JH, Wei L, Zheng QS (2015) Three-dimensional poly-(epsilon-caprolactone) nanofibrous scaffolds directly promote the cardiomyocyte differentiation of murine-induced pluripotent stem cells through Wnt/beta-catenin signaling. BMC Cell Biol 16:22. https://doi.org/10.1186/s12860-015-0067-3
100. Joanne P, Kitsara M, Boitard SE, Naemetalla H, Vanneaux V, Pernot M, Larghero J, Forest P, Chen Y, Menasche P, Agbulut O (2016) Nanofibrous clinical-grade collagen scaffolds seeded with human cardiomyocytes induces cardiac remodeling in dilated cardiomyopathy. Biomaterials 80:157–168. https://doi.org/10.1016/j.biomaterials.2015.11.035
101. Hackelberg S, Tuck SJ, He L, Rastogi A, White C, Liu L, Prieskorn DM, Miller RJ, Chan C, Loomis BR, Corey JM, Miller JM, Duncan RK (2017) Nanofibrous scaffolds for the guidance of stem cell-derived neurons for auditory nerve regeneration. PLoS One 12(7):e0180427. https://doi.org/10.1371/journal.pone.0180427
102. Hyysalo A, Ristola M, Joki T, Honkanen M, Vippola M, Narkilahti S (2017) Aligned poly(epsilon-caprolactone) nanofibers guide the orientation and migration of human pluripotent stem cell-derived neurons, astrocytes, and oligodendrocyte precursor cells in vitro. Macromol Biosci 17(7). https://doi.org/10.1002/mabi.201600517
103. Naskar D, Ghosh AK, Mandal M, Das P, Nandi SK, Kundu SC (2017) Dual growth factor loaded nonmulberry silk fibroin/carbon nanofiber composite 3D scaffolds for in vitro

and in vivo bone regeneration. Biomaterials 136:67–85. https://doi.org/10.1016/j.biomaterials.2017.05.014
104. Wang TY, Bruggeman KF, Kauhausen JA, Rodriguez AL, Nisbet DR, Parish CL (2016) Functionalized composite scaffolds improve the engraftment of transplanted dopaminergic progenitors in a mouse model of Parkinson's disease. Biomaterials 74:89–98. https://doi.org/10.1016/j.biomaterials.2015.09.039
105. Mahmoudi N, Eslahi N, Mehdipour A, Mohammadi M, Akbari M, Samadikuchaksaraei A, Simchi A (2017) Temporary skin grafts based on hybrid graphene oxide-natural biopolymer nanofibers as effective wound healing substitutes: pre-clinical and pathological studies in animal models. J Mater Sci Mater Med 28(5):73. https://doi.org/10.1007/s10856-017-5874-y
106. Hayashi K, Ochiai-Shino H, Shiga T, Onodera S, Saito A, Shibahara T, Azuma T (2016) Transplantation of human-induced pluripotent stem cells carried by self-assembling peptide nanofiber hydrogel improves bone regeneration in rat calvarial bone defects. BDJ Open 2:2. https://doi.org/10.1038/bdjopen.2015.7
107. Agrawal A, Choudhary A (2016) Perspective: materials informatics and big data: realization of the "fourth paradigm" of science in materials science. APL Mater 4(5). https://doi.org/10.1063/1.4946894
108. Rajan K (2005) Materials informatics. Mater Today 8(10):38–45. https://doi.org/10.1016/s1369-7021(05)71123-8
109. Tian B, Liu J, Dvir T, Jin L, Tsui JH, Qing Q, Suo Z, Langer R, Kohane DS, Lieber CM (2012) Macroporous nanowire nanoelectronic scaffolds for synthetic tissues. Nat Mater 11(11):986–994. https://doi.org/10.1038/nmat3404

Correction to: Magnetic Nanoparticles and Alternating Magnetic Field for Cancer Therapy

Harutaka Mekaru, Yuko Ichiyanagi, and Fuyuhiko Tamanoi

Correction to:
Chapter 7 in: D. O. Wang, D. Packwood (eds.), *Cell-Inspired Materials and Engineering*, Fundamental Biomedical Technologies, https://doi.org/10.1007/978-3-030-55924-3_7

The name of one of the author of this chapter should be: Yuko Ichiyanagi which was mistakenly printed as Yuko Ichivanagi. The error has been corrected in this chapter.

The updated online version of this chapter can be found at
https://doi.org/10.1007/978-3-030-55924-3_7

D. O. Wang, D. Packwood (eds.), *Cell-Inspired Materials and Engineering*, Fundamental Biomedical Technologies, https://doi.org/10.1007/978-3-030-55924-3_12

Index

D. O. Wang, D. Packwood (eds.), *Cell-Inspired Materials and Engineering*, Fundamental Biomedical Technologies, https://doi.org/10.1007/978-3-030-55924-3

N

O

P

Q

R

The manufacturer's authorised representative in the EU is Springer Nature Customer Service Centre GmbH, Europaplatz 3, 69115 Heidelberg, Germany. If you have any concerns regarding our products, please contact ProductSafety@springernature.com

Printed and bound by CPI Group (UK) Ltd, Croydon, CR0 4YY
22/09/2026
02221062-0011